LECTURE NOTES ON REGULARITY THEORY FOR THE NAVIER–STOKES EQUATIONS

LECTURE NOTES ON REGULARITY THEORY FOR THE NAVIER–STOKES EQUATIONS

Gregory Seregin

Oxford University, UK

St. Petersburg Department of Steklov Mathematical Institute, RAS, Russia

NEW JERSEY • LONDON • SINGAPORE • BEIJING • SHANGHAI • HONG KONG • TAIPEI • CHENNAI

Published by

World Scientific Publishing Co. Pte. Ltd.
5 Toh Tuck Link, Singapore 596224
USA office: 27 Warren Street, Suite 401-402, Hackensack, NJ 07601
UK office: 57 Shelton Street, Covent Garden, London WC2H 9HE

Library of Congress Cataloging-in-Publication Data
Seregin, Gregory, 1950– author.
[Lecture notes. Selections]
Lecture notes on regularity theory for the Navier-Stokes equations / Gregory Seregin (Oxford University, UK).
pages cm
Includes bibliographical references and index.
ISBN 978-9814623407 (hardcover : alk. paper)
1. Navier-Stokes equations. 2. Fluid dynamics. I. Title.
QA377.S463 2014
515'.353--dc23
2014024553

British Library Cataloguing-in-Publication Data
A catalogue record for this book is available from the British Library.

Printed in Singapore

Preface

The Lecture Notes are based on the TCC (Graduate Taught Course Center) course given by me in Trinity Terms of 2009-2011 at Mathematical Institute of Oxford University. Chapters 1-3 contain material discussed in Trinity Term of 2009 (16 hours in total), Chapters 4-5 contain lectures of 2010 (16 hours), and, finally, lectures of 2011 are covered by Chapter 6 (16 hours).

Chapters 1-5 can be regarded as an Introduction to the Mathematical Theory of the Navier-Stokes equations, relying mainly on the classical PDE's approach. First, the notion of weak solutions is introduced, then their existence is proven (where it is possible), and, afterwards, differentiability properties are analyzed. In other words, we treat the Navier-Stokes equations as a particular case, maybe very difficult, of the theory of nonlinear PDE's. From this point of view, the Lectures Notes do not pretend to be a complete mathematical theory of the Navier-Stokes equations. There are different approaches, for example, more related to harmonic analysis, etc. A corresponding list of references (incomplete, of course) is given at the end of the Lecture Notes.

Finally, Chapters 6 and 7 contain more advanced material, which reflects my scientific interests.

I also would like to thank Tim Shilkin for careful reading of Lecture Notes and for his valuable suggestions.

Contents

Chapter 1

Preliminaries

1.1 Notation

Let us denote by Ω a domain (open connected set) in $\mathbb{R}^n$. Then, $C_0^\infty(\Omega;\mathbb{R}^m)$ is set of all infinitely differentiable functions from Ω into $\mathbb{R}^m$, having a compact support in Ω. If $m = 1$, we use abbreviation $C_0^\infty(\Omega)$. However, even in the case of functions with values in $\mathbb{R}^m$, we shall drop the space $\mathbb{R}^m$ in the notation of the corresponding spaces very often.

A Lebesgue space $L_p(\Omega)$ is endowed with the standard norm

$$\|f\|_{p,\Omega} = \Big(\int\limits_\Omega |f(x)|^p dx\Big)^{\frac{1}{p}}$$

if $1 \leq p < \infty$ and

$$\|f\|_{\infty,\Omega} = \operatorname{ess}\sup_{x\in\Omega} |f(x)|$$

if $p = \infty$.

Lemma 1.1. *Let $1 \leq p < \infty$. Then, $L_p(\Omega) = [C_0^\infty(\Omega)]^{L_p(\Omega)}$, i.e., $L_p(\Omega)$ is the completion of $C_0^\infty(\Omega)$ in $L_p(\Omega)$.*

In what follows, we always assume that the exponent of integrability is finite unless otherwise is specially indicated.

We say that a distribution u, defined in Ω, belongs to the Sobolev space $W_s^k(\Omega)$ if and only if all its weak derivatives up to order k are integrable in Ω with the power s. The norm of this space is defined as

$$\|u\|_{W_s^k(\Omega)} = \sum_{i=0}^k \|\nabla^i u\|_{s,\Omega}.$$

We also let

$$\overset{\circ}{W}{}_s^k(\Omega) = [C_0^\infty(\Omega)]^{W_s^k(\Omega)}.$$

It is said that a distribution u, defined in Ω, belongs to the space $L^k_s(\Omega)$ if and only if all its weak derivatives of order k are integrable in Ω with the power s. This space can be endowed with the usual semi-norm

$$\|u\|_{L^k_s(\Omega)} = \|\nabla^k u\|_{s,\Omega}.$$

Theorem 1.1. $L^k_s(\Omega) \subset L_{s,\mathrm{loc}}(\Omega)$.

PROOF For simplicity, let us consider the case $k = 1$ only. We have a distribution T such that

$$\int\limits_\Omega g(x)\varphi(x)dx = -T(\nabla\varphi) = -<T, \nabla\varphi>$$

for $\varphi \in C^\infty_0(\Omega)$ with $g = (g_i) \in L_s(\Omega)$. Our aim is to show that T is in fact a regular distribution, i.e., there exists a function $u \in L_{s,\mathrm{loc}}(\Omega)$ such that $T = T_u$.

Consider a subdomain $\Omega_0 \Subset \Omega$, i.e., a bounded domain $\Omega_0 \subset \Omega$ such that the closure of Ω_0 belongs to Ω. Let $0 < \varrho < \mathrm{dist}(\Omega_0, \partial\Omega)$. Define a linear functional $l : L_1(\Omega_0) \to \mathbb{R}$ in the following way

$$l(\psi) := <T, \psi_\varrho>$$

for $\psi \in L_1(\Omega_0)$ where

$$\psi_\varrho(x) = \int\limits_\Omega \varpi_\varrho(x - y)\psi(y)dy$$

and ϖ_ϱ is a standard mollifier. Obviously, $\psi_\varrho \in C^\infty_0(\Omega)$. It is easy to check that l is a bounded functional on $L_1(\Omega_0)$ and thus there exists a function $u^\varrho_0 \in L_\infty(\Omega_0)$ such that

$$l(\psi) = \int\limits_{\Omega_0} u^\varrho_0(x)\psi(x)dx$$

for all $\psi \in L_1(\Omega_0)$. Next, for any $\varphi \in C^\infty_0(\Omega_0)$, we have

$$-<T, \nabla\varphi_\rho> = \int\limits_\Omega g\varphi_\rho dx = \int\limits_\Omega g_\varrho\varphi dx.$$

On the other hand, by known properties of mollification we find

$$-<T, \nabla\varphi_\rho> = -<T, (\nabla\varphi)_\rho> = -\int\limits_{\Omega_0} u^\varrho_0 \nabla\varphi dx.$$

The latter means that

$$g_\varrho = \nabla \overline{u}^\varrho_0$$

in Ω_0. Here, $\overline{u}_0^\varrho = u_0^\varrho - [u_0^\varrho]_{\Omega_0}$ and

$$[u_0^\varrho]_{\Omega_0} := \frac{1}{|\Omega_0|} \int\limits_{\Omega_0} u_0^\varrho dx.$$

Assuming that Ω_0 has a sufficiently smooth boundary, we can apply the Poincaré-Sobolev inequality

$$\|\overline{u}_0^\varrho\|_{s,\Omega_0} \leq c(n,s,\Omega_0)\|g_\varrho\|_{s,\Omega_0} \leq c(n,s,\Omega_0)\|g\|_{s,\Omega}.$$

Without loss of generality, we may assume that

$$\overline{u}_0^\varrho \rightharpoonup u_0$$

in $L_s(\Omega_0)$. And thus

$$g = \nabla u_0$$

in Ω_0.

Next, we take a sequence of domains with sufficiently smooth boundary such that $\Omega_k \Subset \Omega$, $\Omega_k \Subset \Omega_{k+1}$ for any natural k and

$$\Omega = \bigcup_{k=0}^{\infty} \Omega_k.$$

Now, a required function u can be defined as follows. We let $w = u_0$ in Ω_0. Then, repeating the above procedure, we find a function $u_1 \in L_s(\Omega_1)$ such that $g = \nabla u_1$ in Ω_1. It is easy to see that $u_1 - w = C_0$ on Ω_0. Then we let $w = u^1 - C_0$ on $\Omega_1 \setminus \Omega_0$. Since $w - u^1 = -C_0$ in Ω_1, $g = \nabla w$ in Ω_1 and $w \in L_s(\Omega_1)$. We can extend function w to Ω_2 in the same way and so on.

So, we have constructed a function $w \in L_{s,\mathrm{loc}}(\Omega)$ such that

$$< T_w, \nabla\varphi >= \int\limits_{\Omega} w(x)\nabla\varphi(x)dx = -\int\limits_{\Omega} g(x)\varphi(x)dx$$

for all $\varphi \in C_0^\infty(\Omega)$ and thus $< T - T_w, \nabla\varphi >= 0$ for the same φ. Hence, there is a constant c such that $T = T_{w+c}$. Letting $u = w + c$, we complete the proof. □

In a similar way, we can show that if $u^m \in C_0^\infty(\Omega)$ is a Cauchy sequence in $L_s^k(\Omega)$, i.e., $\|u^m - u^r\|_{L_s^k(\Omega)} \to 0$ as m and r tend to ∞, then there exists $u \in L_s^k(\Omega)$ such that $\|u^m - u\|_{L_s^k(\Omega)} \to 0$ as m tends ∞. Indeed, supposing for simplicity that $k = 1$, we may assume that ∇u^m converges to $g \in L_s(\Omega)$. Our aim is to find a function $u \in L_{s,\mathrm{loc}}$ such that $g = \nabla u$ in Ω.

Now, let domains Ω_k be as in the proof of the previous statement. By Poincaré-Sobolev inequality, we have

$$\|u^m - u^r - ([u^m]_{\Omega_0} - [u^r]_{\Omega_0})\|_{s,\Omega_0} \leq c(n,s,\Omega_0)\|\nabla(u^m - u^r)\|_{s,\Omega_0} \leq$$

$$\leq c(n,s,\Omega_0)\|\nabla(u^m-u^r)\|_{s,\Omega}\to 0$$

as m and r go to ∞. So,

$$u^m-[u^m]_{\Omega_0}\to u_0$$

in $L_s(\Omega_0)$ and $\nabla u_0=g$ in Ω_0. Repeating this procedure in a bigger domain Ω_1, we finish our proof using the same arguments as in the previous theorem.

Unfortunately, if a sequence $u^m\in C_0^\infty(\Omega)$ converges to $u\in L_s^k(\Omega)$ in $L_s^k(\Omega)$, it converges there to $u+w$ with $\nabla^k w=0$ as well. This tells us that in general we should operate with equivalence classes generated by $u\sim w$ if $\nabla^k(u-w)=0$ in Ω. So, we can introduce a Banach space $\overset{\circ}{L}{}_s^k(\Omega)$ that consists of all equivalence classes containing an element u such that there exists a sequence $u^m\in C_0^\infty(\Omega)$ with $\|\nabla(u^m-u)\|_{s,\Omega}$ as $m\to\infty$.

In many interesting cases, we can get rid of equivalence classes selecting a "good" representative from each of them. This usually happens if we can control a weaker norm. For example, if Ω is bounded, the Friedrichs inequality is valid:

$$\|u\|_{s,\Omega}\leq c\|u\|_{L_s^k(\Omega)}$$

for any $u\in C_0^\infty(\Omega)$ with a positive constant independent of u. So, if $u^m\in C_0^\infty(\Omega)$ converges to $w\in L_s^k(\Omega)$ in $L_s^k(\Omega)$, then we can select a special element $u\in[w]$ such that $u^m\to u$ in $L_s(\Omega)$ and work with it in what follows. So, we have

Proposition 1.2. *For bounded domains Ω,*

$$\overset{\circ}{L}{}_s^k(\Omega)=\overset{\circ}{W}{}_s^k(\Omega).$$

For unbounded domains, things are more complicated. Let us consider, for example, the space $\overset{\circ}{L}{}_2^1(\Omega)$. If $n\geq 3$, then we can use Gagliardo-Nirenberg inequality

$$\|u\|_{p,\Omega}\leq c(n)\|\nabla u\|_{2,\Omega}$$

for any $u\in C_0^\infty(\Omega)$ with $p=2n/(n-2)$. This inequality allows to avoid using equivalence classes for ANY domain Ω in $\mathbb{R}^n$.

For $n=2$, a half-plane $\mathbb{R}^2_+=\{x\in\mathbb{R}^2:\,x_2>0\}$ is still a "good" case thanks to the following inequality:

$$\|u\|_{2,\Pi}\leq\|\nabla u\|_{2,\Pi}$$

for any $u\in C_0^\infty(\mathbb{R}^2_+)$ with $\Pi=\{-\infty<x_1<\infty,\ 0<x_2<1\}$. The proof of it is the same as for the Friedrichs inequality.

1.2 Newtonian Potential

The fundamental solution to the Laplace equation is

$$E(x) = \frac{1}{\omega_n n(n-2)} \frac{1}{|x|^{n-2}}$$

if $n \geq 3$ and

$$E(x) = \frac{1}{2\pi} \ln \frac{1}{|x|}$$

if $n = 2$, where ω_n is the volume of unit ball in $\mathbb{R}^n$.

For a given function $f : \mathbb{R}^n \to \mathbb{R}$, we define the Newtonian potential of f as the following convolution:

$$u = E \star f$$

or

$$u(x) = \int\limits_{\mathbb{R}^n} E(x-y) f(y) dy.$$

In what follows, we are going to use a standard cut-off function $\varphi \in C_0^\infty(\mathbb{R}^n)$, having the following properties:

$$0 \leq \varphi(x) \leq 1 \qquad x \in \mathbb{R}^n,$$

$$\varphi(x) = 1 \quad x \in B(1), \qquad \varphi(x) = 0 \quad x \notin B(2),$$

$$\varphi_R(x) = \varphi(x/R).$$

Here, $B(R)$ is a ball of radius R centred at the origin.

Proposition 2.3. *Let $f \in L_p(\mathbb{R}^n)$ with $1 < p < \infty$ and $u = E \star f$. The following statements are true:*

(*i*)
$$\int\limits_{\mathbb{R}^n} |\nabla^2 u|^p dx \leq c(n,p) \int\limits_{\mathbb{R}^n} |f|^p dx,$$

(*ii*)
$$u \in \overset{\circ}{L}{}^2_p(\mathbb{R}^n),$$

(*iii*)
$$\Delta u = -f$$

in $\mathbb{R}^n$.

PROOF (i) follows from the theory of singular integrals. (iii) follows from (i), (ii), and from the classical PDE theory.

Let us prove (ii), assuming that $n \geq 3$. By Lemma 1.1, there exists a sequence $f_m \in C_0^\infty(\mathbb{R}^n)$ such that $f_m \to f$ in $L_p(\mathbb{R}^n)$. Since $\operatorname{supp} f_m$ is a compact set in $\mathbb{R}^n$,

$$|\nabla^i u_m(x)| \leq \frac{c(m,i)}{|x|^{n-2+i}}$$

for all $x \in \mathbb{R}^n$, for $i = 0, 1, 2$, for all $m = 1, 2, ...$, and for some positive $c(m, i)$. Here, $u_m = E \star f_m$.

Our aim is to show that

$$u_m \in \overset{\circ}{L}{}^2_p(\mathbb{R}^n).$$

Indeed, we have

$$\int\limits_{\mathbb{R}^n} |\nabla^2(\varphi_R u_m - u_m)|^p dx \leq c\Big[\int\limits_{\mathbb{R}^n \setminus B(R)} |\nabla^2 u_m|^p dx +$$

$$+\frac{1}{R^p}\int\limits_{B(2R)\setminus B(R)} |\nabla u_m|^p dx + \frac{1}{R^{2p}}\int\limits_{B(2R)\setminus B(R)} |u_m|^p dx\Big] \leq$$

$$\leq c \int\limits_{\mathbb{R}^n \setminus B(R)} |\nabla^2 u_m|^p dx + C(m)\Big[\frac{1}{R^p}\frac{R^n}{R^{(n-1)p}} + \frac{1}{R^{2p}}\frac{R^n}{R^{(n-2)p}}\Big] \to 0$$

as $R \to \infty$ for each fixed m.

On the other hand, by (i), we have

$$\|\nabla^2 u - \nabla^2 u_m\|_{p,\mathbb{R}^n} \leq c\|f - f_m\|_{p,\mathbb{R}^n} \to 0$$

as $R \to \infty$. This implies (ii). □

Particular cases

1. Let $f(x', x_n) = -f(x', -x_n)$, where $x' = (x_1, x_2, ..., x_{n-1})$. Then

$$u(x', x_n) = -u(x', -x_n)$$

and u, $u_{,n} = \partial u/\partial x_n$, and $u_{,nn}$ are in $L_{p,\mathrm{loc}}(\mathbb{R}^n)$ that implies $u(x', 0) = 0$. So, the Newtonian potential u solves the following Dirichlet problem in half-space:

$$\Delta u = -f \tag{1.2.1}$$

in $\mathbb{R}^n_+ = \{x = (x', x_n) : x_n > 0\}$,

$$u(x', 0) = 0$$

for any x'.

2. The same arguments show that if $f(x', x_n) = f(x', -x_n)$ then u solves the Neumann boundary value problem, i.e., it satisfies (1.2.1) and the Neumann boundary condition

$$u_{,n}(x', 0) = 0$$

for any x'.

1.3 Equation $\operatorname{div} \boldsymbol{u} = \boldsymbol{b}$

We start with the simplest case $\Omega = \mathbb{R}^n$.

Proposition 3.4. *Let* $1 < s < \infty$. *Given* $b \in L_s(\mathbb{R}^n)$, *there exists* $u \in \overset{\circ}{L}{}^1_s(\mathbb{R}^n)$ *with the following properties:*

(i)
$$\operatorname{div} u = b$$
in $\mathbb{R}^n$,

(ii)
$$\|\nabla u\|_{s,\mathbb{R}^n} \leq c(s,n)\|b\|_{s,\mathbb{R}^n}.$$

PROOF We let $h = E \star b$. By Proposition 2.3, $\nabla h \in \overset{\circ}{L}{}^1_s(\mathbb{R}^n)$. If we let $u = -\nabla h$, then, by the same statement,

$$\|\nabla u\|_{s,\mathbb{R}^n} = \|\nabla^2 h\|_{s,\mathbb{R}^n} \leq c(s,n)\|b\|_{s,\mathbb{R}^n}$$

and

$$\operatorname{div} u = -\operatorname{div} \nabla h = -\Delta h = b. \qquad \square$$

In the case of the half-space, i.e., $\Omega = \mathbb{R}^n_+ := \{x = (x', x_n) : x' \in \mathbb{R}^{n-1}, x_n > 0\}$, we have

Proposition 3.5. *Let* $1 < s < \infty$. *Given* $b \in L_s(\mathbb{R}^n_+)$, *there exists* $u \in \overset{\circ}{L}{}^1_s(\mathbb{R}^n_+)$ *with the following properties:*

(i)
$$\operatorname{div} u = b$$
in $\mathbb{R}^n_+$,

(ii)
$$\|\nabla u\|_{s,\mathbb{R}^n_+} \leq c(s,n)\|b\|_{s,\mathbb{R}^n_+}.$$

Remark 1.1. The above vector-valued function u satisfies the homogeneous boundary condition $u|_{x_n=0} = 0$ in the sense of traces in Sobolev spaces.

PROOF OF PROPOSITION 3.5 To show the essence of the matter, let us consider a special case $n = 3$ and $s = 2$.

Let $b \in C^\infty_0(\mathbb{R}^3_+)$ and $\hat{b}$ is the even extension of b to $\mathbb{R}^3$. Clearly, $\hat{b} \in C^\infty_0(\mathbb{R}^3)$. Letting $h = -E \star \hat{b}$, we see that

$$\Delta h = \hat{b} = b \quad \text{in} \quad \mathbb{R}^3_+, \qquad h_{,3}|_{x_3=0} = 0 \tag{1.3.1}$$

and

$$\|\nabla^2 h\|_{2,\mathbb{R}^3_+} \leq c\|b\|_{2,\mathbb{R}^3_+}. \tag{1.3.2}$$

The idea is to look for u in the form

$$u = \nabla h + \operatorname{rot} A,$$

where A is unknown vector field. Obviously,

$$\operatorname{div} u = \Delta h = b \quad \text{in} \quad \mathbb{R}^3_+.$$

Equations for A is coming from condition $u|_{x_3=0} = 0$ that leads to the following relations

$$\operatorname{rot} A = -\nabla h$$

at $x_3 = 0$. We are seeking A, satisfying additional assumptions:

$$A|_{x_3=0} = 0, \qquad A_3 \equiv 0 \quad \text{in} \quad \mathbb{R}^3_+.$$

So, the main equations for $A = (A_1, A_2, 0)$ are:

$$A_{\alpha,3}(x', 0) = B_\alpha(x') \qquad \alpha = 1, 2,$$

where $B_1(x') = h_{,2}(x', 0)$ and $B_2(x') = -h_{,1}(x', 0)$ are known functions.

The theory of traces for functions from Sobolev spaces suggests to seek A in the form:

$$A_\alpha(x', x_3) = x_3 \int\limits_{\mathbb{R}^2} B_\alpha(x' + y'x_3) K(y') dy',$$

where a function $K \in C_0^\infty(\mathbb{R}^2)$ is supposed to obey the following conditions:

$$K(x') = 0 \quad x' \notin B' = \{x' \in \mathbb{R}^2 : |x'| < 1\}, \qquad \int\limits_{\mathbb{R}^2} K(y') dy' = 1.$$

Now, our aim is to show that functions

$$u_1 = h_{,1} + A_{2,3},$$

$$u_2 = h_{,2} - A_{1,3},$$

$$u_3 = h_{,3} + A_{1,2} - A_{2,1},$$

with A described above, satisfy all the requirements.

Indeed, direct calculations gives us:

$$A_{\alpha,3}(x) = x_3 \int\limits_{\mathbb{R}^2} \frac{\partial B_\alpha}{\partial z_\beta}(x' + y'x_3) y_\beta K(y') dy' + \int\limits_{\mathbb{R}^2} B_\alpha(x' + y'x_3) K(y') dy'. \tag{1.3.3}$$

Noticing that

$$\frac{\partial B_\alpha}{\partial z_\beta} = \frac{\partial B_\alpha}{\partial y_\beta}\frac{1}{x_3}$$

and integrating by parts with respect y_β, we can transform the right-hand side of (1.3.3) to the form

$$A_{\alpha,3}(x) = -\int\limits_{\mathbb{R}^2} \Big[B_\alpha(x'+y'x_3) - B_\alpha(x')\Big]\frac{\partial}{\partial y_\beta}(y_\beta K(y'))dy' +$$

$$+\int\limits_{\mathbb{R}^2} B_\alpha(x'+y'x_3)K(y')dy'.$$

By the choice of K, $A_{\alpha,3}(x',0) = B_\alpha(x'), \alpha = 1,2$.

Now, our goal is to show the validity of the estimate

$$\|\nabla^2 A_\alpha\|_{2,\mathbb{R}^3_+} \leq c\|b\|_{2,\mathbb{R}^3_+}, \qquad \alpha = 1,2, \tag{1.3.4}$$

with some universal constant c. To this end, let us discuss a typical statement from the theory of traces in Sobolev spaces.

Lemma 1.2. *For any smooth function $f : \mathbb{R}^3_+ \to \mathbb{R}$, vanishing for sufficiently large $|x|$, the following inequality holds:*

$$\|f\|^2_{L^{\frac{1}{2}}_2(\mathbb{R}^2)} := \int\limits_{\mathbb{R}^2}\int\limits_{\mathbb{R}^2} |f(x',0)-f(y',0)|^2\frac{dx'dy'}{|x'-y'|^3} \leq c\int\limits_{\mathbb{R}^3_+}|\nabla f|^2 dx \tag{1.3.5}$$

with some universal constant c.

PROOF By the shift in variables, we can rewrite the left-hand side of the latter inequality in the following way:

$$\|f\|^2_{L^{\frac{1}{2}}_2(\mathbb{R}^2)} = \int\limits_{\mathbb{R}^2}\frac{dz'}{|z'|^3}\int\limits_{\mathbb{R}^2}|f(x'+z',0)-f(x',0)|^2dx'.$$

Applying the triangle inequality, we find

$$|f(x'+z',0)-f(x',0)| \leq |f(x'+z',|z'|)-f(x',|z'|)| +$$

$$+|f(x'+z',|z'|)-f(x'+z',0)| + |f(x',|z'|)-f(x',0)|. \tag{1.3.6}$$

According to (1.3.6), we should evaluate three integrals. In the first one, the polar coordinates $z' = (\varrho\cos\varphi, \varrho\sin\varphi)$ are used to derive identity:

$$I_1 = \int\limits_{\mathbb{R}^2}\frac{dz'}{|z'|^3}\int\limits_{\mathbb{R}^2}|f(x',|z'|)-f(x',0)|^2dx' =$$

$$= 2\pi \int\limits_{\mathbb{R}^2} dx' \int\limits_0^\infty \frac{1}{\varrho^2} |f(x',\varrho) - f(x',0)|^2 d\varrho.$$

The right-hand side of it can be bounded from above with the help of Hardy's inequality:

$$\int\limits_0^\infty t^{-p}|g(t) - g(0)|^p dt \leq \left(\frac{p}{p-1}\right)^p \int\limits_0^\infty |g'(t)|^p dt$$

with $1 < p < \infty$. So,

$$I_1 \leq 2\pi \int\limits_{\mathbb{R}^2} dx' 4 \int\limits_0^\infty \left|\frac{\partial}{\partial\varrho} f(x',\varrho)\right|^2 d\varrho \leq 8\pi \int\limits_{\mathbb{R}^3_+} |\nabla f|^2 dx.$$

Applying similar arguments to I_3, we show

$$I_2 = \int\limits_{\mathbb{R}^2} \frac{dz'}{|z'|^3} \int\limits_{\mathbb{R}^2} |f(x'+z',|z'|) - f(x'+z',0)|^2 dx' =$$

$$= \int\limits_{\mathbb{R}^2} \frac{dz'}{|z'|^3} \int\limits_{\mathbb{R}^2} |f(y',|z'|) - f(y',0)|^2 dy' \leq 8\pi \int\limits_{\mathbb{R}^3_+} |\nabla f|^2 dx.$$

To estimate the third term, we exploit the following simple inequality

$$|f(x'+z',|z'|) - f(x',|z'|)| = \left|\int\limits_0^1 \frac{\partial f}{\partial t}(x'+tz',|z'|)dt\right| \leq$$

$$\leq |z'| \Big(\int\limits_0^1 |\nabla_x f(x'+tz',|z'|)|^2 dt\Big)^{\frac{1}{2}},$$

which give us

$$I_3 = \int\limits_{\mathbb{R}^2} \frac{dz'}{|z'|^3} \int\limits_{\mathbb{R}^2} |f(x'+z',|z'|) - f(x',|z'|)|^2 dx' \leq$$

$$\leq \int\limits_{\mathbb{R}^2} \frac{dz'}{|z'|} \int\limits_0^1 dt \int\limits_{\mathbb{R}^2} |\nabla_{x'} f(x'+tz',|z'|)|^2 dx' =$$

$$= \int\limits_{\mathbb{R}^2} \frac{dz'}{|z'|} \int\limits_{\mathbb{R}^2} |\nabla_{y'} f(y',|z'|)|^2 dy'.$$

It remains to make the change of variables $z' = (\varrho\cos\varphi, \varrho\sin\varphi)$

$$I_3 \leq 2\pi \int\limits_{\mathbb{R}^3_+} |\nabla f|^2 dx$$

and complete our proof. □

Now, our goal is to show that (1.3.5) holds for any function f of class C^2, having the decay

$$|f(x)| \leq \frac{c}{|x|}, \tag{1.3.7}$$

with a positive constant c.

Denoting by $B'(R)$ the unit disk centered at the origin and letting $f_R = f\varphi_R$, where φ_R is a standard cut-off function, we have

$$\int\limits_{B'(R)}\int\limits_{B'(R)} \frac{|f(x',0) - f(y',0)|^2}{|x'-y'|^3} dx'dy' =$$

$$= \int\limits_{B'(R)}\int\limits_{B'(R)} \frac{|f_R(x',0) - f_R(y',0)|^2}{|x'-y'|^3} dx'dy' \leq$$

$$\leq c\int\limits_{\mathbb{R}^3_+} \varphi_R|\nabla f|^2 dx + \frac{c}{R^2} \int\limits_{(B(2R)\setminus B(R))\cap\mathbb{R}^3_+} |f|^2 dx \leq$$

$$\leq c\int\limits_{\mathbb{R}^3_+} |\nabla f|^2 dx + \frac{c}{R}.$$

Passing to the limit as $R \to \infty$ and using Fatou's lemma, we deduce (1.3.5).

Next, we observe that it is sufficient to show

$$\|\nabla^2 A_\alpha\|_{2,\mathbb{R}^3_+} \leq c\|\nabla h(\cdot,0)\|_{L_2^{\frac{1}{2}}(\mathbb{R}^2)}. \tag{1.3.8}$$

Indeed, since

$$|\nabla h(x)| \leq \frac{c(b)}{|x|^2}$$

for $|x| \gg 1$, one can derive from (1.3.5) and (1.3.2) that

$$\|\nabla h(\cdot,0)\|_{L_2^{\frac{1}{2}}(\mathbb{R}^2)} \leq c\|\nabla^2 h\|_{2,\mathbb{R}^3_+} \leq c\|b\|_{2,\mathbb{R}^3_+}.$$

Then, statement (ii) of Proposition 3.5 for this particular class of b follows.

Now, let us prove (1.3.8), directly working out the second derivatives of A,

$$A_{\alpha,\beta 3}(x',x_3) = -\int\limits_{\mathbb{R}^2} \frac{\partial}{\partial x_\beta} B_\alpha(x'+y'x_3) \frac{\partial}{\partial y_\gamma}(y_\gamma K(y'))dy' +$$

$$+ \int\limits_{\mathbb{R}^2} \frac{\partial}{\partial x_\beta} B_\alpha(x'+y'x_3) K(y')dy'.$$

Obviously,

$$\frac{\partial}{\partial x_\beta} B_\alpha(x'+y'x_3) = \frac{\partial}{\partial y_\beta} B_\alpha(x'+y'x_3)\frac{1}{x_3}.$$

Then

$$S = A_{\alpha,\beta 3}(x) = \frac{1}{x_3}\int\limits_{\mathbb{R}^2} \delta B_\alpha(x',y',x_3) K_\beta(y')dy',$$

where

$$\delta B_\alpha(x',y',x_3) := B_\alpha(x'+y'x_3) - B_\alpha(x')$$

and

$$K_\beta(y') := (y_\gamma K(y'))_{,\beta\gamma} - K_{,\beta}(y').$$

Now, we have

$$\int\limits_{\mathbb{R}^3_+} S^2 dx \leq c\int\limits_0^\infty dx_3 \int\limits_{\mathbb{R}^2} dx' \Big(\frac{1}{x_3}\int\limits_{\mathbb{R}^2} |\delta B_\alpha(x',y',x_3)||K_\beta(y')|dy'\Big)^2$$

and, by Hölder inequality,

$$\int\limits_{\mathbb{R}^3_+} S^2 dx \leq c\int\limits_0^\infty dx_3 \int\limits_{\mathbb{R}^2} dx' \frac{1}{x_3^2}\int\limits_{\mathbb{R}^2} |K_\beta(z')|dz' \int\limits_{\mathbb{R}^2} |\delta B_\alpha(x',y',x_3)|^2 |K_\beta(y')|dy' \leq$$

$$\leq c\int\limits_{\mathbb{R}^2} |K_\beta(y')|dy' \int\limits_0^\infty \frac{dx_3}{x_3^2}\int\limits_{\mathbb{R}^2} |B_\alpha(x'+y'x_3) - B_\alpha(x')|^2 dx'.$$

Introducing polar coordinates $y' = \varrho(\cos\varphi, \sin\varphi)$, we find

$$\int\limits_{\mathbb{R}^3_+} S^2 dx \leq c\int\limits_0^1\int\limits_0^{2\pi} \varrho d\varrho d\varphi \int\limits_0^\infty \frac{dx_3}{x_3^2}\int\limits_{\mathbb{R}^2} |B_\alpha(x'+\varrho(\cos\varphi,\sin\varphi)x_3) - B_\alpha(x')|^2 dx'.$$

If we set $z' = x_3(\cos\varphi, \sin\varphi)$, then

$$\int\limits_{\mathbb{R}^3_+} S^2 dx \le c \int\limits_0^1 \varrho d\varrho \int\limits_{\mathbb{R}^2} \frac{dz'}{|z'|^3} \int\limits_{\mathbb{R}^2} |B_\alpha(x' + z'\varrho) - B_\alpha(x')|^2 dx'.$$

Letting $y' = z'\varrho$, we show

$$\int\limits_{\mathbb{R}^3_+} S^2 dx \le c \int\limits_0^1 d\varrho \int\limits_{\mathbb{R}^2}\int\limits_{\mathbb{R}^2} \frac{|B_\alpha(x' + y') - B_\alpha(x')|^2}{|y'|^3} dx'dy' \le$$

$$\le \int\limits_{\mathbb{R}^2}\int\limits_{\mathbb{R}^2} \frac{|B_\alpha(z') - B_\alpha(y')|^2}{|z' - y'|^3} dz'dy' \le c\|\nabla h(\cdot, 0)\|^2_{L_2^{\frac{1}{2}}(\mathbb{R}^2)}. \tag{1.3.9}$$

With the remaining second derivatives, we proceed as follows:

$$A_{\alpha,33}(x) = \frac{1}{x_3} \int\limits_{\mathbb{R}^2} (B_\alpha(x' + y'x_3) - B_\alpha(x'))\widetilde{K}(y')dy',$$

where

$$\widetilde{K}(y') := (y_\gamma(y_\beta K(y'))_{,\beta}))_{,\gamma} - (y_\gamma K(y'))_{,\gamma},$$

and

$$A_{\alpha,\beta\gamma}(x) = \frac{1}{x_3} \int\limits_{\mathbb{R}^2} (B_\alpha(x' + y'x_3) - B_\alpha(x'))(K(y'))_{,\beta\gamma} dy'.$$

So, similar arguments as above lead to the required bound

$$\int\limits_{\mathbb{R}^3_+} (A^2_{\alpha,33} + A^2_{\alpha,\beta\gamma}) dx \le c\|\nabla h(\cdot, 0)\|^2_{L_2^{\frac{1}{2}}(\mathbb{R}^2)}.$$

Hence, inequality (1.3.8) for smooth compactly supported functions b is proven.

Now, our aim is to show that

$$u \in \overset{\circ}{L}{}^1_2(\mathbb{R}^3_+) \tag{1.3.10}$$

for any $b \in C_0^\infty(\mathbb{R}^3_+)$. The proof of (1.3.10) consists of two parts.

Step 1. First, let us check that $\nabla h \in L_2(\mathbb{R}^3_+)$. Indeed, since $\nabla^2 h \in L_2(\mathbb{R}^3_+)$, an embedding theorem implies $\nabla h \in L_2(B_+(1))$, where $B_+(R) := \{x \in B(R) : \ x_3 > 0\}$. We know that

$$|\nabla h(x)| \le \frac{c(b)}{|x|^2}$$

for $|x| \gg 1$. So,

$$\int\limits_{B_+(R)} |\nabla h|^2 dx \leq \int\limits_{B_+(1)} |\nabla h|^2 dx + \int\limits_{B_+(R)\setminus B_+(1)} |\nabla h|^2 dx \leq ... + c(b) \int\limits_1^R \frac{d\varrho}{\varrho^2} \leq c(b)$$

for any $R > 1$.

STEP 2. Let us show that $\nabla A_\alpha \in L_2(\mathbb{R}^3_+)$. We know

$$A_{\alpha,3}(x) = \int\limits_{\mathbb{R}^2} B_\alpha(x' + y'x_3) K_0(y') dy',$$

where

$$K_0(y') = K(y') - (y_\beta K(y'))_{,\beta}.$$

Let $\alpha = 1$. Then $B_1(x') = h_{,2}(x', 0)$ and

$$B_1(x' + y'x_3) = \frac{\partial}{\partial y_2} h(x' + y'x_3, 0) \frac{1}{x_3}.$$

So,

$$A_{1,3}(x) = -\frac{1}{x_3} \int\limits_{\mathbb{R}^2} (h(x' + y'x_3, 0) - h(x', 0)) \frac{\partial}{\partial y_2} K_0(y') dy'.$$

Repeating the evaluation of $A_{\alpha,\beta 3}$, we find

$$\int\limits_{\mathbb{R}^3_+} A^2_{1,3} dx \leq c \|h(\cdot, 0)\|_{L^{\frac{1}{2}}_2(\mathbb{R}(2)}.$$

Since $|h(x)| \leq c(b)/|x|$ for $|x| \gg 1$, one can derive with the help of Lemma 1.2 the inequality

$$\|h(\cdot, 0)\|_{L^{\frac{1}{2}}_2(\mathbb{R}(2)} \leq c \|\nabla h\|_{2,\mathbb{R}^3_+}.$$

This means that, by Step 1, $A_{1,3} \in L_2(\mathbb{R}^3_+)$. The same holds true for $A_{2,3}$. The proof of the fact that $A_{\alpha,\beta} \in L_2(\mathbb{R}^3_+)$ is an exercise. So, it has been proven that

$$u \in L_2(\mathbb{R}^3_+) \tag{1.3.11}$$

provided $b \in C^\infty_0(\mathbb{R}^3_+)$.

Now, we wish to finish the proof of (1.3.10). Letting $u_R = \varphi_R u$, one can observe that

$$\int\limits_{\mathbb{R}^3_+} |\nabla(u - u_R)|^2 dx \leq c \int\limits_{\mathbb{R}^3_+ \setminus B_+(R)} |\nabla u|^2 dx + c\frac{1}{R^2} \int\limits_{B_+(2R)\setminus B_+(R)} |u|^2 dx \to 0$$

as $R \to \infty$.

The function u_R is not compactly supported in $\mathbb{R}^3_+$ and we need to cut it in the direction of x_3. To this end, let us introduce the following cut-off function: $\chi(t) = 0$ if $-\infty < t \le \varepsilon/2$, $\chi(t) = 2(t-\varepsilon/2)/\varepsilon$ if $\varepsilon/2 < t \le \varepsilon$, and $\chi(t) = 1$ if $t > \varepsilon$. Considering $u_{R,\varepsilon}(x) = u_R(x)\chi(x_3)$, we have

$$\int\limits_{\mathbb{R}^3_+} |\nabla(u_R - u_{R,\varepsilon})|^2 dx \le c \int\limits_0^\varepsilon dx_3 \int\limits_{\mathbb{R}^2} |\nabla u_R|^2 dx' + \frac{1}{\varepsilon^2} \int\limits_{\mathbb{R}^2} dx' \int\limits_0^\varepsilon |u_R|^2 dx_3.$$

The first integrals on the right-hand side of the last inequality tends to zero as $\varepsilon \to 0$. To show that the second term does the same, we are going to use two facts. Firstly, $u_R(x',0) = 0$ and secondly, by the Friedrichs inequality,

$$\int\limits_0^\varepsilon |u_R(x',x_3)|^2 dx_3 \le c\varepsilon^2 \int\limits_0^\varepsilon \Big|\frac{\partial}{\partial x_3} u_R(x',x_3)\Big|^2 dx_3.$$

So, combining the above inequalities, we show that

$$\int\limits_{\mathbb{R}^3_+} |\nabla(u_R - u_{R,\varepsilon})|^2 dx \le c \int\limits_0^\varepsilon \int\limits_{\mathbb{R}^2} |\nabla u_R|^2 dx \to 0$$

as $\varepsilon \to 0$ for each fixed $R > 0$.

It remains to mollify $u_{R,\varepsilon}$. The mollification $(u_{R,\varepsilon})_\tau$ belongs to $C^\infty_0(\mathbb{R}^3_+)$ for $0 < \tau \le \tau(R,\varepsilon)$ and

$$\int\limits_{\mathbb{R}^3_+} |\nabla(u_{R,\varepsilon} - (u_{R,\varepsilon})_\tau)|^2 dx \to 0$$

as $\tau \to 0$ for each fixed R and ε. So, (1.3.10) is proven.

Now, we are going to extend our result to functions $b \in L_2(\mathbb{R}^3_+)$. Given $b \in L_2(\mathbb{R}^3_+)$, there exists $b^{(m)} \in C^\infty_0(\mathbb{R}^3_+)$ such that $\|b^{(m)} - b\|_{2,\mathbb{R}^3_+} \to 0$ as $m \to \infty$. We know that there is $u^{(m)} \in \overset{\circ}{L}{}^1_2(\mathbb{R}^3_+)$ having the properties:

$$\operatorname{div} u^{(m)} = b^{(m)}$$

in $\mathbb{R}^3_+$ and

$$\|\nabla u^{(m)}\|_{2,\mathbb{R}^3_+} \le c\|b^{(m)}\|_{2,\mathbb{R}^3_+}.$$

Moreover, by construction

$$\|\nabla u^{(m)} - \nabla u^{(k)}\|_{2,\mathbb{R}^3_+} \le c\|b^{(m)} - b^{(k)}\|_{2,\mathbb{R}^3_+},$$

which implies that

$$u^{(m)} \to u$$

in $L^1_2(\mathbb{R}^3_+)$. □

Let us mention some consequences and generalizations.

Theorem 3.6. *Let $\Omega \subset \mathbb{R}^n$ be a bounded domain with Lipschitz boundary, $1 < p < \infty$, and let*

$$\bar{L}_p(\Omega) := \{b \in L_p(\Omega) : \int\limits_\Omega b(x)dx = 0\}.$$

Then, for any $b \in \bar{L}_p(\Omega)$, there exists $u \in \overset{\circ}{L}{}^1_2(\Omega)$ with the following properties:

$$\operatorname{div} u = b$$

in Ω and

$$\|\nabla u\|_{p,\Omega} \leq c(p, n, \Omega)\|b\|_{p,\Omega}.$$

Remark 1.2. For bounded domains, we need a restriction on b:

$$\int\limits_\Omega b(x)dx = 0,$$

which is called the compatibility condition.

Remark 1.3. Proof of Theorem 3.6 is based on Propositions 3.4 and 3.5, decomposition of the unity, and changes of coordinates. It is quite involved but does not contain new ideas.

Remark 1.4. There is a different approach to the proof of Theorem 3.6, which is due to Bogovskii. It is simpler than the above proof. But it relies upon the theory of singular integrals.

1.4 Nečas Imbedding Theorem

The main result of this section reads:

Theorem 4.7. *Let $1 < r < \infty$ and let Ω be a domain in $\mathbb{R}^n$. Assume that the gradient of a distribution p, defined in Ω, has the property:*

$$< \nabla p, w > \leq K\|\nabla w\|_{r',\Omega}$$

for any $w \in C_0^\infty(\Omega;\mathbb{R}^n)$.

The following statements are valid:

(i) $p \in L_{r,\mathrm{loc}}(\Omega)$ *and for any* $\Omega' \Subset \Omega$ *there exists a constant* $c(r,n,\Omega',\Omega)$ *such that*

$$\int\limits_{\Omega'} |p-a|^r dx \leq cK$$

for some constant a;

(ii) if $\Omega = \mathbb{R}^n$ *or* $\mathbb{R}^n_+$ *and* $p \in L_r(\Omega)$, *then there exists a constant* $c(r,n)$ *such that*

$$\int\limits_{\Omega} |p|^r dx \leq cK$$

(iii) if Ω *is a bounded Lipschitz domain, then* $p \in L_r(\Omega)$ *and there exists a constant* $c(r,n,\Omega)$ *such that*

$$\int\limits_{\Omega} |p-a|^r dx \leq cK$$

for some constant a.

PROOF

(i) Without loss of generality, we may assume that a bounded domain Ω' has Lipschitz boundary. We claim that there exists a constant $c(r,n,\Omega',\Omega)$ such that

$$|<p,q>| \leq cK\|q\|_{r',\Omega'} \tag{1.4.1}$$

for any $q \in C_0^\infty(\Omega')$ with $[q]_{\Omega'} = 0$. The latter would imply that p is a regular distribution. Here,

$$[q]_\omega = \frac{1}{|\omega|}\int\limits_\omega q(x)dx.$$

By Theorem 3.6, there exists $u \in \overset{\circ}{L}{}^1_{r'}(\Omega')$ such that

$$\operatorname{div} u = q$$

in Ω' and

$$\|\nabla u\|_{r',\Omega'} \leq c(r,n,\Omega')\|q\|_{r',\Omega'}.$$

Functions u and q are supposed to be extended by zero outside Ω'. Let us mollify u in a standard way

$$(u)_\varrho(x) = \int\limits_{\Omega'} \omega_\varrho(x-y)u(y)dy = \int\limits_{\Omega} \omega_\varrho(x-y)u(y)dy = \int\limits_{\mathbb{R}^n} \omega_\varrho(x-y)u(y)dy$$

with the help of a smooth mollifier ω_ϱ. So, $(u)_\varrho \in C_0^\infty(\Omega)$ for $0 < \varrho < \varrho_0(\Omega', \Omega)$. Moreover, we know that

$$\nabla(u)_\varrho = (\nabla u)_\varrho$$

and thus

$$\operatorname{div}(u)_\varrho = (q)_\varrho,$$

$$\|\nabla(u)_\varrho\|_{r',\Omega} \leq \|\nabla u\|_{r',\Omega} \leq c\|q\|_{r',\Omega},$$

by the known mollification properties.

Now, we have (in the sense of distributions)

$$< \nabla p, (u)_\varrho >= - < p, \operatorname{div}(u)_\varrho >= - < p, (q)_\varrho >,$$

which implies

$$| < p, (q)_\varrho > | \leq K\|\nabla(u)_\varrho\|_{r',\Omega} \leq cK\|q\|_{r',\Omega}.$$

It is worthy to notice that there exists a compact K_0 such that $\Omega' \subset K_0 \subset \Omega$, support of $\nabla^k(q)_\varrho$ and support of $\nabla^k q$ belong to K_0, and

$$\nabla^k(q)_\varrho \to \nabla^k q$$

uniformly in K_0 for any $k = 0, 1, ...$ as $\varrho \to 0$ and thus

$$< p, (q)_\varrho > \to < p, q >$$

as $\varrho \to 0$. Tending $\varrho \to 0$, we then find (1.4.1).

It follows from Banach and Riesz theorems that there exist $P \in L_r(\Omega')$ such that $\|P\|_{r,\Omega'} \leq cK$ and

$$< p, q >= \int\limits_{\Omega'} Pq dx$$

for any $q \in C_0^\infty(\Omega')$ with $[q]_{\Omega'} = 0$.

Now, let us test the latter identity with $q = \operatorname{div} u$ for an arbitrary $u \in C_0^\infty(\Omega')$ (it is supposed that all the functions are extended by zero to the whole domain Ω). As a result, we have

$$< \nabla p, u >= - < p, \operatorname{div} u >= - \int\limits_{\Omega'} P \operatorname{div} u dx.$$

This means that $\nabla(p|_{\Omega'} - P) = 0$. And thus, by Theorem 1.1, $p - P =$ constant on Ω'. So, part (i) is proven.
(ii) According to (i), our distribution v is regular and, therefore,

$$< p, q >= \int\limits_{\Omega} p(x)q(x)dx$$

for any $q \in C_0^\infty(\Omega)$.

Given $q \in C_0^\infty(\Omega)$, we find $u \in \overset{\circ}{L}{}^1_{r'}(\Omega)$ such that $\operatorname{div} u = q$ in Ω and

$$\|\nabla u\|_{r',\Omega} \leq c\|q\|_{r',\Omega}.$$

By the definition of $\overset{\circ}{L}{}^1_{r'}(\Omega)$, there exists a sequence $u^{(m)} \in C_0^\infty(\Omega)$ such that

$$\nabla u^{(m)} \to \nabla u \qquad \text{in } L_{r'}(\Omega)$$

and thus

$$q^{(m)} := \operatorname{div} u^{(m)} \to q \qquad \text{in } L_{r'}(\Omega)$$

as $m \to \infty$. Then, as it has been pointed out above, we should have

$$< \nabla p, u^{(m)} >= -\int\limits_{\Omega} p\operatorname{div} u^{(m)}dx = -\int\limits_{\Omega} pq^{(m)}dx \leq cK\|\nabla u^{(m)}\|_{r',\Omega}.$$

Passing to the limit, we find the estimate

$$-\int\limits_{\Omega} pqdx \leq cK\|q\|_{r',\Omega}, \tag{1.4.2}$$

which allows us to state that

$$\|p\|_{r,\Omega} \leq cK.$$

(iii) Here, it is enough to repeat the same arguments as in (i), replacing Ω' with Ω, under the additional restriction on q that is $[q]_\Omega = 0$. As a result, we get estimate (1.4.2) that holds for any $q \in C_0^\infty(\Omega)$ provided $[q]_\Omega = 0$. Repeating arguments, used at the end of the proof of the statement (i), we conclude that there exists $P \in L_r(\Omega)$ such that $\|P\|_{r,\Omega} \leq cK$ and $P = p - a$ for some constant a. □

We end up this section with recollecting known facts related to duality between function spaces. For a given Banach space V, let V' be its dual one, i.e., the space of all bounded linear functionals on V. Very often, we need to identify V' with a particular function space and such a choice as a rule depends on the problem under consideration. There is a relatively general construction that is very popular in the theory of evolution problems. To describe it, let us state the corresponding standing assumptions. We are given a reflexive Banach space V with the norm $\|\cdot\|_V$ and a Hilbert space with the scalar product $(\cdot,\cdot)$. It is supposed that V is continuously imbedded into H, i.e., there exists a constant c such that $\|v\|_H = \sqrt{(v,v)} \leq c\|v\|_V$ for any $v \in V$ and let V be dense in H.

As usual, we identify H' with H itself, i.e., $H' = H$ (in the known functional analysis sense). Now, let us fix $f \in H$, then $v \mapsto (f,v)$ is a bounded linear functional on V and thus there exists $v'_f \in V'$ with the properties:

$$< v'_f, v >= (f,v) \qquad \forall v \in V$$

and

$$\|v'_f\|_{V'} \leq c\|f\|_H \qquad \forall f \in H.$$

So, we have a bounded linear operator $\tau : H \to V'$ (one-to-one by density) defined by the identity $\tau f = v'_f$ for $f \in H$.

Obviously, $\tau(H)$ is a linear manifold of V'. Moreover, it is dense there. To see that, assume it is not, i.e., there exists $v'_0 \in V'$ but $v'_0 \notin [\tau(H)]^{V'}$. By the Hahn-Banach theorem, there exists $v'' \in V'' := (V')'$ with the properties $< v'', v'_0 >= 1$ and $< v'', v' >= 0$ for any $v' \in \tau(H)$. Since V is reflexive, there should be $v \in V$ so that $< v'', v' >=< v', v >$ for any $v' \in V'$. This gives us: $< v'_0, v >= 1$ and $< v'_f, v >= (f,v) = 0$ for any $f \in H$. Therefore, $v = 0$ and we get a contradiction. The latter allows us to identify V' with the closure of $\tau(H)$ in V'. But we can go further and identify duality relation between V and V' with the scalar product $(\cdot,\cdot)$ on H. Very often, we call such an identification of V' the space dual to V relative to the Hilbert space H.

So, under our standing assumptions, $v' \in V'$ means that there exists a sequence sequence $f_m \in H$ such that

$$\sup\{|(f_k - f_n, v)| : \ \|v\|_V = 1\} \to 0$$

as $k, n \to \infty$ and (v', v) is just notation for $\lim\limits_{k\to\infty} (f_k, v)$ that exists for all $v \in V$. Moreover,

$$\|v'\|_{V'} = \sup\{|(v', v)| : \ \|v\|_V = 1\}.$$

If a domain Ω is such that the space $\overset{\circ}{L}{}^1_{r'}(\Omega)$ is continuously imbedded into the space $L_2(\Omega)$, all standing assumptions with the particular choice of spaces $V = \overset{\circ}{L}{}^1_{r'}(\Omega)$ and $H = L_2(\Omega)$ hold. In this case, we shall denote by $L_r^{-1}(\Omega)$ an identification of the space $\left(\overset{\circ}{L}{}^1_{r'}(\Omega)\right)'$ according to the aforesaid scheme and the main assumption of Theorem 4.7 can be replaced with the following one:

$$\nabla p \in L_r^{-1}(\Omega).$$

1.5 Spaces of Solenoidal Vector Fields

First, let us introduce the set of all smooth divergence free vector fields compactly supported in Ω:

$$C^\infty_{0,0}(\Omega) := \{v \in C^\infty_0(\Omega) : \ \operatorname{div} v = 0 \text{ in}\,\Omega\}.$$

Next, for $1 \leq r < \infty$, we define the following "energy" spaces

$$\overset{\circ}{J}{}^1_r(\Omega) := [C^\infty_{0,0}(\Omega)]^{L^1_r(\Omega)}$$

and

$$\hat{J}^1_r(\Omega) := \{v \in \overset{\circ}{L}{}^1_r(\Omega) : \ \operatorname{div} v = 0 \text{ in}\,\Omega\}.$$

In general,

$$\hat{J}^1_r(\Omega) \supseteq \overset{\circ}{J}{}^1_r(\Omega).$$

For $r = 2$, we use abbreviations:

$$V(\Omega) := \overset{\circ}{J}{}^1_2(\Omega), \qquad \hat{V}(\Omega) := \hat{J}^1_2(\Omega).$$

Here, it is an example of a domain in $\mathbb{R}^3$

$$\Omega_\star = \mathbb{R}^3 \setminus \{x = (0, x_2, x_3), x_2^2 + x_3^2 \geq 1\},$$

for which

$$\hat{V}(\Omega_*) \setminus V(\Omega_*) \neq \emptyset.$$

This example is due to J. Heywood.

There is a wide class of domains for which the above spaces coincide. For example, we have

Theorem 5.8. *Let $1 < m < \infty$ and let Ω be $\mathbb{R}^n$, or $\mathbb{R}^n_+$, or a bounded domain with Lipschitz boundary. Then*

$$\overset{\circ}{J}{}^1_m(\Omega) = \hat{J}^1_m(\Omega).$$

PROOF See next section.

1.6 Linear Functionals Vanishing on Divergence Free Vector Fields

Proposition 6.9. *Let $\Omega = \mathbb{R}^n$ or $\mathbb{R}^n_+$ or be a bounded Lipschitz domain. Assume that $1 < s < \infty$. Let, further, $l : \overset{\circ}{L}{}^1_s(\Omega) \to \mathbb{R}$ be a linear functional having the following properties:*

$$|l(v)| \leq c\|\nabla v\|_{s,\Omega}$$

for any $v \in \overset{\circ}{L}{}^1_s(\Omega)$ and

$$l(v) = 0$$

for any $v \in \hat{J}^1_s(\Omega)$.

Then there exists a function $p \in L_{s'}(\Omega)$, $s' = s/(s-1)$, such that

$$l(v) = \int\limits_\Omega p \operatorname{div} v dx$$

for any $v \in \overset{\circ}{L}{}^1_s(\Omega)$.

PROOF Let us consider case $\Omega = \mathbb{R}^n$ or $\mathbb{R}^n_+$.

We define a linear functional $G : L_s(\Omega) \to \mathbb{R}$ as follows. Given $q \in L_s(\Omega)$, take any $u \in \overset{\circ}{L}{}^1_s(\Omega)$ such that $\operatorname{div} u = q$ and let $G(q) = l(u)$. By Proposition 3.5, there is at least one function u with this property. Next, one should show that functional G is well-defined, i.e., for any $v \in \overset{\circ}{L}{}^1_s(\Omega)$ with $\operatorname{div} v = q$, we have $l(u) = l(v)$. Indeed, $u - v \in \hat{J}^1_s(\Omega)$ and by our assumptions $l(u - v) = 0 = l(u) - l(v)$.

It is not a difficult exercise to verify that G is a linear functional.

Now, we can select a special vector-valued function $u \in \overset{\circ}{L}{}^1_s(\Omega)$, for which we have the identity $\operatorname{div} u = q$ and the estimate

$$\|\nabla u\|_{s,\Omega} \leq c\|q\|_{s,\Omega}.$$

The latter implies

$$G(q) = l(u) \leq c\|\nabla u\|_{s,\Omega} \leq c\|q\|_{s,\Omega}$$

for any $q \in L_s(\Omega)$. So, the functional G is bounded on $L_s(\Omega)$ and by Riesz theorem, there exists $p \in L_{s'}(\Omega)$ such that

$$G(q) = \int\limits_\Omega pq dx$$

for any $q \in L_s(\Omega)$. Now, for each $u \in \overset{\circ}{L}{}^1_s(\Omega)$, we have the identity

$$l(u) = G(\operatorname{div} u) = \int\limits_\Omega p \operatorname{div} u dx.$$

For bounded Lipschitz domains, one should replace the space $L_s(\Omega)$ with its subspace

$$\bar{L}_s(\Omega) = \{q \in L_s(\Omega) : \ [q]_\Omega = 0\}$$

and use the same arguments as above. □

However, we can assume that our functional vanish on $\overset{\circ}{J}{}^1_s(\Omega)$ only.

Theorem 6.10. *Let $\Omega = \mathbb{R}^n$ or $\mathbb{R}^n_+$ or be a bounded Lipschitz domain. Assume that $1 < s < \infty$. Let, further, $l : \overset{\circ}{L}{}^1_s(\Omega) \to \mathbb{R}$ be a linear functional having the following properties:*

$$|l(v)| \leq c\|\nabla v\|_{s,\Omega}$$

for any $v \in \overset{\circ}{L}{}^1_s(\Omega)$ and

$$l(v) = 0$$

for any $v \in \overset{\circ}{J}{}^1_s(\Omega)$.

Then there exists a function $p \in L_{s'}(\Omega)$, $s' = s/(s-1)$, such that

$$l(v) = \int\limits_\Omega p \operatorname{div} v dx$$

for any $v \in \overset{\circ}{L}{}^1_s(\Omega)$.

Proof We start with bounded domains. Let us consider a sequence of bounded smooth domains Ω_m, $m = 1, 2, ...$, with the following properties:

$$\Omega_m \subset \Omega_{m+1}$$

and

$$\Omega = \bigcup_{m=1}^{\infty} \Omega_m.$$

Given $v \in \overset{\circ}{L}{}^1_s(\Omega_m)$, define $v^m = v$ in Ω_m and $v^m = 0$ outside Ω_m. Obviously, $v^m \in \overset{\circ}{L}{}^1_s(\Omega)$. We also define a linear functional $l_m : \overset{\circ}{L}{}^1_s(\Omega_m) \to \mathbb{R}$ as follows:

$$l_m(v) := l(v^m)$$

for any $v \in \overset{\circ}{L}{}^1_s(\Omega_m)$. It is bounded and, moreover,

$$|l_m(v)| \leq c\|\nabla v\|_{s,\Omega_m}$$

with a constant c independent of m.

Using standard properties of mollification, one can show the following fact: if $v \in \hat{J}^1_s(\Omega_m)$, then $v^m \in \overset{\circ}{J}{}^1_s(\Omega)$. This immediately implies that $l_m(v) = 0$ for any $v \in \hat{J}^1_s(\Omega_m)$. According to Proposition 6.9, there exists $p_m \in L_{s'}(\Omega_m)$ such that

$$l_m(v) = \int\limits_{\Omega_m} p_m \mathrm{div}\, v dx$$

for any $v \in \overset{\circ}{L}{}^1_s(\Omega_m)$. Obviously, p_m is defined up to an arbitrary constant. Moreover, $p_{m+1} - p_m = c(m) = constant$ in Ω_m. So, we can change p_{m+1} adding a constant to achieve the identity $p_{m+1} = p_m$ in Ω_m that makes it possible to introduce a function $p \in L_{s',\mathrm{loc}}(\Omega)$ so that $p = p_m$ on Ω_m. By construction, it satisfies identity

$$l(v) = \int\limits_{\Omega} p \mathrm{div}\, v dx, \tag{1.6.1}$$

and the inequality

$$l(v) \leq c\|\nabla v\|_{s,\Omega}$$

for every $v \in C^\infty_0(\Omega)$ and, as it follows from Theorem 4.7, $p \in L_{s'}(\Omega)$. If so, identity (1.6.1) can be extended to all functions $v \in \overset{\circ}{L}{}^1_s(\Omega)$ by density arguments.

Let us consider the case $\Omega = \mathbb{R}^n_+$. The case $\Omega = \mathbb{R}^3$ can be treated in the same and even easier. Our arguments are similar to previous ones. Let $\Omega_m = \Omega + e_n \frac{1}{m}$. To show that $v \in \hat{J}^1_s(\Omega_m)$ implies $v^m \in \overset{\circ}{J}{}^1_s(\Omega)$, we find a sequence $v^{(k)} \in C^\infty_0(\Omega_m)$ such that $\|\nabla v^{(k)} - \nabla v\|_{s,\Omega_m} \to 0$ as $k \to \infty$. Let $\mathrm{supp} v^{(k)} \subset B_+(e_n \frac{1}{m}, R_k) = B_+(R_k) + e_n \frac{1}{m}$ for some $R_k > 0$. By scaling arguments, we can find a function $w^{(k)} \in \overset{\circ}{L}{}^1_s(B_+(e_n \frac{1}{m}, R_k))$ with the following properties:

$$\mathrm{div} w^{(k)} = \mathrm{div} v^{(k)}$$

in $B_+(e_n \frac{1}{m}, R_k)$ and

$$\|\nabla w^{(k)}\|_{s,B_+(e_n \frac{1}{m}, R_k)} \leq c\|\mathrm{div} v^{(k)}\|_{s,B_+(e_n \frac{1}{m}, R_k)}.$$

One should emphasize that a constant in the above inequality is independent of k.

We let further $u^{(k)} = v^{(k)} - w^{(k)}$ so that $u^{(k)} \in \overset{\circ}{L}{}^1_s(B_+(e_n\frac{1}{m}, R_k))$, $\mathrm{div} u^{(k)m} = 0$ in Ω, and

$$\|\nabla u^{(k)m} - \nabla v^m\|_{2,\Omega} \leq \|\nabla v^{(k)m} - \nabla v^m\|_{2,\Omega} + \|\nabla w^{(k)m}\|_{2,\Omega} \to 0$$

as $k \to \infty$. Obviously,

$$(u^{(k)m})_\varrho \in C^\infty_{0,0}(\Omega)$$

for fixed k and sufficiently small $\varrho > 0$ and

$$\nabla (u^{(k)m})_\varrho = (\nabla u^{(k)m})_\varrho \to \nabla u^{(k)m}$$

in $L_s(\Omega)$ as $\varrho \to 0$. So, the required implication has been proven.

By Proposition 6.9, there exists $p_m \in L_{s'}(\Omega_m)$. Obviously, $p_{m+1} = p_m$ in Ω_m. So, we may define a function p so that $p = p_m$ in Ω_m. Next, we have

$$l_m(v) = \int\limits_{\Omega_m} p_m \mathrm{div} v dx \leq C\|\nabla v\|_{s,\Omega_m}$$

for any $v \in C^\infty_0(\Omega_m)$ and thus

$$< \nabla p_m, v > \leq C\|\nabla v\|_{s,\Omega_m}$$

for any $v \in C^\infty_0(\Omega_m)$. From Theorem 4.7 it follows that

$$\|p\|_{s',\Omega_m} \leq cC$$

for any natural number m. This certainly implies that $p \in L_{s'}(\Omega)$. □

PROOF OF THEOREM 5.8 Indeed, assume that there exists $v_* \in \hat{J}^1_s(\Omega)$ but $v_* \notin \overset{\circ}{J}{}^1_s(\Omega)$. By Banach theorem, there exists a functional

$$l_* \in \left(\overset{\circ}{L}{}^1_s(\Omega)\right)'$$

with the following properties:

$$l_*(v_*) = 1$$

and

$$l_*(v) = 0$$

for any $v \in \overset{\circ}{J}{}^1_s(\Omega)$. By Theorem 6.10, there exists $p \in L_{s'}(\Omega)$ such that

$$l_*(v) = \int\limits_\Omega \mathrm{pdiv}\, v dx$$

for all $v \in \overset{\circ}{L}{}^1_s(\Omega)$. However,

$$l_*(v_*) = \int\limits_\Omega \mathrm{pdiv}\, v_* dx = 0$$

since $v_* \in \hat{J}^1_s(\Omega)$. This is a contradiction. □

1.7 Helmholtz-Weyl Decomposition

Let

$$\overset{\circ}{J}(\Omega) := [C^{\infty}_{0,0}(\Omega)]^{L_2(\Omega)}$$

and

$$G(\Omega) := \{v \in L_2(\Omega; \mathbb{R}^n) : \ v = \nabla p \text{ for some distribution } p\}.$$

Remark 1.5. We know that if a distribution $p \in G(\Omega)$, then in general $p \in L_{2,\mathrm{loc}}(\Omega)$. However, if Ω is a bounded Lipschitz domain, then in fact $p \in L_2(\Omega)$.

Theorem 7.11. *(Ladyzhenskaya) For any domain $\Omega \in \mathbb{R}^n$,*

$$L_2(\Omega) := \overset{\circ}{J}(\Omega) \oplus G(\Omega).$$

PROOF Obviously, our statement is equivalent to the following identity

$$G(\Omega) = (\overset{\circ}{J}(\Omega))^{\perp}.$$

STEP 1 Let Ω be a bounded Lipschitz domain. It is easy to see that

$$G(\Omega) \subseteq (\overset{\circ}{J}(\Omega))^{\perp},$$

since

$$\int\limits_{\Omega} v \cdot \nabla p dx = 0$$

for any $p \in G(\Omega)$ and for any $v \in C^{\infty}_{0,0}(\Omega)$. Now, assume

$$u \in (\overset{\circ}{J}(\Omega))^{\perp},$$

i.e., $u \in L_2(\Omega)$ and

$$\int\limits_{\Omega} u \cdot v dx = 0$$

for any $v \in C^{\infty}_{0,0}(\Omega)$. By Poincaré inequality,

$$l(v) = \int\limits_{\Omega} u \cdot v dx \leq \Big(\int\limits_{\Omega} |u|^2\Big)^{\frac{1}{2}} \Big(\int\limits_{\Omega} |v|^2\Big)^{\frac{1}{2}} \leq$$

$$\leq c(\Omega)\|u\|_{2,\Omega}\|\nabla v\|_{2,\Omega}$$

for any $v \in \overset{\circ}{L}{}^1_2(\Omega)$. So, $l : \overset{\circ}{L}{}^1_2(\Omega) = \overset{\circ}{W}{}^1_2(\Omega) \to \mathbb{R}$ is bounded and $l(v) = 0$ for any $v \in \overset{\circ}{J}{}^1_2(\Omega) =: V(\Omega)$.

By Theorem 6.10, there exists $p \in L_2(\Omega)$ such that

$$l(v) = \int_\Omega p \operatorname{div} v dx$$

for any $v \in \overset{\circ}{L}{}^1_2(\Omega)$. Therefore, $u = \nabla p$ and thus $p \in G(\Omega)$ and

$$(\overset{\circ}{J}(\Omega))^\perp \subseteq G(\Omega).$$

STEP 2 We proceed in a similar way as in the proof of Theorem 6.10. Consider a sequence of domains Ω_j with the properties: $\Omega_j \subset \Omega_{j+1}$ and

$$\Omega = \bigcup_{j=1}^{\infty} \Omega_j,$$

where Ω_j is a bounded Lipschitz domain.

Since $v \in L_2(\Omega) \Rightarrow v \in L_2(\Omega_j)$, we can state that, for any j,

$$v = u^{(j)} + \nabla p^{(j)},$$

where

$$u^{(j)} \in \overset{\circ}{J}(\Omega_j), \qquad p^{(j)} \in W^1_2(\Omega_j).$$

We know that $p^{(j)}$ is defined up to a constant, which can be fixed by the condition

$$\int_{B_*} p^{(j)} dx = 0,$$

where B_* is a fixed ball belonging to Ω_1.

Here, we are going to make use of the following version of Poincaré's inequality

$$\int_{\tilde{\Omega}} |q|^2 dx \leq c(n, \tilde{\Omega}, B_*) \Big[\int_{\tilde{\Omega}} |\nabla q|^2 dx + \Big| \int_{B_*} q dx \Big|^2 \Big] \qquad (1.7.1)$$

that holds in a bounded Lipschitz domain $\tilde{\Omega}$ containing the ball B_*. A proof of (1.7.1) is based on standard compactness arguments and can be regarded as a good exercise.

We further let $\tilde{u}^{(j)} = u^{(j)}$ in Ω_j and $\tilde{u}^{(j)} = 0$ outside Ω_j. It is easy to check that $\tilde{u}^{(j)} \in \overset{\circ}{J}(\Omega)$ and

$$\|\tilde{u}^{(j)}\|_{2,\Omega} = \|u^{(j)}\|_{2,\Omega_j} \leq \|v\|_{2,\Omega_j} \leq \|v\|_{2,\Omega}$$

and, hence, without loss of generality, we may assume that

$$\tilde{u}^{(j)} \rightharpoonup u$$

in $L_2(\Omega)$ and

$$u \in \overset{\circ}{J}(\Omega).$$

Next, by (1.7.1), we have for $j \geq s$

$$\int\limits_{\Omega_s} |p^{(j)}|^2 dx \leq c(s) \int\limits_{\Omega_s} |\nabla p^{(j)}|^2 dx \leq c(s) \|v\|^2_{2,\Omega}.$$

Letting $s = 1$, we find a subsequence $\{j^1_k\}_{k=1}^\infty$ so that

$$p^{(j^1_k)} \rightharpoonup p_1, \qquad \nabla p^{(j^1_k)} \rightharpoonup \nabla p_1,$$

in $L_2(\Omega_1)$. Then we let $s = 2$ and select a subsequence $\{j^2_k\}_{k=1}^\infty$ of $\{j^1_k\}_{k=1}^\infty$ such that

$$p^{(j^2_k)} \rightharpoonup p_2, \qquad \nabla p^{(j^2_k)} \rightharpoonup \nabla p_2$$

in $L_2(\Omega_2)$. Obviously, $p_2 = p_1$ in Ω_1. Proceeding in the same way, we find a subsequence $\{j^l_k\}_{k=1}^\infty$ of $\{j^{l-1}_k\}_{k=1}^\infty$ such that

$$p^{(j^l_k)} \rightharpoonup p_l, \qquad \nabla p^{(j^l_k)} \rightharpoonup \nabla p_l$$

in $L_2(\Omega_l)$. For the same reason, $p_l = p_{l-1}$ in Ω_{l-1}. Hence, the function p, defined

$$p = p_l$$

in Ω_l, is well-defined. Using the celebrated diagonal Cantor process, we find a subsequence $p^{(j_s)}$ such that

$$p^{(j_s)} \rightharpoonup p, \qquad \nabla p^{(j_s)} \rightharpoonup \nabla p$$

in $L_2(\omega)$ for each $\omega \Subset \Omega$. Moreover, we have the estimate

$$\int\limits_{\omega} |\nabla p|^2 dx \leq \int\limits_{\Omega} |v|^2 dx$$

for any $\omega \Subset \Omega$. So, it is easy to deduce from here that $p \in G(\Omega)$.

Now, fix $w \in C^\infty_0(\Omega)$. We have

$$\int\limits_{\Omega_{j_s}} v \cdot w dx = \int\limits_{\Omega_{j_s}} \nabla p^{(j_s)} \cdot w dx + \int\limits_{\Omega_{j_s}} u^{(j_s)} \cdot w dx.$$

For sufficiently large s_0, $\operatorname{supp} w \subset \Omega_{j_{s_0}}$ and thus for $s > s_0$

$$\int\limits_{\Omega_{j_{s_0}}} v \cdot w dx = \int\limits_{\Omega_{j_{s_0}}} \nabla p^{(j_s)} \cdot w dx + \int\limits_{\Omega_{j_{s_0}}} u^{(j_s)} \cdot w dx.$$

Passing $s \to \infty$, we show that $v = u + \nabla p$. Orthogonality and uniqueness can be proven in a standard way (exercise). □

1.8 Comments

The main goal for writing up Chapter 1 is to show author's preferences how the theory of function spaces related to the Navier-Stokes equations can be developed. In our approach, the basic things are estimates of certain solutions to the equation $\operatorname{div} u = f$ and their applications to the derivation of the Nečas embedding theorem. Each part of this theory can be given in either more compact way or even in a different way. For example, in Section 3, one could apply very nice Bogovskii's approach, see [Bogovskii (1980)], based on the theory of singular integrals. For more generic and detailed investigation of spaces arising in the Navier-Stokes theory, we refer the reader to monographs [Ladyzhenskaya (1970)], [Temam (2010)], and [Galdi (2000)].

Chapter 2

Linear Stationary Problem

2.1 Existence and Uniqueness of Weak Solutions

Let us consider the Dirichlet problem for the Stokes system

$$\begin{cases} -\Delta u + \nabla p = f \\ \\ \operatorname{div} u = 0 \end{cases} \qquad \text{in } \Omega, \tag{2.1.1}$$

$$u|_{\partial\Omega} = 0 \tag{2.1.2}$$

and if $n = 3$ and Ω is unbounded then $u(x) \to u_0$ as $|x| \to \infty$.

In what follows, we always consider the simplest case

$$u_0 = 0.$$

Let

$$(f, g) := \int\limits_{\Omega} f(x)g(x)dx.$$

If u and p are smooth, then, for any $v \in C^\infty_{0,0}(\Omega)$, integration by parts gives the following identity:

$$\int\limits_{\Omega}(-\Delta u + \nabla p)\cdot v dx = \int\limits_{\Omega} \nabla u : \nabla v dx = (\nabla u, \nabla v) = (f, v),$$

which shows how weak solutions can be defined.

Let us list our standing assumptions: $n = 2$ or 3 and

$$f \in (L^1_2(\Omega))'.$$

For example, the above condition holds if $f = \operatorname{div} F$ with $F \in L_2(\Omega; \mathbb{M}^{n\times n})$, where $\mathbb{M}^{n\times n}$ is the space of real-valued $n \times n$-matrices.

Definition 2.1. A function $u \in \hat{V}(\Omega)$ is called a weak solution to boundary value problem (2.1.1) and (2.1.2) if and only if

$$(\nabla u, \nabla v) = < f, v >$$

for any $v \in C^\infty_{0,0}(\Omega)$.

Remark 2.1. If the domain Ω is such that Poincaré inequality holds in it:

$$\|w\|_{2,\Omega} \leq c\|\nabla w\|_{2,\Omega}$$

for any $w \in C^\infty_0(\Omega)$, then the space $(L^1_2(\Omega))'$ can be identified with the space $L^{-1}_2(\Omega)$ and $< f, v >= (f, v)$, see Section 4 of Chapter 1 for details.

Remark 2.2. Boundary conditions are understood in the sense of traces, see the definition of spaces $\overset{\circ}{L}{}^1_2(\Omega)$ and $\hat{V}(\Omega)$ in Sections 1 and 5 of Chapter 1. If Ω is unbounded and $n = 3$, then condition $u(x) \to 0$ as $|x| \to \infty$ holds in the following sense:

$$\Big(\int\limits_\Omega |u|^6 dx\Big)^{\frac{1}{6}} \leq c\Big(\int\limits_\Omega |\nabla u|^2 dx\Big)^{\frac{1}{2}}.$$

Lemma 2.1. *(Existence). Assume that the domain Ω is such that if $v \in \overset{\circ}{L}{}^1_2(\Omega)$ and $\|\nabla v\|_{2,\Omega} = 0$, then $v = 0$. Given f, there exists at least one weak solution to boundary value problem (2.1.1) and (2.1.2) that satisfies the estimate*

$$\|\nabla u\|_{2,\Omega} \leq \|f\|_{(L^1_2(\Omega))'}.$$

PROOF Indeed, $[u, v] = (\nabla u, \nabla v)$ is a scalar product in $\hat{V}(\Omega)$. On the other hand, $l(v) =< f, v >$ defines the linear functional on $C^\infty_{0,0}(\Omega)$ that is bounded on $V(\Omega)$:

$$|l(v)| \leq \|f\|_{(L^1_2(\Omega))'}\|v\|_{\overset{\circ}{L}{}^1_2(\Omega)}.$$

Now, the required existence is an easy consequence of the Banach extension theorem and Riesz representation theorem. □

If the assumption of the lemma does not hold, one should work with equivalence classes.

Lemma 2.2. *(Uniqueness). Assume that Ω either $\mathbb{R}^3$ or $\mathbb{R}^n_+$ or bounded domain in $\mathbb{R}^n$ with Lipschitz boundary. Then problem (2.1.1) and (2.1.2) has a unique weak solution.*

PROOF As we know from Chapter 1, in this case,

$$\hat{V}(\Omega) = V(\Omega). \tag{2.1.3}$$

Assume that there are two different solutions u^1 and u^2. Then

$$(\nabla(u^1 - u^2), \nabla v) = 0$$

for every $v \in C^\infty_{0,0}(\Omega)$ and, by (2.1.3),

$$\|\nabla(u^1 - u^2)\|_{2,\Omega} = 0.$$

This immediately implies $u^1 = u^2$. □

If $n = 2$ and $\Omega = \mathbb{R}^2$, the uniqueness takes place in the equivalence classes, i.e., $u^1 - u^2 \in [0]$. The equivalence class $[0]$ consists of functions that are constant in $\mathbb{R}^2$.

To recover the pressure, let us assume that Ω satisfies conditions of Theorem 6.10 of Chapter 1 and consider the following linear functional

$$l(v) = (\nabla u, \nabla v) - < f, v > .$$

It is bounded in $\overset{\circ}{L}{}^1_2(\Omega)$ and vanishes in $V(\Omega)$. By Theorem 6.10 of Chapter 1, there exists a function $p \in L_2(\Omega)$ such that

$$(\nabla u, \nabla v) - < f, v > = (p, \operatorname{div} v)$$

for any $v \in \overset{\circ}{L}{}^1_2(\Omega)$. In other words, functions u and p satisfy the Stokes system in the sense of distributions. □

2.2 Coercive Estimates

Proposition 2.1. *Let Ω be a domain with smooth boundary ($\Omega = \mathbb{R}^n$ or $\mathbb{R}^n_+$ or bounded domain). Let functions*

$$f \in L_2(\Omega), \quad g \in W^1_2(\Omega), \quad u \in V(\Omega), \quad p \in L_2(\Omega),$$

with $[g]_\Omega = 0$ if Ω is bounded, satisfy the nonhomogeneous Stokes system

$$\begin{cases} -\Delta u + \nabla p = f & \\ & \text{in } \Omega, \\ \operatorname{div} u = g & \end{cases}$$

in the sense of distributions.

Then $\nabla^2 u, \nabla p \in L_2(\Omega)$ and the coercive estimate

$$\|\nabla^2 u\|_{2,\Omega} + \|\nabla p\|_{2,\Omega} \le c(n, \Omega)(\|\nabla g\|_{2,\Omega} + \|f\|_{2,\Omega})$$

holds.

PROOF To demonstrate the essence of the matter, we restrict ourselves to the case $\Omega = \mathbb{R}^3_+$.

STEP 1 Here, we are going to estimate tangential derivatives of u, i.e., derivatives with respect to x_α, $\alpha = 1,2$. Let $h = (h_1, h_2, 0)$ be a vector in $\mathbb{R}^3$ and $\Delta_h f(x) := f(x+h) - f(x)$. We have

$$\begin{cases} -\Delta \Delta_h u + \nabla \Delta_h p = \Delta_h f \\ \\ \operatorname{div} \Delta_h u = \Delta_h g \end{cases}$$

with $\Delta_h u \in V(\mathbb{R}^3_+)$. According to Proposition 3.5 of Chapter 1, there exists $w_h \in V(\mathbb{R}^3_+)$ such that $\operatorname{div} w_h = \Delta_h g$ and

$$\|\nabla w_h\|_{2,\mathbb{R}^3_+} \leq c\|\Delta_h g\|_{2,\mathbb{R}^3_+}$$

with a constant c independent of h. Then the previous system can be transformed to the following form:

$$\begin{cases} -\Delta(\Delta_h u - w_h) + \nabla \Delta_h p = \Delta_h f + \Delta w_h \\ \\ \operatorname{div}(\Delta_h u - w_h) = 0 \end{cases} \quad \text{in } \mathbb{R}^3_+.$$

Let us denote $\Delta_h u - w_h$ by v. We know that $v \in V(\mathbb{R}^3_+)$ and, therefore, there exists a sequence $v^k \in C^\infty_0(\mathbb{R}^3_+)$ such that $\|\nabla(v^k - v)\|_{2,\mathbb{R}^3_+} \to 0$ as $k \to \infty$.

Testing the above system with v^k, we find

$$I_k = (\nabla v, \nabla v^k) = (\Delta_h f, v^k) - (\nabla w_h, \nabla v^k).$$

Our aim is to estimate the first term of the right-hand side in the above identity. Indeed, we have

$$(\Delta_h f, v^k) = -\int\limits_{\mathbb{R}^3_+} f(x+h) \cdot \Delta_h v(x) dx \leq \|f\|_{2,\mathbb{R}^3_+} \Big(\int\limits_{\mathbb{R}^3_+} |\Delta_h v|^2 dx \Big)^{\frac{1}{2}}.$$

Since $h \cdot e_3 = 0$, it is not difficult to show that

$$(\Delta_h f, v^k) \leq |h| \|f\|_{2,\mathbb{R}^3_+} \|\nabla v^k\|_{2,\mathbb{R}^3_+}.$$

Passing to the limit as $k \to \infty$, we find

$$\|\nabla v\|_{2,\mathbb{R}^3_+} \leq |h| \|f\|_{2,\mathbb{R}^3_+} + \|\nabla w_h\|_{2,\mathbb{R}^3_+}.$$

So, we have

$$\frac{1}{|h|}\|\nabla(\Delta_h u)\|_{2,\mathbb{R}^3_+} \leq c\Big[\|f\|_{2,\mathbb{R}^3_+} + \frac{1}{|h|}\|\Delta_h g\|_{2,\mathbb{R}^3_+}\Big].$$

Tending h to zero, we find the bound for tangential derivatives of v

$$\|\nabla u_{,\alpha}\|_{2,\mathbb{R}^3_+} \le cI,$$

where $\alpha = 1, 2$ and

$$I = \|f\|_{2,\mathbb{R}^3_+} + \|\nabla g\|_{2,\mathbb{R}^3_+}.$$

To estimate tangential derivatives of the pressure, we come back to the system at the beginning of Step 1. The first equation there can be re-written as follows:

$$\nabla \triangle_h p = \triangle \triangle_h u + \triangle_h f$$

with $h = (h_1, h_2, 0)$. We know that there exists a function $w_h \in \overset{\circ}{L}{}^1_2(\mathbb{R}^3_+)$ such that $\mathrm{div} w_h = \triangle_h p$ in $\mathbb{R}^3_+$ and

$$\|\nabla w_h\|_{2,\mathbb{R}^3_+} \le c\|\triangle_h p\|_{2,\mathbb{R}^3_+}.$$

We also can find a sequence $w_h^k \in C_0^\infty(\mathbb{R}^3_+)$ such that $\nabla w_h^k \to \nabla w_h$ in $L_2(\mathbb{R}^3_+)$ and

$$\int\limits_{\mathbb{R}^3_+} \triangle_h p \mathrm{div} w_h^k dx = \int\limits_{\mathbb{R}^3_+} \nabla(\triangle_h u) : \nabla w_h^k dx + \int\limits_{\mathbb{R}^3_+} \triangle_h f \cdot w_h^k dx.$$

The last term on the right-hand side can be treated as above and, as a result, we have

$$(\triangle_h f, w_h^k) \le h\|f\|_{2,\mathbb{R}^3_+}\|\nabla w_h^k\|_{2,\mathbb{R}^3_+}.$$

Applying Hölder inequality, the above estimate, and passing to the limit as $k \to \infty$, we show that

$$\|\triangle_h p\|_{2,\mathbb{R}^3_+} \le c(\|\nabla \triangle_h u\|_{2,\mathbb{R}^3_+} + h\|f\|_{2,\mathbb{R}^3_+})$$

and, hence,

$$\|p_{,\alpha}\|_{2,\mathbb{R}^3_+} \le cI.$$

STEP 2 Let us start with evaluation of terms $u_{3,33}$ and $p_{,3}$. $u_{3,33}$ can be estimated simply with the help of the equation $\mathrm{div}\, u = u_{\alpha,\alpha} + u_{3,3} = g$. This gives us $u_{3,33} = g_{,3} - u_{\alpha,3\alpha}$ and thus

$$\|u_{3,33}\|_{2,\mathbb{R}^3_+} \le cI.$$

As to the second term, the above estimate, the equation $p_{,3} = f_3 + \triangle u_3$, and bounds for tangential derivatives lead to the inequality

$$\|p_{,3}\|_{2,\mathbb{R}^3_+} \le cI.$$

So, the full gradient of the pressure p obeys the estimate

$$\|\nabla p\|_{2,\mathbb{R}^3_+} \leq cI.$$

The remaining part of the second derivatives can be estimated with the help of the equation $u_{\alpha,33} = -u_{\alpha,\beta\beta} + p_{,\alpha} - f_\alpha$ and previous bounds. So, we have

$$\|\nabla^2 u\|_{2,\mathbb{R}^3_+} \leq cI$$

and this completes the proof. □

Remark 2.3. The Stokes system holds a.e. in Ω provided assumptions of Proposition 2.1 are satisfied.

In fact, we have more general statement, which is called Cattabriga-Solonnikov estimates.

Theorem 2.2. *Assume that all assumptions of Proposition 2.1 are fulfilled. Let Ω be a bounded domain with sufficiently smooth boundary. In addition, assume that*

$$f \in W^k_r(\Omega), \qquad g \in W^{k+1}_r(\Omega)$$

with $[g]_\Omega = 0$ and with integer k. Then

$$\|\nabla^2 u\|_{W^k_r(\Omega)} + \|\nabla p\|_{W^k_r(\Omega)} \leq c(n,r,k,\Omega)\Big[\|f\|_{W^k_r(\Omega)} + \|\nabla g\|_{W^k_r(\Omega)}\Big].$$

2.3 Local Regularity

Proposition 3.3. *Assume that we are given functions*

$$v \in W^1_2(B_+), \qquad q \in L_2(B_+), \qquad f \in L_2(B_+), \qquad g \in W^1_2(B_+),$$

satisfying the Stokes system

$$\begin{cases} -\Delta v + \nabla q = f \\ \qquad\qquad\qquad\qquad \text{in } B_+ \\ \operatorname{div} v = g \end{cases}$$

and the boundary condition

$$v|_{x_3=0} = 0.$$

Then, for any $\tau \in]0,1[$,

$$\nabla^2 v, \nabla q \in L_2(B_+(\tau))$$

and the following estimate is valid:

$$\|\nabla^2 v\|_{2,B_+(\tau)} + \|\nabla q\|_{2,B_+(\tau)} \leq c(\tau)\Big[\|f\|_{2,B_+} + \|q\|_{2,B_+} + \\ + \|g\|_{W^1_2(B_+)} + \|v\|_{W^1_2(B_+)}\Big].$$

PROOF Let a cut-off function $\varphi \in C_0^\infty(\mathbb{R}^3)$ possess the properties: $0 \leq \varphi \leq 1$, $\varphi \equiv 1$ in $B(\tau)$, and $\varphi \equiv 0$ outside $B(1)$. Introducing new functions $u = \varphi v$ and $p = \varphi q$, we can verify that they satisfy the following system:

$$\left\{\begin{array}{l} -\Delta u + \nabla p = \tilde{f} = \varphi f - 2\nabla v \nabla \varphi - v\Delta\varphi + q\nabla\varphi \in L_2(\mathbb{R}^3_+) \\ \qquad \operatorname{div} v = \tilde{g} = \varphi g + v \cdot \nabla\varphi \in W^1_2(\mathbb{R}^3_+) \end{array}\right. \quad \text{in } \mathbb{R}^3_+.$$

By assumptions, $u \in \overset{\circ}{L}{}^1_2(\mathbb{R}^3_+)$ and $p \in L_2(\mathbb{R}^3_+)$ and thus we are in a position to apply Proposition 2.1, which reads that $\nabla^2 u$, $\nabla p \in L_2(\mathbb{R}^3_+)$ and

$$\|\nabla^2 u\|_{2,\mathbb{R}^3_+} + \|\nabla p\|_{2,\mathbb{R}^3_+} \leq c\Big[\|\nabla\tilde{g}\|_{2,\mathbb{R}^3_+} + \|\tilde{f}\|_{2,\mathbb{R}^3_+}\Big].$$

Then all the statements of Proposition 3.3 follow. □

The statement below can be proven in the same way as Proposition 3.3.

Proposition 3.4. *Assume that we are given functions*

$$v \in W^1_2(B), \qquad q \in L_2(B), \qquad f \in L_2(B), \qquad g \in W^1_2(B),$$

satisfying

$$\left\{\begin{array}{l} -\Delta v + \nabla q = f \\ \quad \operatorname{div} v = g \end{array}\right. \quad \text{in } B.$$

Then, for any $\tau \in]0,1[$,

$$\nabla^2 v, \nabla q \in L_2(B(\tau))$$

and the following estimate is valid:

$$\|\nabla^2 v\|_{2,B(\tau)} + \|\nabla q\|_{2,B(\tau)} \leq c(\tau)\Big[\|f\|_{2,B} + \|q\|_{2,B} +$$
$$+ \|g\|_{W^1_2(B)} + \|v\|_{W^1_2(B)}\Big].$$

2.4 Further Local Regularity Results, $n = 2, 3$

Proposition 4.5. *Assume that a divergence free vector field* $v \in W^1_2(B)$ *obeys the identity*

$$\int\limits_B \nabla v : \nabla w dx = 0$$

for any $w \in C^\infty_{0,0}(B)$. *Then*

$$\sup_{x \in B(1/2)} |\nabla v(x)|^2 \leq c(n) \int\limits_B |\nabla v|^2 dx.$$

PROOF As it has been explained, one can introduce the pressure $q \in L_2(B)$ with $[q]_B = 0$ such that

$$\left\{\begin{array}{l} -\Delta v + \nabla q = 0 \\ \\ \operatorname{div} v = 0 \end{array}\right. \qquad \text{in } B$$

in the sense of distributions and

$$\|q\|_{2,B} \leq c\|\nabla v\|_{2,B}. \tag{2.4.1}$$

STEP 1 Let $\bar{v} = v - [v]_B$ and fix $1/2 < \tau_1 < 1$. By previous results, see Proposition 3.4,

$$\int\limits_{B(\tau_1)} (|\nabla^2 v|^2 + |\nabla q|^2)dx \leq c(\tau_1, n)\Big[\int\limits_B |\bar{v}|^2 dx + \int\limits_B |\nabla v|^2 dx + \int\limits_B |q|^2 dx\Big].$$

According to Poincaré's inequality

$$\int\limits_B |\bar{v}|^2 dx \leq c(n) \int\limits_B |\nabla v|^2 dx$$

and estimate (2.4.1), one can state that

$$\int\limits_{B(\tau_1)} (|\nabla^2 v|^2 + |\nabla q|^2)dx \leq c(\tau_1, n) \int\limits_B |\nabla v|^2 dx \equiv cI.$$

STEP 2 Now, obviously, functions $v_{,k}$ and $q_{,k}$ obey the system

$$\left\{\begin{array}{l} -\Delta v_{,k} + \nabla q_{,k} = 0 \\ \\ \operatorname{div} v_{,k} = 0 \end{array}\right. \qquad \text{in } B(\tau_1)$$

in the sense of distributions. Repeating the previous arguments in two balls $B(\tau_2)$ and $B(\tau_1)$ with $1/2 < \tau_2 < \tau_1$, we find

$$\int\limits_{B(\tau_2)} (|\nabla^2 v_{,k}|^2 + |\nabla q_{,k}|^2)dx \leq c(\tau_2, \tau_1, n) \int\limits_{B(\tau_1)} (|\nabla^2 v|^2 + |\nabla q|^2)dx \leq c(\tau_2, \tau_1, n)I.$$

As a result,

$$\int\limits_{B(1/2)} \Big(\sum_{i=1}^{l} |\nabla^i v|^2 + \sum_{i=1}^{l-1} |\nabla^i q|^2\Big) dx \leq c(l, n)I.$$

Taking $l = 3$ and using Sobolev's imbedding theorem, we show

$$\sup_{x \in B(1/2)} (|\nabla v(x)|^2 + |q(x)|^2) \leq c(n)I. \qquad \square$$

The proof of the following statement is slightly more complicated but still can be made along similar lines.

Proposition 4.6. *Assume that a divergence free vector field $v \in W^1_2(B_+)$ satisfies the boundary condition*

$$v|_{x_3=0} = 0$$

and the identity

$$\int\limits_{B_+} \nabla v : \nabla w dx = 0$$

for any $w \in C^\infty_{0,0}(B_+)$. Then

$$\sup_{x \in B_+(1/2)} |\nabla v(x)|^2 \leq c(n) \int\limits_{B_+} |\nabla v|^2 dx.$$

PROOF First, we recover the pressure $q \in L_2(B_+)$ with $[q]_{B_+} = 0$ such that

$$\left\{ \begin{array}{l} -\Delta v + \nabla q = 0 \\ \\ \operatorname{div} v = 0 \end{array} \right. \qquad \text{in } B_+$$

in the sense of distributions with the estimate

$$\|q\|_{2,B_+} \leq c \|\nabla v\|_{2,B_+}. \tag{2.4.2}$$

Fix $1/2 < \tau_1 < 1$. By Proposition 3.3, we have additional regularity so that

$$\int\limits_{B_+(\tau_1)} (|\nabla^2 v|^2 + |\nabla q|^2) dx \leq c(\tau_1, n) \int\limits_{B_+} \Big[|v|^2 + |\nabla v|^2 + |q|^2 \Big] dx.$$

Since $v|_{x_3=0} = 0$, Poincaré type inequality ensures the bound:

$$\int\limits_{B_+} |v|^2 dx \leq c(n) \int\limits_{B_+} |\nabla v|^2 dx$$

and, by (2.4.2),

$$\int\limits_{B_+(\tau_1)} (|\nabla^2 v|^2 + |\nabla q|^2) dx \leq c(\tau_1, n) \int\limits_{B_+} |\nabla v|^2 dx \equiv cI.$$

Tangential derivatives of v and q satisfy the same equations and boundary conditions:

$$\left\{ \begin{array}{l} -\Delta v_{,\alpha} + \nabla q_{,\alpha} = 0 \\ \\ \operatorname{div} v_{,\alpha} = 0 \end{array} \right. \qquad \text{in } B_+$$

and

$$v_{,\alpha}|_{x_3=0} = 0.$$

Assume that $n = 3$ ($n = 2$ is an exercise). We have for $1/2 < \tau_2 < \tau_1$

$$\int\limits_{B_+(\tau_2)} (|\nabla^2 v_{,\alpha}|^2 + |\nabla q_{,\alpha}|^2) dx \leq c(\tau_1, n) \int\limits_{B_+} |\nabla v|^2 dx \equiv cI.$$

It remains to evaluate $v_{i,333}$ and $q_{,33}$. To this end, we are going to exploit the incompressibility condition: $v_{3,333} = -v_{\alpha,\alpha 33} \in L_2(B_+(\tau_2))$, which gives us the bound

$$\int\limits_{B_+(\tau_2)} |\nabla^3 v_3|^2 dx \leq cI.$$

To estimate $v_{\alpha,333}$, $\alpha = 1, 2$, one can make use of the identity

$$q_{,3i} = \Delta v_{3,i}$$

and conclude that

$$\int\limits_{B_+(\tau_2)} |\nabla q_{,3}|^2 dx \leq cI.$$

Now, exploiting the equations $-\Delta v_{\alpha,3} + q_{,\alpha 3} = 0$ one more time, we find

$$v_{\alpha,333} = -v_{\alpha,\beta\beta 3} + q_{\alpha 3} \in L_2(B_+(\tau_2)).$$

The latter implies

$$\int\limits_{B_+(\tau_2)} |v_{\alpha,333}|^2 dx \leq cI.$$

So, the final estimate

$$\int\limits_{B_+(\tau_2)} (|\nabla^3 v|^2 + |\nabla^2 v|^2 + |\nabla v|^2 + |\nabla^2 q|^2 + |\nabla q|^2 + |q|^2) dx \leq cI$$

comes out and it implies

$$\sup_{x \in B_+(1/2)} |\nabla v(x)|^2 \leq c(n) I. \qquad \square$$

2.5 Stokes Operator in Bounded Domains

In this section, we always assume that a bounded domain $\Omega \in \mathbb{R}^n$ has smooth boundary, and $n = 2$ or 3.

By Ladyzhenskaya's theorem, given $f \in L_2(\Omega)$, there exists a unique $f_1 \in \overset{\circ}{J}(\Omega)$ such that

$$f = f_1 + \nabla q$$

with $q \in W^1_2(\Omega)$. We let $Pf := f_1$. The operator $P : L_2(\Omega) \to L_2(\Omega)$ is called the Leray projector.

It is worthy to notice that the Dirichlet problem

$$\begin{cases} -\Delta u + \nabla p = f \in L_2(\Omega) \\ \\ \operatorname{div} u = 0 \end{cases} \qquad \text{in } \Omega,$$

$$u|_{\partial\Omega} = 0$$

can be transformed into the equivalent one

$$\begin{cases} -\Delta u + \nabla p_1 = f_1 \in \overset{\circ}{J}(\Omega) \\ \\ \operatorname{div} u = 0 \end{cases} \qquad \text{in } \Omega,$$

$$u|_{\partial\Omega} = 0,$$

where $f_1 = Pf$ and $p_1 = p - q$. So, without loss of generality, we always may assume that the right-hand side in the Stokes system belongs to $\overset{\circ}{J}(\Omega)$.

We know that

$$\|\nabla^2 u\|_{2,\Omega} + \|\nabla p\|_{2,\Omega} \leq c\|f\|_{2,\Omega}.$$

We can also re-write the Dirichlet problem in the operator form

$$\widetilde{\Delta} u = f,$$

where

$$\widetilde{\Delta} := P\Delta : \overset{\circ}{J}(\Omega) \to \overset{\circ}{J}(\Omega)$$

is a unbounded operator with the domain of the definition

$$\operatorname{dom} \widetilde{\Delta} := \{u \in W^2_2(\Omega) : \ \operatorname{div} u = 0, \ u|_{\partial\Omega} = 0\} = \overset{\circ}{J}{}^1_2(\Omega) \cap W^2_2(\Omega).$$

It is called the Stokes operator.

The Stokes operator $\widetilde{\Delta}$ has the similar properties as the Laplace operator under the Dirichlet boundary conditions. Let us list these properties:

(i) the Stokes operator has a discrete spectrum

$$-\widetilde{\Delta}u = \lambda u, \qquad u \in \overset{\circ}{J}(\Omega), \qquad u \neq 0,$$

$$0 < \lambda_1 < \lambda_2 < ... < \lambda_m < ..., \qquad \lambda_m \to \infty,$$

(ii) $\dim \ker(-\widetilde{\Delta} - \lambda_k \mathbb{I})$ is finite for each $k \in \mathbb{N}$,

(iii) the set $\{\varphi_k\}_{k=1}^\infty$ of eigenvectors (eigenfunctions) of the Stokes operator is an orthogonal basis in $\overset{\circ}{J}(\Omega)$ so that $(\varphi_k, \varphi_j) = \delta_{ij}$,

(iv) the set $\{\varphi_k\}_{k=1}^\infty$ is an orthogonal system in $\overset{\circ}{J}{}^1_2(\Omega)$ as well as in $\operatorname{dom}\widetilde{\Delta}$ so that $\lambda_k = \|\nabla\varphi_k\|^2_{2,\Omega} = \|\widetilde{\Delta}\varphi_k\|_{2,\Omega}$,

(v) if $f \in \overset{\circ}{J}(\Omega)$, then $\|f\|^2_{2,\Omega} = \sum_{k=1}^\infty |c_k|^2 < \infty$, where $c_k = (f, \varphi_k)$, and the series $\sum_{k=1}^\infty c_k\varphi_k$ converges to f in $L_2(\Omega)$,

if $f \in \overset{\circ}{J}{}^1_2(\Omega)$, then $\|\nabla f\|^2_{2,\Omega} = \sum_{k=1}^\infty |c_k|^2\lambda_k < \infty$ and series $\sum_{k=1}^\infty c_k\varphi_k$ converges to f in $W^1_2(\Omega)$,

if $f \in \operatorname{dom}\widetilde{\Delta}$, then $\|\widetilde{\Delta}f\|^2_{2,\Omega} = \sum_{k=1}^\infty |c_k|^2\lambda_k^2 < \infty$ and series $\sum_{k=1}^\infty c_k\varphi_k$ converges to f in $W^2_2(\Omega)$.

The proof of all above statements is based on the Hilbert-Schmidt theorem and the compactness of the embedding of $W^1_2(\Omega)$ into $L_2(\Omega)$.

Let us describe extension of $\widetilde{\Delta}$ to $\overset{\circ}{J}{}^1_2(\Omega)$. We know that

$$\widetilde{\Delta} : \overset{\circ}{J}{}^1_2(\Omega) \cap W^2_2(\Omega) \to \overset{\circ}{J}(\Omega)$$

is a bijection. Given $u \in \overset{\circ}{J}{}^1_2(\Omega) \cap W^2_2(\Omega)$, we have

$$(-\widetilde{\Delta}u, v) = (-\Delta u + \nabla p, v) = (-\Delta u, v) = (\nabla u, \nabla v)$$

for any $v \in C^\infty_{0,0}(\Omega)$. From the latter identity, we immediately derive the following estimate

$$\|\widetilde{\Delta}u\|_{(\overset{\circ}{J}{}^1_2(\Omega))'} \leq \|\nabla u\|_{2,\Omega} = \|u\|_{\overset{\circ}{J}{}^1_2(\Omega)}.$$

Here, we use the identification of the dual space $(\overset{\circ}{J}{}^1_2(\Omega))'$ described in Section 4 of Chapter 1 with $V = \overset{\circ}{J}{}^1_2(\Omega)$ and $H = \overset{\circ}{J}(\Omega)$ and in what follows we are not going to introduce any special notation for this particular identification. Since the space $\overset{\circ}{J}{}^1_2(\Omega) \cap W^2_2(\Omega)$ is dense in $\overset{\circ}{J}{}^1_2(\Omega)$, there exists a unique extension of the Stokes operator $\widetilde{\Delta}$ (denoted again by $\widetilde{\Delta}$) from $\overset{\circ}{J}{}_2^1(\Omega) \cap W_2^2(\Omega)$ to $\overset{\circ}{J}{}_2^1(\Omega)$. Moreover, we have the following statement:

Proposition 5.7. *(i) The extension $\widetilde{\Delta} : \overset{\circ}{J}{}_2^1(\Omega) \to (\overset{\circ}{J}{}_2^1(\Omega))'$ is a bijection.*

(ii) If $f \in (\overset{\circ}{J}{}_2^1(\Omega))'$, then

$$\|f\|^2_{(\overset{\circ}{J}{}_2^1(\Omega))'} = \sum_{k=1}^{\infty} f_k^2/\lambda_k,$$

where $f_k = (f, \varphi_k)$.

Proof of Proposition 5.7 Obviously, $\widetilde{\Delta} : \overset{\circ}{J}{}_2^1(\Omega) \to \widetilde{\Delta}(\overset{\circ}{J}{}_2^1(\Omega))$ is a bijection. Our aim is to show that

$$\widetilde{\Delta}(\overset{\circ}{J}{}_2^1(\Omega)) = (\overset{\circ}{J}{}_2^1(\Omega))'. \tag{2.5.1}$$

Lemma 2.3.
(i) for $f \in (\overset{\circ}{J}{}_2^1(\Omega))'$, we have

$$\|f\|^2_{(\overset{\circ}{J}{}_2^1(\Omega))'} \leq \sum_{k=1}^{\infty} f_k^2/\lambda_k.$$

(ii) if

$$\sum_{k=1}^{\infty} f_k^2/\lambda_k < \infty,$$

then the series $\sum_{k=1}^{\infty} f_k\varphi_k$ converges to f in $(\overset{\circ}{J}{}_2^1(\Omega))'$, $f \in \widetilde{\Delta}(\overset{\circ}{J}{}_2^1(\Omega))$, and

$$\|f\|^2_{(\overset{\circ}{J}{}_2^1(\Omega))'} = \sum_{k=1}^{\infty} f_k^2/\lambda_k.$$

Proof Fix an arbitrary function $a \in \overset{\circ}{J}{}_2^1(\Omega))$, then

$$a^N = \sum_{k=1}^{N} a_k\varphi_k \to a$$

in $\overset{\circ}{J}{}_2^1(\Omega))$. So,

$$(f, a) = \lim_{N\to\infty} (f, a^N) = \lim_{N\to\infty} \sum_{k=1}^{N} f_k a_k \leq$$

$$\leq \Big(\sum_{k=1}^{N} f_k^2/\lambda_k\Big)^{\frac{1}{2}}\Big(\sum_{k=1}^{N} a_k^2\lambda_k\Big)^{\frac{1}{2}} \leq \Big(\sum_{k=1}^{\infty} f_k^2/\lambda_k\Big)^{\frac{1}{2}}\|\nabla a\|_{2,\Omega}.$$

The latter certainly implies (i).

(ii) First, let us show that the series $\sum_{k=1}^{\infty} f_k\varphi_k$ converges in $(\overset{\circ}{J}{}_2^1(\Omega))'$. Indeed, let $f_N = \sum_{k=1}^{N} f_k\varphi_k$ and then, by (i),

$$\|f_N - f_M\|^2_{(\overset{\circ}{J}{}_2^1(\Omega))'} \leq \sum_{k=M+1}^{N} f_k^2/\lambda_k \to 0$$

as $M, N \to 0$.

We denote by $f \in (\overset{\circ}{J}{}_2^1(\Omega))'$ the sum of our series. Then, by (i),

$$\|f - f_N\|^2_{(\overset{\circ}{J}{}_2^1(\Omega))'} \leq \sum_{k=N+1}^{\infty} f_k^2/\lambda_k \to 0$$

and thus

$$\|f_N\|_{(\overset{\circ}{J}{}_2^1(\Omega))'} \to \|f\|_{(\overset{\circ}{J}{}_2^1(\Omega))'}.$$

Now, the goal is to prove that $f \in \widetilde{\Delta}(\overset{\circ}{J}{}_2^1(\Omega))$. Indeed, we have

$$f_N = \sum_{k=1}^{N} f_k\varphi_k = \widetilde{\Delta}\Big(\sum_{k=1}^{N} f_k\varphi_k/\lambda_k\Big) = \widetilde{\Delta}u_N,$$

where

$$u_N = \sum_{k=1}^{N} f_k\varphi_k/\lambda_k \in \overset{\circ}{J}{}_2^1(\Omega) \cap W_2^2(\Omega).$$

By direct calculations,

$$\|\nabla u_N - \nabla u_M\|^2_{2,\Omega} = \sum_{k=M+1}^{N} f_k^2/\lambda_k \to 0.$$

Then, by definition of the extension of $\widetilde{\Delta}$,

$$\widetilde{\Delta}u_N \to \widetilde{\Delta}u = f.$$

Next, we have

$$\|f_N\|^2_{(\overset{\circ}{J}{}_2^1(\Omega))'} = \|\widetilde{\Delta}u_N\|^2_{(\overset{\circ}{J}{}_2^1(\Omega))'} = \|\nabla u_N\|^2_{2,\Omega} = \sum_{k=1}^{N} f_k^2/\lambda_k \to \|f\|^2_{(\overset{\circ}{J}{}_2^1(\Omega))'}. \square$$

Lemma 2.4.

$$\widetilde{\Delta}(\overset{\circ}{J}{}_2^1(\Omega)) = \{f \in (\overset{\circ}{J}{}_2^1(\Omega))' : \sum_{k=1}^{\infty} f_k^2/\lambda_k < \infty\} =: U.$$

PROOF According to Lemma 2.3 (ii), we have

$$U \subseteq \widetilde{\Delta}(\overset{\circ}{J}{}^1_2(\Omega)).$$

Now, assume that $f \subseteq \widetilde{\Delta}(\overset{\circ}{J}{}^1_2(\Omega))$, i.e., $f = \widetilde{\Delta} u$ for some $u \in \overset{\circ}{J}{}^1_2(\Omega)$. Then we have

$$f_k = (f, \varphi_k) = (\widetilde{\Delta}\, u, \varphi_k) = (u, \widetilde{\Delta}\, \varphi_k) = \lambda_k u_k.$$

Since

$$\|\nabla u\|^2_{2,\Omega} = \sum_{k=1}^{\infty} u_k^2 \lambda_k < \infty,$$

we find

$$\sum_{k=1}^{\infty} f_k^2/\lambda_k < \infty.$$

So, $f \in U$ and thus $U \in \widetilde{\Delta}(\overset{\circ}{J}{}^1_2(\Omega))$. □

Now, we proceed with the proof of Proposition 5.7. We are done, if the implication

$$f \in (\overset{\circ}{J}{}^1_2(\Omega))' \Rightarrow \sum_{k=1}^{\infty} f_k^2/\lambda_k < \infty$$

is shown. To this end, we let

$$a_k = f_k/\lambda_k, \qquad a^N = \sum_{k=1}^{N} a_k \varphi_k.$$

Then

$$\|\nabla a^N\|^2_{2,\Omega} = \sum_{k=1}^{N} |a_k|^2 \|\nabla \varphi_k\|^2_{2,\Omega} = \sum_{k=1}^{N} f_k^2/\lambda_k.$$

So, we have

$$(f, a^N) = \sum_{k=1}^{N} f_k^2/\lambda_k \leq \|f\|_{(\overset{\circ}{J}{}^1_2(\Omega))'} \|\nabla a^N\|_{2,\Omega} = \|f\|_{(\overset{\circ}{J}{}^1_2(\Omega))'} \Big(\sum_{k=1}^{N} f_k^2/\lambda_k\Big)^{\frac{1}{2}},$$

which implies

$$\sum_{k=1}^{N} f_k^2/\lambda_k \leq \|f\|^2_{(\overset{\circ}{J}{}^1_2(\Omega))'}$$

for any natural number N. This completes the proof of Proposition 5.7. □

2.6 Comments

Chapter 2 contains standard results on linear stationary Stokes system including the notion of Stokes operator in smooth bounded domains. In addition, various global and local interior and boundary regularity results are discussed.

Chapter 3

Non-Linear Stationary Problem

3.1 Existence of Weak Solutions

Consider the Dirichlet boundary value problem for the classical stationary Navier-Stokes system

$$\begin{cases} -\nu\triangle u + u \cdot \nabla u + \nabla p = f \\ \\ \operatorname{div} u = 0 \end{cases} \quad \text{in } \Omega, \tag{3.1.1}$$

$$u|_{\partial\Omega} = 0 \tag{3.1.2}$$

and if $n = 3$ and Ω is unbounded then $u(x) \to 0$ as $|x| \to \infty$. Here, ν is a positive parameter called viscosity. We always assume that

$$f \in (L^1_2(\Omega))'.$$

Definition 3.1. A function $u \in \hat{V}(\Omega)$ is called a weak solution to boundary value problem (3.1.1) and (3.1.2) if

$$\nu(\nabla u, \nabla v) = (u \otimes u, \nabla v) + < f, v >$$

for any $v \in C^\infty_{0,0}(\Omega)$.

For $n = 2$ or 3, the imbedding theorems ensure that

$$u \in L_{4,\mathrm{loc}}(\Omega).$$

So, the first term on the right-hand side in the identity of Definition 3.1 is well-defined.

If domain Ω is bounded and has Lipschitz boundary, then

$$u \in L_4(\Omega).$$

Proposition 1.1. *Let Ω be a bounded Lipschitz domain. Then boundary value problem (3.1.1) and (3.1.2) has at least one weak solution.*

PROOF Let us reduce our boundary value problem to a fixed point problem and try to apply the celebrated Leray-Schauder principle.

Theorem 1.2. *(Leray-Schauder principle) Let X be a separable Banach space, $B : X \to X$ be a continuous operator. Assume that the operator B has the following additional properties:*

(i) B is a compact operator, i.e., it maps bounded sets of X into pre-compact sets of X. In other words, B is a completely continuous operator;

(ii) all possible solutions to the equation

$$u = \lambda B(u)$$

satisfy the inequality $\|u\|_X < R$ with R independent of $\lambda \in [0, 1]$.

Then operator B has at least one fixed point u, i.e., $u = B(u)$.

We define, as usual, $[u, v] := (\nabla u, \nabla v)$ a scalar product on $V(\Omega)$ that coincides with $\hat{V}(\Omega)$ under assumptions of the proposition. It is not difficult to show that, for any $w \in V(\Omega)$,

$$\operatorname{div} w \otimes w \in L_2^{-1}(\Omega).$$

As it has been pointed out in Chapter 2, Section 1, for bounded domains, we can identify the space $(\overset{\circ}{L}{}_2^1(\Omega))'$ with the space $L_2^{-1}(\Omega)$ and replace $< \cdot, \cdot >$ with $(\cdot, \cdot)$.

According to statements of Chapter 2, given $w \in V(\Omega)$, there exists a unique $u \in V(\Omega)$ such that

$$\nu(\nabla u, \nabla v) = (w \otimes w, \nabla v) + (f, v)$$

for any $v \in V(\Omega)$. By Riesz representation theorem, we can define an operator $A : V(\Omega) \to V(\Omega)$ so that

$$[A(w), v] := (w \otimes w, \nabla v)$$

and

$$[F, v] := (f, v).$$

So, the previous identity can be re-written in the operator form

$$u = \frac{1}{\nu}(A(w) + F).$$

Then, the existence of weak solutions is equivalent to the existence of fixed point of the above operator equation.

First, let us show that A is a completely continuous operator. To this end, we take an arbitrary weakly converging sequence such that

$$w^{(k)} \rightharpoonup w$$

in $V(\Omega)$. Then, the compactness of the imbedding of $V(\Omega)$ into $L_4(\Omega)$ gives us:

$$w^{(k)} \otimes w^{(k)} \to w \otimes w$$

in $L_2(\Omega)$. From the main identity, it follows that

$$\nu[u^{(k)} - u^{(m)}, v] = (w^{(k)} \otimes w^{(k)} - w^{(m)} \otimes w^{(m)}, \nabla v) = 0$$

for any $v \in V(\Omega)$. It remains to insert $v = u^{(k)} - u^{(m)}$ into the above relation and make use of the fact that

$$\|w^{(k)} \otimes w^{(k)} - w^{(m)} \otimes w^{(m)}\|_{2,\Omega} \to 0$$

as $k, m \to \infty$. So, complete continuity of A has been proven.

Now, we need to get estimates of all possible solutions to the equation

$$\nu u_\lambda = \lambda A(u_\lambda) + F,$$

depending on a parameter $\lambda \in [0,1]$. Since $(u_\lambda \otimes u_\lambda, \nabla u_\lambda) = 0$, we have

$$\nu[u_\lambda, u_\lambda] = (u_\lambda \otimes u_\lambda, \nabla u_\lambda) + (f, u_\lambda) \leq \|f\|_{L_2^{-1}(\Omega)} \|\nabla u_\lambda\|_{2,\Omega}$$

and thus

$$\|\nabla u_\lambda\|_{2,\Omega} \leq \frac{1}{\nu}\|f\|_{L_2^{-1}(\Omega)}.$$

The right-hand side of the above inequality is independent of λ and thus the existence of at least one fixed point follows from the Leray-Schauder principle. □

Regarding the uniqueness of weak solutions, we have the following statement.

Lemma 3.1. *Assume that all assumptions of Proposition 1.1 hold. Let in addition*

$$\frac{c_0^2(n,\Omega)}{\nu^2}\|f\|_{L_2^{-1}(\Omega)} < 1,$$

where $c_0(n,\Omega)$ is a constant in the inequality

$$\|v\|_{4,\Omega} \leq c_0(n,\Omega)\|\nabla v\|_{2,\Omega} \tag{3.1.3}$$

for any $v \in \overset{\circ}{L}{}_2^1(\Omega)$.

Then, our boundary value problem (3.1.1) and (3.1.2) has a unique weak solution.

PROOF Let u^1 and u^2 be two different solutions to boundary value problem (3.1.1), (3.1.2). Then, we have

$$\nu[u^1 - u^2, u^1 - u^2] = (u^1 \otimes u^1 - u^2 \otimes u^2, \nabla(u^1 - u^2)) =$$

$$= (u^1 \otimes (u^1 - u^2), \nabla(u^1 - u^2)) + ((u^1 - u^2) \otimes u^2, \nabla(u^1 - u^2)) =$$

$$= (u^1 \otimes (u^1 - u^2), \nabla(u^1 - u^2)) \leq \|u^1\|_{4,\Omega} \|u^1 - u^2\|_{4,\Omega} \|\nabla(u^1 - u^2)\|_{2,\Omega}.$$

Applying inequality (3.1.3) twice and taking into account the last estimate in the proof of Proposition 1.1 for u^1, i.e.,

$$\|\nabla u^1\|_{2,\Omega} \leq \frac{1}{\nu} \|f\|_{L_2^{-1}(\Omega)},$$

we find

$$\nu \|\nabla(u^1 - u^2)\|^2_{2,\Omega} \leq c_0^2 \|\nabla u^1\|^2_{2,\Omega} \|\nabla(u^1 - u^2)\|^2_{2,\Omega} \leq$$

$$\leq \frac{c_0^2}{\nu} \|\nabla(u^1 - u^2)\|^2_{2,\Omega} \|f\|_{L_2^{-1}(\Omega)}.$$

This, by a contradiction, implies the statement of the lemma. □

Proposition 1.3. *Assume that unbounded domain Ω is either $\mathbb{R}^3$ or $\mathbb{R}^n_+$, $n = 2, 3$. Then problem (3.1.1) and (3.1.2) has at least one weak solution satisfying the estimate*

$$\|\nabla u\|_{2,\Omega} \leq \frac{1}{\nu} \|f\|_{(L_2^1(\Omega))'}.$$

PROOF Let $R \gg 1$. Consider problem (3.1.1), (3.1.2) in $\Omega_R := B(R) \cap \Omega$. By Proposition 1.1, there exists $u_R \in V(\Omega_R)$, satisfying the identity

$$\nu(\nabla u_R, \nabla v)_{\Omega_R} = (u_R \otimes u_R, \nabla v)_{\Omega_R} + (f, v)_{\Omega_R}$$

for any $v \in C^\infty_{0,0}(\Omega_R)$. Extending u_R by zero to the whole domain Ω, we notice that

$$\|\nabla u_R\|_{2,\Omega} = \|\nabla u_R\|_{2,\Omega_R} \leq \frac{1}{\nu} \|f\|_{L_2^{-1}(\Omega_R)} \leq \frac{1}{\nu} \|f\|_{(L_2^1(\Omega))'}.$$

The latter allows us to select a subsequence, still denoted by u_R, with the following properties:

$$\nabla u_R \rightharpoonup \nabla u$$

in $L_2(\Omega)$ and

$$u_R \to u$$

in $L_{2,\mathrm{loc}}(\Omega)$. For $n = 3$, this follows from Hölder inequality, boundedness in $L_6(\Omega)$, and compactness of the imbedding of $W_2^1(B(R))$ into $L_2(B(R))$ for any fixed $R > 1$. If $n = 2$, we can use the inequality

$$\|w\|_{2,\mathbb{R}\times]o,a[} \leq c(a)\|\nabla w\|_{\mathbb{R}^2_+}$$

that is valid for any function $w \in C_0^\infty(\mathbb{R}^2_+)$ and for all $a > 1$.

It remains to pass to the limits as $R \to \infty$ in the identity for u_R and show that

$$\nu(\nabla u, \nabla v) = (u \otimes u, \nabla v) + < f, v >$$

for any $v \in C_{0,0}^\infty(\Omega)$, which means that u is a required weak solution. □

Now, the question is whether we can recover the pressure? We shall consider two cases.

CASE 1 Here, we assume that Ω is a bounded domain with Lipschitz boundary. Since, for $v \in C_0^\infty(\Omega)$,

$$l(v) := \nu(\nabla u, \nabla v) - (u \otimes u, \nabla v) - (f, v) \leq$$

$$\leq C\|\nabla v\|_{2,\Omega},$$

with a positive constant $C = C(\nu, \|\nabla u\|_{2,\Omega}, \|u\|_{4,\Omega}, \|f\|_{L_2^{-1}(\Omega)})$, and $l(v) = 0$ for any $v \in C_{0,0}^\infty(\Omega)$, we can use the same arguments as before to recover the pressure. According to them, there exists $p \in L_2(\Omega)$ such that

$$\nu(\nabla u, \nabla v) = (u \otimes u, \nabla v) + (f, v) + (p, \operatorname{div} v)$$

for any $v \in C_0^\infty(\Omega)$.

CASE 2 Here, we can use a similar procedure, described in Section 1, where

$$\Omega = \bigcup_{m=1}^{\infty} \Omega_m, \qquad \Omega_m \subset \Omega_{m+1},$$

and Ω_m is a bounded Lipschitz domain. Since $u \in L_{4,\mathrm{loc}}(\Omega)$ implies $u \in L_4(\Omega_m)$, one can state that there exists $p_m \in L_2(\Omega_m)$ such that

$$\nu(\nabla u, \nabla v) = (u \otimes u, \nabla v) + < f, v > + (p_m, \operatorname{div} v)$$

for any $v \in C_0^\infty(\Omega_m)$. Moreover, we can fix p_m so that $p_m = p_{m+1}$ in Ω_m. So, now, if we introduce a function p, letting $p = p_m$ in Ω_m, then $p \in L_{2,\mathrm{loc}}(\Omega)$ and the following identity is valid:

$$\nu(\nabla u, \nabla v) = (u \otimes u, \nabla v) + < f, v > + (p, \operatorname{div} v)$$

for any $v \in C_0^\infty(\Omega)$.

3.2 Regularity of Weak Solutions

We need the following auxiliary statement.

Lemma 3.2. *Let a non-decreasing function* $\Phi :]0, R_0] \to \mathbb{R}_+$ *satisfy the following condition:*

$$\Phi(\varrho) \leq c\Big(\Big(\frac{\varrho}{R}\Big)^m + \varepsilon\Big)\Phi(R) + CR^s \tag{3.2.1}$$

for any $0 < \varrho < R \leq R_0$*, for some positive constants* $c, C, \varepsilon > 0$*, and for some* $m > s > 0$*.*

There exist positive numbers $\varepsilon_0 = \varepsilon_0(m, s, c)$ *and* $c_1 = c_1(m, s, c)$ *such that if* $\varepsilon < \varepsilon_0$*, then*

$$\Phi(\varrho) \leq c_1\Big[\Big(\frac{\varrho}{R_0}\Big)^s + C\varrho^s\Big] \tag{3.2.2}$$

for any $0 < \varrho \leq R_0$*.*

PROOF Let $\varrho = \tau R$, $0 < \tau < 1$, $\varepsilon_0 = \tau^m$. So, if $\varepsilon < \varepsilon_0$, then

$$\Phi(\tau R) \leq 2c\tau^m \Phi(R) + CR^s = 2c\tau^{\frac{m-s}{2}}\tau^{\frac{m+s}{2}}\Phi(R) + CR^s \leq$$

$$\leq \tau^{\frac{m+s}{2}}\Phi(R) + CR^s.$$

If we select ε_0 so that $2c\tau^{\frac{m-s}{2}} \leq 1$, then, after iterations, we have

$$\Phi(\tau^k R_0) \leq \tau^{k\frac{m+s}{2}}\Phi(R_0) + C\tau^s R_0^s(1 + \tau^{\frac{m+s}{2}} + ... + \tau^{\frac{m+s}{2}(k-1)}) \leq$$

$$\leq \tau^{k\frac{m+s}{2}}\Phi(R_0) + C\tau^s R_0^s \frac{1}{1 - \tau^{\frac{m-s}{2}}}.$$

Given $0 < \varrho \leq R_0$, we find an integer number k such that

$$R_0\tau^{k+1} < \varrho \leq R_0\tau^k.$$

Then

$$\Phi(\varrho) \leq \Phi(\tau^k R_0) \leq \Big(\frac{1}{\tau}\frac{\varrho}{R}\Big)^s \Phi(R_0) + C\Big(\frac{\varrho}{\tau}\Big)^s \frac{1}{1 - \tau^{\frac{m-s}{2}}}. \qquad \square$$

We are going to prove the following local estimates for weak solutions to the non-linear stationary Navier-Stokes system.

Lemma 3.3. *Let a divergence free vector-valued function* $u \in W_2^1(B(R))$ *and a tensor-valued function* $F \in L_r(B(R))$*, with* $r > n = 3$*, satisfy the identity*

$$\int\limits_{B(R)} \nabla u : \nabla v dx = \int\limits_{B(R)} u \otimes u : \nabla v dx + \int\limits_{B(R)} F : \nabla v dx$$

for any $v \in C^\infty_{0,0}(B(R))$.

Then,

$$\int\limits_{B(\varrho)} |\nabla u|^2 dx \leq c\Big(\Big(\frac{\varrho}{R}\Big)^3 + R\Big(\int\limits_{B(R)} |u|^6 dx\Big)^{\frac{1}{3}}\Big) \int\limits_{B(R)} |\nabla u|^2 dx +$$

$$+cR^{3(1-\frac{2}{r})}\Big(\int\limits_{B(R)} |F|^r dx\Big)^{\frac{2}{r}}$$

for $0 < \varrho \leq R$. *Here,* c *is a universal positive constant.*

Lemma 3.4. *Let a divergence free vector-valued function* $u \in W^1_2(B_+(R))$, *with* $u|_{x_3=0} = 0$, *and a tensor-valued function* $F \in L_r(B_+(R))$, *with* $r > n = 3$, *satisfy the identity*

$$\int\limits_{B_+(R)} \nabla u : \nabla v dx = \int\limits_{B_+(R)} u \otimes u : \nabla v dx + \int\limits_{B_+(R)} F : \nabla v dx$$

for any $v \in C^\infty_{0,0}(B_+(R))$.

Then,

$$\int\limits_{B_+(\varrho)} |\nabla u|^2 dx \leq c\Big(\Big(\frac{\varrho}{R}\Big)^3 + R\Big(\int\limits_{B_+(R)} |u|^6 dx\Big)^{\frac{1}{3}}\Big) \int\limits_{B_+(R)} |\nabla u|^2 dx +$$

$$+cR^{3(1-\frac{2}{r})}\Big(\int\limits_{B_+(R)} |F|^r dx\Big)^{\frac{2}{r}}$$

for $0 < \varrho \leq R$. *Here,* c *is a universal positive constant.*

Proof of Lemma 3.3 We know that $\operatorname{div} u \otimes u \in L^{-1}_2(B(R))$. Hence, there exist

$$\tilde{u}_R \in \overset{\circ}{J}{}^1_2(B(R)), \qquad \tilde{p}_R \in L_2(B(R)),$$

with $[\tilde{p}_R]_{B(R))} = 0$, so that

$$\begin{cases} -\nu\triangle\tilde{u}_R + \nabla\tilde{p}_R = -\operatorname{div} u \otimes u - \operatorname{div} F \\ \qquad\qquad \operatorname{div}\tilde{u}_R = 0 \end{cases} \quad \text{in } B(R). \qquad (3.2.3)$$

Multiplying the first equation in (3.2.3) by $\tilde{u}_R$ and integrating the product by parts, we find

$$\int\limits_{B(R)} |\nabla\tilde{u}_R|^2 dx = \int\limits_{B(R)} (u \otimes u - [u \otimes u]_{B(R)}) : \nabla\tilde{u}_R dx + \int\limits_{B(R)} F : \tilde{u}_R dx$$

and, therefore, after application of the Cauchy-Schwartz inequality, we get

$$\int\limits_{B(R)} |\nabla \tilde{u}_R|^2 dx \leq 2\Big(\int\limits_{B(R)} |u\otimes u - [u\otimes u]_{B(R)}|^2 dx + \int\limits_{B(R)} |F|^2 dx\Big).$$

Next, we treat the first term on the right-hand side of the latter relation with the help of Gagliardo-Nirenberg inequality and, then, with the help of Hölder inequality. As a result, we have

$$\int\limits_{B(R)} |u\otimes u - [u\otimes u]_{B(R)}|^2 dx \leq c\Big(\int\limits_{B(R)} |\nabla (u\otimes u)|^{\frac{6}{5}} dx\Big)^{\frac{5}{3}} \leq$$

$$\leq c\Big(\int\limits_{B(R)} |u|^{\frac{6}{5}} |\nabla u|^{\frac{6}{5}} dx\Big)^{\frac{5}{3}} \leq c\Big(\int\limits_{B(R)} |u|^6 dx\Big)^{\frac{1}{3}} \Big(\int\limits_{B(R)} |\nabla u|^{\frac{3}{2}} dx\Big)^{\frac{4}{3}} \leq$$

$$\leq c\Big(\int\limits_{B(R)} |u|^6 dx\Big)^{\frac{1}{3}} R \int\limits_{B(R)} |\nabla u|^2 dx,$$

where c is a universal constant.

Let $u_R = u - \tilde{u}_R$. This function satisfies the identity

$$\int\limits_{B(R)} \nabla u_R : \nabla v dx = 0$$

for any $v \in C^\infty_{0,0}(B(R))$. By the results of Chapter 2, see Section 2, we have the following estimate

$$\int\limits_{B(\varrho)} |\nabla u_R|^2 dx \leq c\Big(\frac{\varrho}{R}\Big)^3 \int\limits_{B(R)} |\nabla u_R|^2 dx,$$

which, in turn, implies another one:

$$\int\limits_{B(\varrho)} |\nabla u|^2 dx \leq c\Big(\frac{\varrho}{R}\Big)^3 \int\limits_{B(R)} |\nabla u|^2 dx + c \int\limits_{B(R)} |\nabla \tilde{u}_R|^2 dx.$$

At first, we apply our earlier estimates for

$$\int\limits_{B(R)} |\nabla \tilde{u}_R|^2 dx$$

and, then, Hölder's inequality for the term, containing F, in order to get the estimate of Lemma 3.3.

Lemma 3.5. *(Ch.-B. Morrey) Let $u \in W^1_m(\Omega)$ satisfy the condition*

$$\int\limits_{B(\varrho)} |\nabla u|^m dx \leq K\varrho^{n-m+m\alpha}$$

for some $0 < \alpha < 1$ and for any $B(x_0, \varrho) \subset \Omega \subset \mathbb{R}^n$ such that $0 < \varrho < \varrho_0$ with two positive constants K and ϱ_0.

Then $u \in C^\alpha_{\rm loc}(\Omega)$, i.e., $u \in C^\alpha(\overline{\Omega}_1)$ for any subdomain $\Omega_1 \Subset \Omega$.

Here, $C^\alpha(\overline{\omega})$ is a Hölder space with the norm $\|u\|_{C^\alpha(\overline{\omega})} := \|u\|_{C(\overline{\omega})} + [u]_{\alpha,\overline{\omega}}$, where

$$[u]_{\alpha,\overline{\omega}} := \sup\Big\{\frac{|u(x) - u(y)|}{|x-y|^\alpha} \,:\, x, y \in \overline{\omega},\, x \neq y\Big\}.$$

Lemma 3.6. *Assume that all assumptions of Lemma 3.3 hold with $R = a$. Then*

$$u \in C^{1-\frac{3}{r}}_{\rm loc}(B(a)).$$

PROOF We remind that the case $n = 3$ is considered only. Fix $\Omega_1 \Subset B(a)$ and find Ω such that $\Omega_1 \Subset \Omega \Subset B(a)$. By shift, we have, for any $B(x_0, R) \subset B(a)$, the following estimate

$$\Phi(x_0, \varrho) \leq c\Big[\Big(\Big(\frac{\varrho}{R}\Big)^3 + RA\Big)\Phi(x_0, R) + CR^{3-2+2\alpha}\Big],$$

with $\alpha = 1 - 3/R$,

$$\Phi(x_0, R) := \int\limits_{B(x_0,R)} |\nabla u|^2 dx, \quad A := \Big(\int\limits_{B(a)} |u|^6 dx\Big)^{\frac{1}{3}}, \quad C := \Big(\int\limits_{B(a)} |F|^r dx\Big)^{\frac{2}{r}}.$$

Now, we apply Lemma 3.2 with $m = 3$, $s = 3 - 2 + 2\alpha = 1 + 2\alpha$. If we let

$$R_0 := \frac{1}{2}\min\Big\{\text{dist}\,(\partial B(a), \Omega), \frac{\varepsilon_0}{A}\Big\},$$

then $B(x_0, R) \subset B(a)$ for any $x_0 \in \Omega$ and $AR < \varepsilon_0$ as long as $0 < R < R_0$. Hence,

$$\Phi(x_0, \varrho) \leq c_1\Big[\Big(\frac{\varrho}{R_0}\Big)^{3-2+2\alpha}\Phi(x_0, R) + C\varrho^{3-2+2\alpha}\Big]$$

for any $x_0 \in \Omega$ and for any $0 < \varrho \leq R_0$. So, we have

$$\int\limits_{B(x_0,\varrho)} |\nabla u|^2 dx \leq K\varrho^{3-2+2\alpha}$$

for any $0 < \varrho \leq R_0$, where $K = K(r, \|u\|_{W^1_2(B(a))}, \|F\|_{r,B(a)}, R_0)$ provided $B(x_0, \varrho) \subset \Omega$. $\square$

Lemma 3.7. *Assume that all assumptions of Lemma 3.4 hold with $R = a$. Then*

$$u \in C^\alpha(\bar{B}_+(b))$$

for any $0 < b < a$ with $\alpha = 1 - 3/r$.

PROOF We have two types of estimates. The first one is so-called "interior". For any $b_1 \in]b, a[$, the following estimate is valid:

$$\Phi(x_0, \varrho) \leq K\varrho^{1+2\alpha} \tag{3.2.4}$$

for $x_0 \in B_+(b_1)$, $x_{30} \geq \frac{1}{2}(a - b_1)$, and $0 < \varrho \leq R_0 = \frac{1}{2}\min\{a - b_1, \varrho_0\}$ with $\varrho_0 = \varepsilon_0/A$. Here, K depends on r, R_0, $\|u\|_{W^1_2(B_+(a))}$, and $\|F\|_{r,B_+(a)}$.

The second estimate is "boundary" one:

$$\Phi_+(x_0, \varrho) := \int\limits_{B_+(x_0,\varrho)} |\nabla u|^2 dx \leq K_+\varrho^{1+2\alpha} \tag{3.2.5}$$

for $x_0 = (x_0', 0)$, $|x_0'| < \frac{1}{2}(a - b_1)$, $0 < \varrho \leq R_0$, and K_+ depends on the same arguments as K.

Now, let us denote by $\tilde{u}$ extension of u to the whole ball $B(a)$ by zero and let

$$\tilde{\Phi}(x_0, \varrho) := \int\limits_{B(x_0,\varrho)} |\nabla \tilde{u}|^2 dx$$

with $0 < \varrho \leq R_0$ and $x_0 \in B(b_1)$.

Consider two cases: $x_{30} \geq \frac{1}{2}(a - b_1)$ and $x_{30} < \frac{1}{2}(a - b_1)$. In the first case, we may use our "interior" estimate (3.2.4) and the definition of $\tilde{u}$. As a result, we arrive at the inequality

$$\tilde{\Phi}(x_0, \varrho) \leq K\varrho^{1+2\alpha}. \tag{3.2.6}$$

In the second case, we first assume that $x_{30} > 0$ and if $x_{30} \geq \varrho$, we still have estimate (3.2.6). Now, suppose that $x_{30} < \varrho$. Then, by (3.2.5), we have

$$\tilde{\Phi}(x_0, \varrho) = \int\limits_{B(x_0,\varrho)\cap B_+(a)} |\nabla u|^2 dx \leq \int\limits_{B_+((x_0',0),\varrho + x_{30})} |\nabla u|^2 dx \leq$$

$$\leq K_+(\varrho + x_{30})^{1+2\alpha} \leq 2^{1+2\alpha} K_+ \varrho^{1+2\alpha}.$$

Now, assume that $x_{30} < 0$. If $|x_{30}| \geq \varrho$, then, obviously, $\tilde{\Phi}(x_0, \varrho) = 0$. So, let us suppose that $-x_{30} < \varrho$. Here,

$$\tilde{\Phi}(x_0, \varrho) = \int\limits_{B(x_0,\varrho)\cap B_+(a)} |\nabla u|^2 dx \leq \int\limits_{B_+((x_0',0),\varrho)} |\nabla u|^2 dx \leq$$

$$\leq K_+ \varrho^{1+2\alpha}.$$

So, the statement of the lemma follows from Morrey's condition on Hölder continuity, see Lemma 3.5. □

Proposition 2.4. *Let $u \in W^1_2(B(2a))$ be a divergence free function and satisfy the identity*

$$\int\limits_{B(2a)} (\nabla u : \nabla v - u \otimes u : \nabla v) dx = \int\limits_{B(2a)} f \cdot v dx$$

for any $v \in C^\infty_{0,0}(B(2a))$. If f is of class C^∞ in $B(2a)$, then u is of class C^∞ in $B(a)$.

Proof It is not difficult to check that there exists a tensor-valued function F of class C^∞ such that $f = -\operatorname{div} F$. Then, the identity from the statement of the proposition can be re-written in the following way

$$\int\limits_{B(2a)} \nabla u : \nabla v dx = \int\limits_{B(2a)} u \otimes u : \nabla v + \int\limits_{B(2a)} F : \nabla v dx$$

for any $v \in C^\infty_{0,0}(B(2a))$. From Lemma 3.6, it follows that u belongs, at least, to $C(\overline{B}(3a/2))$. Using the same arguments as in Section 1, we can recover a pressure $p \in L_2(B(2a))$ (exercise) so that

$$\left\{ \begin{array}{l} -\Delta u + \nabla p = -\operatorname{div} G := -\operatorname{div}(u \otimes u + F) \\ \\ \operatorname{div} u = 0 \end{array} \right. \qquad \text{in } B(2a).$$

Since $\operatorname{div} G \in L_2(B(3a/2))$, we can apply results of Chapter II on properties of solutions to the Stokes system and find

$$\nabla^2 u \in L_2(B(a_1)) \Rightarrow \nabla u \in L_6(B(a_1)), \quad \nabla p \in L_2(B(a_1))$$

for any $a < a_1 < \frac{3}{2}a$.

Next, we know that, for $k = 1, 2, 3$, functions $u_{,k} \in W^1_2(B(a_1))$ and $p_{,k} \in L_2(B(a_1))$ satisfy the system

$$\left\{ \begin{array}{l} -\Delta u_{,k} + \nabla p_{,k} = -\operatorname{div} G_{,k} \\ \\ \operatorname{div} u_{,k} = 0 \end{array} \right. \qquad \text{in } B(a_1).$$

Since $\nabla^2 G \in L_2(B(a_1))$, we can use the linear theory one more time and get:

$$\nabla^3 u \in L_2(B(a_2)) \Rightarrow \nabla^2 u \in L_6(B(a_2)), \quad \nabla^2 p \in L_2(B(a_2))$$

for any $a < a_2 < a_1$.

Then, for $k, s = 1, 2, 3$, functions $u_{,ks} \in W^1_2(B(a_1))$ and $p_{,ks} \in L_2(B(a_1))$ satisfy the system

$$\left\{ \begin{array}{l} -\Delta u_{,ks} + \nabla p_{,ks} = -\operatorname{div} G_{,ks} \\ \\ \operatorname{div} u_{,ks} = 0 \end{array} \right. \qquad \text{in } B(a_3),$$

with $\nabla^3 G \in L_2(B(a_2))$. The similar arguments allow us to deduce that

$$\nabla^4 u \in L_2(B(a_3)) \Rightarrow \nabla^3 u \in L_6(B(a_3)), \quad \nabla^3 p \in L_2(B(a_3))$$

for any $a < a_3 < a_2$. Proceeding, further, in the same way, we complete the proof of the lemma. □

Proposition 2.5. *Let $u \in W^1_2(B(2a))$ be a divergence free function and satisfy the conditions: $u|_{x_3=0} = 0$ and*

$$\int\limits_{B_+(2a)} (\nabla u : \nabla v - u \otimes u : \nabla v) dx = \int\limits_{B_+(2a)} f \cdot v dx$$

for any $v \in C^\infty_{0,0}(B_+(2a))$. If f is of class C^∞ in $B(2a) \cap \{x_3 \geq 0\}$, then u is of class C^∞ in $B(a) \cap \{x_3 \geq 0\}$.

PROOF We start with our proof in a way similar to the proof of the previous proposition, i.e., we find F of class C^∞ in $B(2a) \cap \{x_3 \geq 0\}$ so that $f = -\operatorname{div} F$. Then, we recover the pressure $p \in L_2(B_+(2a))$, which gives us:

$$\left\{ \begin{array}{l} -\Delta u + \nabla p = -\operatorname{div} G := -\operatorname{div}(u \otimes u + F) \\ \\ \operatorname{div} u = 0 \end{array} \right. \qquad \text{in } B_+(2a),$$

$$u|_{x_3=0} = 0.$$

By Lemma 3.7, $u \in C(\overline{B}_+(3a/2))$ and, by the linear theory,

$$\nabla^2 u \in L_2(B_+(a_1)) \Rightarrow \nabla u \in L_6(B_+(a_1)), \quad \nabla p \in L_2(B_+(a_1))$$

for any $a < a_1 < \frac{3}{2}a$.

Next, for $\alpha = 1,2$ and for $k = 1,2,3$, we have functions $u_{,k} \in W^1_2(B_+(a_1))$ and $p_{,k} \in L_2(B_+(a_1))$ satisfying the system

$$\left\{\begin{array}{l} -\Delta u_{,k} + \nabla p_{,k} = -\operatorname{div} G_{,k} \\ \\ \qquad \operatorname{div} u_{,k} = 0 \end{array}\right. \qquad \text{in } B_+(a_1)$$

and the boundary condition

$$u_{,\alpha}|_{x_3=0} = 0,$$

where $\nabla^2 G \in L_2(B_+(a_1))$. Then, again, we apply the linear theory and conclude that

$$\nabla^2 u_{,\alpha} \in L_2(B_+(a_2)) \quad \nabla p_{,\alpha} \in L_2(B_+(a_2))$$

for any $a < a_2 < a_1$. We need to establish the same properties for $\nabla^2 u_{,3}$ and $\nabla p_{,3}$. To achieve this goal, it is sufficient to evaluate $u_{k,333}$ and $p_{,33}$ for $k = 1,2,3$, which is, in fact, not so difficult. Indeed, denoting $g_{ik} := -G_{ij,jk}$, we first use the incompressibility condition:

$$u_{3,333} = -u_{\alpha,\alpha 33} \in L_2(B_+(a_2)).$$

For other derivatives, we use the equations:

$$p_{,33} = g_{33} + u_{3,kk3} \in L_2(B_+(a_2))$$

and

$$u_{\alpha,333} = -g_{\alpha 3} + p_{,\alpha 3} - u_{\alpha,\beta\beta 3} \in L_2(B_+(a_2)).$$

So, we can state

$$\nabla^3 u \in L_2(B_+(a_2)) \Rightarrow \nabla^2 u \in L_6(B_+(a_2)), \quad \nabla^2 p \in L_2(B_+(a_2)).$$

Next, for $\alpha, \beta = 1,2$ and for $k, j = 1,2,3$, we have functions

$$u_{,kj} \in W^1_2(B_+(a_2)) \qquad p_{,kj} \in L_2(B_+(a_2))$$

satisfying the conditions

$$\left\{\begin{array}{l} -\Delta u_{,kj} + \nabla p_{,kj} = -\operatorname{div} G_{,kj} \\ \\ \qquad \operatorname{div} u_{,kj} = 0 \end{array}\right. \qquad \text{in } B_+(a_2),$$

$$u_{,\alpha\beta}|_{x_3=0} = 0,$$

where $\nabla^3 G \in L_2(B_+(a_2))$. Here, we are going to proceed as in the case of the third derivatives. We let

$$h_{ijk} = -G_{im,mjk}.$$

From the linear theory, one can deduce that

$$\nabla^2 u_{,\alpha\beta} \in L_2(B_+(a_3)), \qquad \nabla p_{,\alpha\beta} \in L_2(B_+(a_3))$$

for $a < a_3 < a_2$. We start again with the incompressibility condition:

$$u_{3,333\alpha} = -u_{\beta,\beta 3\alpha} \in L_2(B_+(a_3)).$$

So,

$$\nabla^3 u_{3,\alpha} \in L_2(B_+(a_3)).$$

Then,

$$p_{,33\alpha} = h_{33\alpha} + u_{3,jj3} \in L_2(B_+(a_3))$$

and thus

$$\nabla^2 p_{,\alpha} \in L_2(B_+(a_3)).$$

Next,

$$u_{\beta,333\alpha} = -u_{\beta,\gamma\gamma 3\alpha} + p_{,\beta 3\alpha} - h_{\beta\alpha 3} \in L_2(B_+(a_3)).$$

So, we have

$$\nabla^3 u_{,\alpha} \in L_2(B_+(a_3)).$$

Now, let us go back to the incompressibility condition:

$$u_{3,3333} = -u_{\beta,\beta 333} \Rightarrow \nabla^4 u_3 \in L_2(B_+(a_3)).$$

For the pressure, we have

$$p_{,333} = h_{333} + \Delta u_{3,33} \Rightarrow \nabla^3 p \in L_2(B_+(a_3)).$$

Finally,

$$u_{\alpha,3333} = -u_{\alpha,\beta\beta 33} + p_{,\alpha 33} - h_{\alpha 33} \Rightarrow \nabla^4 u_\alpha \in L_2(B_+(a_3)).$$

Proceeding in a similar further, we complete the proof of the proposition. □

Theorem 2.6. *Let Ω be $\mathbb{R}^n$, or $\mathbb{R}^n_+$, or a bounded with smooth boundary. Let $u \in \hat{V}(\Omega)$ be a weak solution to the stationary Navier-Stokes equations, see Definition 3.1. Assume that the right hand side in these equations is of class C^∞ in the closure of the domain Ω. Then u is also of class C^∞ in the closure of the domain Ω.*

PROOF For $\Omega = \mathbb{R}^n$ or $\mathbb{R}^n_+$, the statement follows from Propositions 2.4 and 2.5. □

3.3 Comments

Chapter 3 contains standard results on the existence and regularity of solutions to the non-linear stationary boundary value problem. The main point of the chapter is the local regularity technique, which differs a bit from the technique developed for standard elliptic systems.

Chapter 4

Linear Non-Stationary Problem

4.1 Derivative in Time

Let us recall some definitions from the theory of distributions. $\mathcal{D}(\Omega)$ is a vector space that consists of all elements, belonging to $C_0^\infty(\Omega)$, where the convergence of a sequence of functions $\varphi_k \in C_0^\infty(\Omega)$ to a function $\varphi \in C_0^\infty(\Omega)$ is understood in the following sense. There exists a compact $K \subset \Omega$ such that $\mathrm{supp}\varphi_k, \mathrm{supp}\varphi \subset K$ and $\nabla^m\varphi_k \to \nabla^m\varphi$ uniformly on K for any $m \geq 0$. The space of all linear functionals on $\mathcal{D}(\Omega)$, being continuous with respect to the above convergence in $\mathcal{D}(\Omega)$, is denoted by $\mathcal{D}'(\Omega)$. Elements of $\mathcal{D}'(\Omega)$ are called distributions.

We may consider the space $\mathcal{D}'(a,b;\mathcal{D}'(\Omega))$. Given $T \in \mathcal{D}'(a,b;\mathcal{D}'(\Omega))$, let us denote by $\partial_t T$ or even by $\frac{d}{dt}T$ the following distribution

$$(\partial_t T(\varphi))(\chi) = -T(\varphi)(\partial_t \chi)$$

for any $\varphi \in \mathcal{D}(\Omega)$ and for any $\chi \in \mathcal{D}(a,b)$.

It is a too general definition for our purposes and we are going to use somewhat more specific. Let V be a Banach space, V^* be its dual space with duality relation $< v^*, v >$.

Definition 4.1. Let $v^* \in L_{1,\mathrm{loc}}(a,b;V^*)$ ($t \mapsto v^*(\cdot,t) \in V^*$ is measurable and $t \mapsto \|v^*(\cdot,t)\|_{V^*}$ is in $L_{1,\mathrm{loc}}(a,b)$). We call $u^* \in \mathcal{D}'(a,b;V^*)$ derivative of v^* in t if and only if

$$< u^*, v > (\chi) = -\int\limits_a^b < v^*(\cdot,t), v(\cdot) > \partial_t \chi(t)dt$$

for any $v \in V$ and for any $\chi \in C_0^\infty(a,b)$. We let $u^* = \partial_t v^*$.

As usual, the left-hand side of the above identity is written in the same way as the right-hand side, i.e.,

$$\int\limits_a^b < \partial_t v^*(\cdot,t), v(\cdot) > \chi(t)dt = -\int\limits_a^b < v^*(\cdot,t), v(\cdot) > \partial_t\chi(t)dt$$

for any $v \in V$ and for any $\chi \in C_0^\infty(a,b)$, although the left-hand side might make no sense as Lebesgue's integral.

Let us discuss the relationship between introduced notion of the derivative in time and the Sobolev derivatives. Assume that

$$\begin{aligned} V, V^* \in L_{1,\text{loc}}(\Omega), \quad C_0^\infty(\Omega) \subset V, \quad < v^*, v >= \int\limits_\Omega v^* v dx, \\ v^* \in L_{1,\text{loc}}(a,b; L_{1,\text{loc}}(\Omega)) = L_{1,\text{loc}}(\Omega \times]a,b[), \\ \partial_t v^* \in L_{1,\text{loc}}(\Omega \times]a,b[). \end{aligned} \tag{4.1.1}$$

Then $\partial_t v^*$ is a usual Sobolev derivative of v^* in the domain $\Omega\times]a,b[$. To understand why, we are going to use the following simple statement.

Lemma 4.1. *Given $\varepsilon > 0$ and $\varphi \in C_0^\infty(\Omega\times]a,b[)$, there exist positive integer number N and functions $\varphi_k \in C_0^\infty(\Omega)$, $\chi_k \in C_0^\infty(a,b)$, $k = 1, 2, ..., N$ such that*

$$\|\varphi - \sum_{k=1}^N \varphi_k \chi_k\|_{C^1(\overline{\Omega}\times[a,b])} < \varepsilon.$$

Let us assume that Lemma 4.1 has been proved. Suppose that $\partial_t v^*$ is the derivative in the sense of Definition 4.1 and satisfies assumptions (4.1.1). Our aim is to show that it is Sobolev's derivative as well. Take an arbitrary $\varepsilon > 0$ and an arbitrary function $\varphi \in C_0^\infty(\Omega\times]a,b[)$ and fix them. Clearly, $\varphi \in C_0^\infty(\Omega'\times]a',b'[)$ for some $\Omega' \Subset \Omega$ and for some $a < a'$ and $b' < b$. Let a natural number $N(\varepsilon)$ and functions φ_k and χ_k be from Lemma 4.1 in the domain $\Omega' \times]a', b'[$. Then we have

$$\Big| -\int_{a'}^{b'} \int_{\Omega'} v^*(x,t)\partial_t \varphi(x,t)dxdt - \int_{a'}^{b'} \int_{\Omega'} \partial_t v^*(x,t)\varphi(x,t)dxdt\Big| \leq$$

$$\leq \Big| \int_{a'}^{b'} \int_{\Omega'} v^*(x,t)\Big(\partial_t\varphi(x,t) - \partial_t \sum_{k=1}^{N(\varepsilon)} \varphi_k(x)\chi_k(t)\Big)dxdt\Big| +$$

$$+ \Big| \int_{a'}^{b'} \int_{\Omega'} \partial_t v^*(x,t)\Big(\varphi(x,t) - \sum_{k=1}^{N(\varepsilon)} \varphi_k(x)\chi_k(t)\Big)dxdt\Big| \leq$$

$$\leq c\|\varphi - \sum_{k=1}^{N(\varepsilon)} \varphi_k\chi_k\|_{C^1(\overline{\Omega}'\times[a',b'])}\Big(\|v^*\|_{L_1(\Omega'\times]a',b'[)} + \|\partial_t v^*\|_{L_1(\Omega'\times]a',b'[)}\Big)$$

$$\leq c\varepsilon\Big(\|v^*\|_{L_1(\Omega'\times]a',b'[)} + \|\partial_t v^*\|_{L_1(\Omega'\times]a',b'[)}\Big).$$

Tending ε to zero, we get

$$\int_{a'}^{b'} \int_{\Omega'} \partial_t v^*(x,t)\varphi(x,t)dxdt = -\int_{a'}^{b'} \int_{\Omega'} v^*(x,t)\partial_t\varphi(x,t)dxdt \tag{4.1.2}$$

for any $\varphi \in C_0^\infty(\Omega\times]a,b[)$. So, $\partial_t v^*$ is Sobolev's derivative as well.

Regarding the inverse statement, we argue as follows. Suppose that $\partial_t v^*$ is Sobolev's derivative, i.e., it satisfies identity (4.1.2) with v^* and $\partial_t v^*$ from $L_{1,\text{loc}}(\Omega\times]a,b[)$. If we assume in addition that $C_0^\infty(\Omega)$ is dense in V, then $\partial_t v^*$ is a derivative of v^* in the sense of Definition 4.1.

Proof of Lemma 4.1 We may extend φ by zero to the whole $\mathbb{R}^n \times \mathbb{R}$ ($\Omega \subset \mathbb{R}^n$). Take a cube $C_l\times]-l,l[$ so that $C_l\times]-l,l[\supset \operatorname{supp}\varphi$. Here, $C_l = \{x \in \mathbb{R}^n : |x_i| < l,\ i = 1,2,...,n\}$. Then we can expand φ as the Fourier series in spatial variable x

$$\varphi(x,t) = \sum_{k=0}^{\infty} \sum_{|m|=k} c_m(t)e^{i\pi\frac{x\cdot m}{l}},$$

where

$$c_m(t) = \frac{1}{(2l)^n}\int_{C_l} \varphi(x,t)e^{-i\pi\frac{x\cdot m}{l}}dx.$$

The Fourier series converges very well. So, after taking real and imaginary parts, given $\varepsilon > 0$, we find the number $N(\varepsilon)$ such that

$$\|\varphi - \Phi_{N(\varepsilon)}\|_{C^1(\overline{C}_l\times[-l,l])} < \varepsilon,$$

where $\Phi_{N(\varepsilon)}(x,t) = \sum_{k=1}^{N(\varepsilon)} \varphi_k(x)\chi_k(t)$. Assume that there exist functions $\varphi_0 \in C_0^\infty(\Omega)$, $\chi_0 \in C_0^\infty(]a,b[)$ with the following property

$$\varphi_0(x)\chi_0(t) = 1 \tag{4.1.3}$$

if $(x,t) \in \mathrm{supp}\varphi$. We may let then

$$\widetilde{\Phi}_{N(\varepsilon)} = \Phi_{N(\varepsilon)}\varphi_0\chi_0$$

and show

$$\|\varphi - \widetilde{\Phi}_{N(\varepsilon)}\|_{C^1(\overline{\Omega}\times[a,b])} = \|(\varphi - \Phi_{N(\varepsilon)})\varphi_0\chi_0\|_{C^1(\overline{\Omega}\times[a,b])} \leq c(\Omega,a,b,l)\varepsilon.$$

To justify (4.1.3), let us introduce the following sets

$$(\mathrm{supp}\varphi)_t = \{x \in \Omega : \ (x,t) \in \mathrm{supp}\,\varphi\},$$

$$\Lambda = \{t \in [a,b] : \ (\mathrm{supp}\,\varphi)_t \neq \emptyset\}.$$

Let $t_1 = \inf\limits_{t\in\Lambda} t$ and $t_2 = \sup\limits_{t\in\Lambda} t$. We claim that $a < t_1 \leq t_2 < b$. Assume that $t_2 = b$. Then, by the definition, there exists a sequence $(x_k, t_k) \in \mathrm{supp}\,\varphi$ with $t_k \to b$ as $k \to \infty$. Selecting if necessary a subsequence, we have a contradiction for the limit point $(x,b) \in \mathrm{supp}\,\varphi$. Now, let us show that

$$K = \cup_{t_1 \leq t \leq t_2}(\mathrm{supp}\,\varphi)_t$$

is a closed set of $\mathbb{R}^3$. Assume that $x_k \in K$ and $x_k \to x$ as $k \to \infty$. For each k, one can find $t_k \in [t_1,t_2]$ such that $(x_k,t_k) \in \mathrm{supp}\,\varphi$. We may assume that $t_k \to t \in [t_1,t_2]$ and then, by the definition of the support, $(x,t) \in \mathrm{supp}\,\varphi$. So, $x \in K$ and thus K is closed. It remains to find an open set $\Omega_1 \Subset \Omega$ such that $K \subset \Omega_1$. So, $\mathrm{supp}\,\varphi \subset \Omega_1 \times [t_1,t_2]$. The rest of the proof is easy.
□

4.2 Explicit Solution

Consider a bounded domain $\Omega \subset \mathbb{R}^n$ with smooth boundary and the following initial-boundary value problem

$$\begin{aligned} \partial_t u - \Delta u = f - \nabla p \quad \text{and} \quad \mathrm{div}\, u = 0 \quad \text{in} \quad Q_T = \Omega \times]0,T[,& \\ u = 0 \quad \text{on} \quad \partial\Omega \times [0,T],& \quad (4.2.1) \\ u(x,0) = a(x) \quad x \in \Omega.& \end{aligned}$$

Assume that

$$a \in \overset{\circ}{J}(\Omega). \tag{4.2.2}$$

This problem can be written in the operator form

$$\partial_t u - \widetilde{\Delta} u = f \in L_2(0,T;(\overset{\circ}{J}{}^1_2(\Omega))'),$$
$$u|_{t=0} = a \in \overset{\circ}{J}(\Omega), \tag{4.2.3}$$

see notation for the Stokes operator $\widetilde{\Delta}$ and for the dual space in the last section of Chapter 2.

Our task is to construct an explicit solution provided eigenvalues and eigenfunctions of the Stokes operator $\widetilde{\Delta}$ in the domain Ω are known. So, we are given:

$$-\widetilde{\Delta}\varphi_k = \lambda_k \varphi_k \quad \text{in} \quad \Omega,$$
$$\varphi_k = 0 \quad \text{on} \quad \Omega, \tag{4.2.4}$$

where $k = 1, 2, \ldots$.

First, let us expand functions f and a as Fourier series, using eigenfunctions of the Stokes operator,

$$f(x,t) = \sum_{k=1}^{\infty} f_k(t)\varphi_k(x),$$

where

$$f_k(t) = (f(\cdot,t), \varphi_k(\cdot))$$

and

$$a(x) = \sum_{k=1}^{\infty} a_k \varphi_k(x), \qquad a_k = (a, \varphi_k).$$

By our assumptions,

$$\|f\|^2_{L_2(0,T;(\overset{\circ}{J}{}^1_2(\Omega))')} = \int_0^T \sum_{k=1}^{\infty} \frac{1}{\lambda_k} |f_k(t)|^2 dt < \infty,$$
$$\|a\|^2_{2,\Omega} = \sum_{k=1}^{\infty} a_k^2 < \infty. \tag{4.2.5}$$

We are looking for a solution to (4.2.3) of the form

$$u(x,t) = \sum_{k=1}^{\infty} c_k(t)\varphi_k(x). \tag{4.2.6}$$

Assume that

$$c_k(0) = a_k, \qquad k = 1, 2, \ldots. \tag{4.2.7}$$

Our further calculations are going to be formal. Later on, we will explain in what sense the formal solution is a solution to problem (4.2.3). So, if we insert (4.2.6) into (4.2.3), then the identity

$$\sum_{k=1}^{\infty} c'_k\varphi_k + \lambda_k c_k\varphi_k = \sum_{k=1}^{\infty} f_k\varphi_k$$

comes out, which holds if one lets

$$c'_k(t) + \lambda_k c_k(t) = f_k(t), \qquad c_k(0) = a_k, \tag{4.2.8}$$

where $k = 1, 2, ...$ System (4.2.8) has a unique solution

$$c_k(t) = e^{-\lambda_k t}\Big(a_k + \int_0^t e^{\lambda_k\tau} f_k(\tau)d\tau\Big). \tag{4.2.9}$$

Let us analyze properties of the formal solution (4.2.6) and (4.2.9). We have

$$c_k^2(t) \leq 2e^{-2\lambda_k t}a_k^2 + 2\Big|\int_0^t e^{-\lambda_k(t-\tau)} f_k(\tau)d\tau\Big|^2$$

$$\leq 2e^{-2\lambda_k t}a_k^2 + 2\int_0^t e^{-2\lambda_k(t-\tau)}d\tau \int_0^t f_k^2(\tau)d\tau$$

$$\leq 2e^{-2\lambda_k t}a_k^2 + \frac{1}{\lambda_k}\int_0^t f_k^2(\tau)d\tau - \frac{1}{\lambda_k}e^{-2\lambda_k t}\int_0^t f_k^2(\tau)d\tau.$$

So, finally,

$$c_k^2(t) \leq 2e^{-2\lambda_k t}a_k^2 + \frac{1}{\lambda_k}\int_0^t f_k^2(\tau)d\tau. \tag{4.2.10}$$

Summing up the above inequalities, we establish the estimate

$$\|u(\cdot,t)\|_2^2 = \sum_{k=1}^{\infty} c_k^2(t) \leq 2e^{-2\lambda_1 t}\sum_{k=1}^{\infty} a_k^2 + \sum_{k=1}^{\infty}\frac{1}{\lambda_k}\int_0^t f_k^2(\tau)d\tau$$
$$\leq 2e^{-2\lambda_1 t}\|a\|_2^2 + \|f\|^2_{L_2(0,T;(\overset{\circ}{J}{}^1_2(\Omega))')}, \tag{4.2.11}$$

which implies

$$\|u\|^2_{L_\infty(0,T;L_2(\Omega))} \leq 2\|a\|_2^2 + \|f\|^2_{L_2(0,T;(\overset{\circ}{J}{}^1_2(\Omega))')}. \tag{4.2.12}$$

To get the second estimate, we multiply the first equation in (4.2.8) by $c_k(t)$ and apply Young's inequality

$$\begin{aligned} c_k'(t)c_k(t) + \lambda_k c_k^2(t) &= f_k(t)c_k(t) \\ &\leq \frac{1}{2}\frac{f_k^2(t)}{\lambda_k} + \frac{1}{2}\lambda_k c_k^2(t). \end{aligned}$$

So,

$$(c_k^2(t))' + \lambda_k c_k^2(t) \leq \frac{f_k^2(t)}{\lambda_k}.$$

Integration in t gives us:

$$\begin{aligned} c_k^2(T) + \lambda_k \int_0^T c_k^2(t)dt &\leq c_k^2(0) + \int_0^T \frac{f_k^2(t)}{\lambda_k}dt \\ &= a_k^2 + \int_0^T \frac{f_k^2(t)}{\lambda_k}dt. \end{aligned}$$

Then, after summation, we arrive at the second estimate

$$\begin{aligned} \|\nabla u\|^2_{L_2(0,T;L_2(\Omega))} &= \int_0^T \sum_{k=1}^{\infty} \lambda_k c_k^2(t)dt \\ &\leq \|a\|_2^2 + \|f\|^2_{L_2(0,T;(\overset{\circ}{J}{}_2^1(\Omega))')}. \end{aligned} \tag{4.2.13}$$

Bounds (4.2.12) and (4.2.13) are called *energy* estimates.

The final estimate will be derived from (4.2.8) in the following way

$$\begin{aligned} \|\partial_t u\|^2_{L_2(0,T;(\overset{\circ}{J}{}_2^1(\Omega))')} &= \int_0^T \sum_{k=1}^{\infty} \frac{|c_k'(t)|^2}{\lambda_k}dt \\ &\leq 2\int_0^T \sum_{k=1}^{\infty} \lambda_k c_k^2(t)dt + 2\int_0^T \sum_{k=1}^{\infty} \frac{f_k^2(t)}{\lambda_k}dt \\ &\leq 2\|\nabla u\|^2_{L_2(0,T;L_2(\Omega))} + 2\|f\|^2_{L_2(0,T;(\overset{\circ}{J}{}_2^1(\Omega))')}. \end{aligned}$$

So, applying (4.2.13), we find the third estimate

$$\|\partial_t u\|^2_{L_2(0,T;(\overset{\circ}{J}{}_2^1(\Omega))')} \leq 2\|a\|_2^2 + 4\|f\|^2_{L_2(0,T;(\overset{\circ}{J}{}_2^1(\Omega))')}. \tag{4.2.14}$$

Now, we wish to figure out in which sense (4.2.3) holds. Let us take an arbitrary function $w \in L_2(0,T;\overset{\circ}{J}{}_2^1(\Omega))$ and expand it as a Fourier series

$$w(x,t) = \sum_{k=1}^{\infty} d_k(t)\varphi_k(x).$$

Obviously,

$$\|\nabla w\|^2_{L_2(0,T;L_2(\Omega))} = \int\limits_0^T \sum_{k=1}^\infty \lambda_k d_k^2(t)dt < \infty.$$

Hence,

$$\int\limits_0^T\int\limits_\Omega \partial_t u \cdot w dxdt = \int\limits_0^T \sum_{k=1}^\infty c_k'(t) d_k(t)dt,$$

$$\int\limits_0^T\int\limits_\Omega \nabla u : \nabla w dxdt = \int\limits_0^T\int\limits_\Omega \sum_{k=1}^\infty c_k(t)d_k(t)|\nabla \varphi_k|^2 dxdt =$$

$$= \int\limits_0^T \sum_{k=1}^\infty \lambda_k c_k(t) d_k(t)dt,$$

$$\int\limits_0^T\int\limits_\Omega f \cdot w dxdt = \int\limits_0^T \sum_{k=1}^\infty f_k(t) d_k(t)dt,$$

and, by (4.2.8),

$$\int\limits_0^T\int\limits_\Omega \Big(\partial_t u \cdot w + \nabla u : \nabla w - f \cdot w\Big) dxdt =$$

$$= \int\limits_0^T \sum_{k=1}^\infty \Big(c_k'(t) + \lambda_k c_k(t) - f_k(t)\Big) d_k(t)dt = 0.$$

Taking $w(x,t) = \chi(t)v(x)$ with $v \in \overset{\circ}{J}{}^1_2(\Omega)$ and $\chi \in C^1_0(0,T)$, we get that, for a.a. $t \in]0,T[$, the identity

$$\int\limits_\Omega \Big(\partial_t u(x,t) \cdot v(x) + \nabla u(x,t) : \nabla v(x)\Big) dx = \int\limits_\Omega f(x,t) \cdot v(x) dx \qquad (4.2.15)$$

holds for all $v \in \overset{\circ}{J}{}^1_2(\Omega)$. To be more precise, (4.2.15) is fulfilled at all Lebesgue's points of the following functions $t \mapsto \partial_t u(\cdot,t)$, $t \mapsto \nabla u(\cdot,t)$, and $t \mapsto f(\cdot,t)$. Identity (4.2.15) is called the weak form of the first equation in (4.2.3).

It remains to establish in what sense the initial data in (4.2.3) are satisfied.

Lemma 4.2. *Function* $t \mapsto u(\cdot,t) \in (\overset{\circ}{J}{}^1_2(\Omega))'$ *can be modified on a zero-measure subset of* $[0,T]$ *so that, for each* $v \in \overset{\circ}{J}{}^1_2(\Omega)$, *the function*

$$t \mapsto \int\limits_\Omega u(\cdot,t)\cdot v(\cdot)dx$$

is continuous on $[0,T]$.

Proof Since $u \in L_2(0,T;(\overset{\circ}{J}{}^1_2(\Omega))')$, a.a. points $t_0 \in [0,T]$ are Lebesgue's points of $t \mapsto u(\cdot,t)$ in the following sense

$$\frac{1}{2\varepsilon}\int\limits_{t_0-\varepsilon}^{t_0+\varepsilon} \|u(\cdot,t)-u(\cdot,t_0)\|_{(\overset{\circ}{J}{}^1_2(\Omega))'}dt \to 0$$

as $\varepsilon \to 0$.

Denote by S the set of al Lebesgue's points of $t \mapsto u(\cdot,t)$. We know that $|S| = T$. Let $t_0 < t_1$ be two points from S. By the definition of the derivative $\partial_t u$,

$$\int\limits_0^T\int\limits_\Omega \partial_t u(x,t)\cdot v(x)\chi(t)dxdt = -\int\limits_0^T\int\limits_\Omega u(x,t)\cdot v(x)\partial_t\chi(t)dxdt$$

for any $v \in \overset{\circ}{J}{}^1_2(\Omega)$ and for any $\chi \in C^1_0(0,T)$. We can easily extend the latter identity to functions $\chi \in \overset{\circ}{W}{}^1_2(0,T)$. Pick up a test function $\chi = \chi_\varepsilon$ so that $\chi_\varepsilon(t) = 0$ if $0 < t \le t_0 - \varepsilon$ or $t_1 + \varepsilon \le t < T$, $\chi_\varepsilon(t) = 1$ if $t_0+\varepsilon \le t \le t_1 - \varepsilon$, $\chi_\varepsilon(t) = (t - t_0 + \varepsilon)/(2\varepsilon)$ if $t_0 - \varepsilon < t < t_0 + \varepsilon$, and $\chi_\varepsilon(t) = (t_1 + \varepsilon - t)/(2\varepsilon)$ if $t_1 - \varepsilon < t < t_1 + \varepsilon$. Then, we have

$$\int\limits_0^T\int\limits_\Omega \partial_t u(x,t)\cdot v(x)\chi_\varepsilon(t)dxdt = \frac{1}{2\varepsilon}\int\limits_{t_1-\varepsilon}^{t_1+\varepsilon}\int\limits_\Omega u(x,t)\cdot v(x)dxdt$$
$$-\frac{1}{2\varepsilon}\int\limits_{t_0-\varepsilon}^{t_0+\varepsilon}\int\limits_\Omega u(x,t)\cdot v(x)dxdt. \quad (4.2.16)$$

Obviously,

$$\Big|\frac{1}{2\varepsilon}\int\limits_{t_0-\varepsilon}^{t_0+\varepsilon}\int\limits_\Omega \Big(u(x,t)-u(x,t_0)\Big)\cdot v(x)dxdt\Big|$$

$$\le \frac{1}{2\varepsilon}\int\limits_{t_0-\varepsilon}^{t_0+\varepsilon} \|u(\cdot,t)-u(\cdot,t_0)\|_{(\overset{\circ}{J}{}^1_2(\Omega))'}dt\|v(\cdot)\|_{\overset{\circ}{J}{}^1_2(\Omega)} \to 0$$

as $\varepsilon \to 0$. So, after taking the limit, we find

$$\int_\Omega u(x,t_1)\cdot v(x)dx = \int_\Omega u(x,t_0)\cdot v(x)dx + \int_{t_0}^{t_1}\int_\Omega \partial_t u(x,\tau)\cdot v(x)dxd\tau$$

for a.a. $t_1 \in [0,T]$. Since the right-hand side of the above identity is a continuous function with respect to t_1, the left-hand side is continuous in t_1 as well. □

Now, coming back to our function u, we notice that $u \in L_\infty(0,T;L_2(\Omega))$. Therefore, we can state that

$$t \mapsto \int_\Omega u(x,t)\cdot v(x)dx \quad \textit{is continuous in } t \textit{ on } [0,T]$$

for each function $v \in \overset{\circ}{J}(\Omega)$ and even for each function $v \in L_2(\Omega)$. The latter follows from the fact that, for any $v \in L_2(\Omega)$, the Helmholtz-Weyl decomposition holds in the Ladyzhenskaya form so that $v = v_1 + \nabla p$, where $v_1 \in \overset{\circ}{J}(\Omega)$ and $p \in W^1_2(\Omega)$, and thus

$$\int_\Omega u(x,t)\cdot v(x)dx = \int_\Omega u(x,t)\cdot v_1(x)dx,$$

since $u(\cdot,t) \in \overset{\circ}{J}(\Omega)$.

Since $u(\cdot,0) = a(\cdot)$ by construction of u, our initial data are satisfied at least in the following sense

$$\lim_{t\to+0}\int_\Omega u(x,t)\cdot v(x)dx = \int_\Omega a(x)\cdot v(x)dx$$

for any $v \in L_2(\Omega)$.

However, in our particular case, we can gain even more.

Theorem 2.1. *Assume that*

$$u \in L_2(0,T;\overset{\circ}{J}{}^1_2(\Omega)), \qquad \partial_t u \in L_2(0,T;(\overset{\circ}{J}{}^1_2(\Omega))').$$

Then $u \in C([0,T];L_2(\Omega))$ *and*

$$\int_{t_1}^{t}\int_\Omega \partial_t u\cdot u dxdt = \frac{1}{2}\|u(\cdot,t)\|^2_{2,\Omega} - \frac{1}{2}\|u(\cdot,t_1)\|^2_{2,\Omega} \tag{4.2.17}$$

for all $t,t_1 \in [0,T]$.

PROOF Indeed, for

$$u(x,t) = \sum_{k=1}^{\infty} d_k(t)\varphi_k(x),$$

we have

$$\|u\|^2_{L_2(0,T;\overset{\circ}{J}{}^1_2(\Omega))} = \int_0^T \sum_{k=1}^{\infty} \lambda_k d_k^2(t)dt$$

and

$$\|\partial_t u\|^2_{L_2(0,T;(\overset{\circ}{J}{}^1_2(\Omega))')} = \int_0^T \sum_{k=1}^{\infty} \frac{1}{\lambda_k}(d_k'(t))^2 dt.$$

In a view of Lemma 4.2, it is sufficient to show that the function $t \mapsto \|u(\cdot,t)\|_{2,\Omega}$ is continuous. We know that functions $t \mapsto d_k(t)$ are continuous on $[0,T]$. Therefore, the function $t \mapsto g_N(t) = \sum_{k=1}^N d_k^2(t)$ is continuous on $[0,T]$ as well. We know also that $g_N(t) \to \|u(\cdot,t)\|^2_{2,\Omega}$ as $N \to \infty$. So, we need to show that the sequence $g_N(t)$ is uniformly bounded and the convergence is uniform.

First, we show uniform boundedness. Indeed,

$$g_N(t) - g_N(t_1) = 2\int_{t_1}^{t} \sum_{k=1}^{N} d_k'(\tau)d_k(\tau)d\tau \tag{4.2.18}$$

for any $t, t_1 \in [0,T]$ and thus

$$g_N(t) \leq g_N(t_1) + 2\Big(\int_0^T \sum_{k=1}^{\infty} \frac{1}{\lambda_k}(d_k'(t))^2 dt\Big)^{\frac{1}{2}} \Big(\int_0^T \sum_{k=1}^{\infty} \lambda_k d_k^2(t)dt\Big)^{\frac{1}{2}}$$
$$\leq g_N(t_1) + 2\|\nabla u\|_{L_2(0,T;L_2(\Omega))}\|\partial_t u\|_{L_2(0,T;(\overset{\circ}{J}{}^1_2(\Omega))')}.$$

The above inequality can be integrated with respect to t_1

$$g_N(t) \leq \frac{1}{T}\int_0^T g_N(t_1)dt_1 + 2\|\nabla u\|_{L_2(0,T;L_2(\Omega))}\|\partial_t u\|_{L_2(0,T;(\overset{\circ}{J}{}^1_2(\Omega))')}$$

$$\leq \frac{1}{T\lambda_1}\|\nabla u\|^2_{L_2(0,T;L_2(\Omega))} + 2\|\nabla u\|_{L_2(0,T;L_2(\Omega))}\|\partial_t u\|_{L_2(0,T;(\overset{\circ}{J}{}^1_2(\Omega))')}.$$

So, the uniform boundedness follows.

From (4.2.18), one can deduce uniform continuity. Indeed,

$$|g_N(t) - g_N(t_1)| \leq 2\Big(\int_{t_1}^{t} \|\nabla u(\cdot,t)\|^2_{2,\Omega}dt\Big)^{\frac{1}{2}} \|\partial_t u\|_{L_2(0,T;(\overset{\circ}{J}{}^1_2(\Omega))')}.$$

Hence, $g_N(t)$ converges to $\|u(\cdot,t)\|^2_{2,\Omega}$ uniformly, which means that the function $t \mapsto \|u(\cdot,t)\|^2_{2,\Omega}$ is continuous on $[0,T]$.

Finally, (4.2.17) follows directly from (4.2.18) if $N \to +\infty$. □

Actually, an abstract analogue of Theorem 2.1 takes places:

Theorem 2.2. *Let H be a Hilbert space, V be a reflexive Banach space, and V is continuously imbedded into H. Let V contain a countable set S which is dense in V and in H, i.e.,*

$$V = [S]^V, \qquad H = [S]^H.$$

Let V^* be a dual space to V with respect to scalar product in H with the norm

$$\|v^*\|_{V^*} = \sup\{(v^*, v)_H : \ v \in V, \ \|v\|_V = 1\}.$$

Assume that $v \in L_p(0,T;V) \cap L_2(0,T;H)$ and $\partial_t v \in L_{p'}(0,T;V^*)$ with $p' = p/(p-1)$ and $p > 1$.

Then, $v \in C([0,T];H)$ and

$$\|v(\cdot,t)\|^2_H - \|v(\cdot,t_1)\|^2_H = 2\int\limits_{t_1}^{t} (\partial_t v(\cdot,\tau), v(\cdot,\tau))_H d\tau$$

for any $t, t_1 \in [0,T]$.

PROOF We start with some general facts. Let $t \mapsto v(\cdot,t)$, where $v \in L_p(0,T;V)$. It is supposed that v is extended by zero outside $[0,T]$. The first fact is the integral continuity: for any $\varepsilon > 0$, there is a number $\delta(\varepsilon) > 0$ such that

$$\int\limits_0^T \|v(\cdot,t+h) - v(\cdot,t)\|^p_V dt < \varepsilon$$

whenever $|h| < \delta(\varepsilon)$. This property provides the following. Let

$$v_\varepsilon(\cdot,t) = \int\limits_0^T \omega_\varepsilon(t-\tau) v(\cdot,\tau) d\tau,$$

where ω_ε is a standard mollifying kernel. We then have

$$\int_0^T \|v_\varepsilon(\cdot,t) - v(\cdot,t)\|_V^p dt \leq$$

$$\leq \int_{-\infty}^{\infty} \|v_\varepsilon(\cdot,t) - v(\cdot,t)\|_V^p dt =$$

$$= \int_{-\infty}^{\infty} \| \int_{-\infty}^{\infty} \omega_\varepsilon(t-\tau)(v(\cdot,\tau) - v(\cdot,t))d\tau\|_V^p dt \leq$$

$$\leq \int_{-\infty}^{\infty}\int_{-\infty}^{\infty} \omega_\varepsilon(t-\tau)\|v(\cdot,\tau) - v(\cdot,t))\|_V^p d\tau dt =$$

$$= \int_{-\infty}^{\infty}\int_{-\infty}^{\infty} \omega_\varepsilon(t_1)\|v(\cdot,t_1+t) - v(\cdot,t))\|_V^p d\tau dt \leq \gamma$$

provided $\varepsilon < \frac{1}{2}\delta(\gamma)$. This means that $v_\varepsilon \to v$ in $L_p(0,T;V)$.

It can be shown (exercise) that

$$\partial_t v_\varepsilon(\cdot,t) = (\partial_t v)_\varepsilon(\cdot,t)$$

provided $0 < \varepsilon \leq t \leq T - \varepsilon$. So, we can claim

$$\partial_t v_\varepsilon \to \partial_t v \qquad \text{in } L_{p',\mathrm{loc}}(0,T;V^*).$$

Further, we can use the same trick as in the case of star-shaped domains. Without loss of generality, we may replace the interval $]0,T[$ with $]-1,1[$. We take $\lambda > 1$ and define

$$v^\lambda(\cdot,t) = v(\cdot,\frac{t}{\lambda}), \qquad |t| \leq \lambda$$

and thus

$$\partial_t v^\lambda(\cdot,t) = \frac{1}{\lambda}\partial_s v(\cdot,s)|_{s=\frac{t}{\lambda}}.$$

Here, the crucial things are as follows:

$$\|v^\lambda - v\|_{L_p(-1,1;V)} + \|v^\lambda - v\|_{L_2(-1,1;H)} \to 0$$

and

$$\|\partial_t v^\lambda - \partial_t v\|_{L_{p'}(-1,1;V^*)} \to 0$$

as $\lambda \to 1$. Moreover, for fixed $\lambda > 1$,

$$\|v^\lambda - (v^\lambda)_\varepsilon\|_{L_p(-1,1;V)} + \|v^\lambda - (v^\lambda)_\varepsilon\|_{L_2(-1,1;H)} \to 0$$

and

$$\|\partial_t v^\lambda - \partial_t(v^\lambda)_\varepsilon\|_{L_{p'}(-1,1;V^*)} \to 0$$

as $\varepsilon \to 0$. Summarizing these two properties, we may construct a sequence $v^{(k)}$ that is differentiable in t and satisfies:

$$\|v^{(k)} - v\|_{L_p(-1,1;V)} + \|v^{(k)} - v\|_{L_2(-1,1;H)} \to 0$$

and

$$\|\partial_t v^{(k)} - \partial_t v\|_{L_{p'}(-1,1;V^*)} \to 0$$

as $k \to \infty$.

Now, let $u = v^{(k)} - v^{(m)}$, we have the identity

$$\|u(\cdot,t)\|_H^2 = 2\int_{t_1}^{t} (\partial_t u(\cdot,\tau), u(\cdot,\tau))_H d\tau + \|u(\cdot,t_1)\|_H^2, \tag{4.2.19}$$

which implies the bound

$$\sup_{-1<t<1} \|u(\cdot,t)\|_H^2 \le \frac{1}{2}\Big(2\|\partial_t u\|_{L_{p'}(-1,1;V^*)}\|u\|_{L_p(-1,1;V)} + \|u\|^2_{L_2(-1,1;H)}\Big).$$

In turn, the above inequality yields that the $v^{(k)}$ is a Cauchy sequence in $C([0,T];H)$ and thus $v^{(k)}$ converges to v in $C([0,T];H)$. The identity of Theorem 2.2 can be derived from (4.2.19) with $u = v^{(k)}$ and $k \to \infty$. □

Theorem 2.3. *Assume that $a \in \overset{\circ}{J}(\Omega)$ and $f \in L_2(0,T;(\overset{\circ}{J}{}^1_2(\Omega))')$. There exists a unique function u called a weak solution to (4.2.1) such that:*

$$u \in L_2(0,T;\overset{\circ}{J}{}^1_2(\Omega)), \quad \partial_t u \in L_2(0,T;(\overset{\circ}{J}{}^1_2(\Omega))'); \tag{4.2.20}$$

for a.a. $t \in [0,T]$,

$$\int_\Omega \Big[\partial_t u(x,t)\cdot v(x) + \nabla u(x,t) : \nabla v(x)\Big] dx = \int_\Omega f(x,t)\cdot v(x) dx \tag{4.2.21}$$

for any $v \in \overset{\circ}{J}{}^1_2(\Omega)$;

$$u(\cdot,0) = a(\cdot) \tag{4.2.22}$$

and (4.2.22) is fulfilled in the L_2-sense, i.e., $\|u(\cdot,t) - a(\cdot)\|_{2,\Omega} \to 0$ as $t \to +0$. Moreover,

$$u \in C([0,T];L_2(\Omega)).$$

PROOF Existence has been already proven. It remains to show uniqueness. Assume that u^1 is another solution satisfying (4.2.20)–(4.2.22). Then for $w = u - u^1$ we have

$$\int_\Omega \partial_t w(x,t) \cdot v(x)dx + \int_\Omega \nabla w(x,t) : \nabla v(x)dx = 0$$

for a.a. $t \in [0,T]$ and for any $v \in \overset{\circ}{J}{}^1_2(\Omega)$ and thus

$$\int_\Omega \partial_t w(x,t) \cdot w(x,t)dx + \int_\Omega |\nabla w(x,t)|^2 dx = 0.$$

Integrating the latter identity with respect to t in $[0,t_0]$, we get, by Theorem 2.1,

$$\|w(\cdot,t_0)\|^2_{2,\Omega} \leq \|w(\cdot,0)\|^2_{2,\Omega} = 0$$

for any $t_0 \in [0,T]$. □

4.3 Cauchy Problem

Here, we assume that $\Omega = \mathbb{R}^n$ and consider the following initial value problem

$$\partial_t u - \Delta u = f - \nabla p \quad \text{and} \quad \operatorname{div} u = 0 \quad \text{in} \quad Q_T = \mathbb{R}^n \times]0,T[,$$
$$u(x,0) = a(x) \quad x \in \mathbb{R}^n. \tag{4.3.1}$$

Assume that

$$a \in \overset{\circ}{J} \equiv \overset{\circ}{J}(\mathbb{R}^n). \tag{4.3.2}$$

It is supposed also that

$$f \in L_2(0,T;\overset{\circ}{J}). \tag{4.3.3}$$

In this case, the Cauchy problem can be reduced to the Cauchy problem for the heat equation

$$\partial_t u - \Delta u = f \qquad \text{in} \quad Q_T,$$
$$u(x,0) = a(x) \quad x \in \mathbb{R}^n. \tag{4.3.4}$$

Indeed, assume that u is a solution to the Cauchy problem (4.3.4). Take the divergence of equations in (4.3.4). Then we have

$$\partial_t \mathrm{div} u - \Delta \mathrm{div} u = 0 \qquad \text{in} \quad Q_T,$$
$$\mathrm{div} u(x,0) = 0 \quad x \in \mathbb{R}^n.$$

By the unique solvability of the Cauchy problem for the heat equation, one can state that $\mathrm{div} u = 0$ in Q_T. The pressure field is an arbitrary function of t.

Solution to (4.3.4) can be given in an explicit form with the help of the fundamental solution to the heat equation:

$$u(x,t) = \int\limits_{\mathbb{R}^n} \Gamma(x-y,t)a(x)dx + \int\limits_0^t \int\limits_{\mathbb{R}^n} \Gamma(x-y,t-\tau)f(y,\tau)dyd\tau,$$

where

$$\Gamma(a,t) = \frac{1}{(4\pi t)^{\frac{n}{2}}} e^{-\frac{|x|^2}{4t}}$$

for $x \in \mathbb{R}^n$ and $t > 0$. This formula is a good source for understanding properties of solutions to (4.3.1).

4.4 Pressure Field. Regularity

Let us go back to initial-boundary value problem (4.2.1) and its functional formulation (4.2.2)

$$\partial_t u - \widetilde{\Delta} u = f \in L_2(0,T; (\overset{\circ}{J}{}^1_2(\Omega))'),$$
$$u|_{t=0} = a \in \overset{\circ}{J}(\Omega). \tag{4.4.1}$$

Assuming that our domain Ω is bounded and has sufficiently smooth boundary, we have constructed a weak solution to (4.4.1) with the help of eigenfunctions φ_k of the Stokes operator in Ω, namely, in the form:

$$u(x,t) = \sum_{k=1}^{\infty} c_k(t)\varphi_k(x).$$

For unknown coefficient $c_k(t)$, the following system of equations

$$c_k'(t) + \lambda_k c_k(t) = f_k(t),$$
$$c_k(0) = a_k \tag{4.4.2}$$

holds, where $k = 1, 2, ...$ and where

$$f_k(t) = \int\limits_{\Omega} f(x,t) \cdot \varphi_k(x)dx = (f(\cdot,t), \varphi_k(\cdot)), \qquad a_k = (a, \varphi_k).$$

Now, we are going to assume additionally that

$$a \in \overset{\circ}{J}{}^1_2(\Omega), \qquad f \in L_2(0,T; \overset{\circ}{J}(\Omega)). \tag{4.4.3}$$

Then,

$$\|f\|^2_{L_2(0,T;\overset{\circ}{J}(\Omega))} = \|f\|^2_{2,Q_T} = \int\limits_0^T \sum_{k=1}^{\infty} f_k^2(t)dt < \infty,$$

$$\|\nabla a\|^2_{2,\Omega} = \sum_{k=1}^{\infty} \lambda_k a_k^2 < \infty. \qquad (4.4.4)$$

Next, let us multiply the first equation in (4.4.2) by c_k, sum up the result from 1 to N, integrate the sum in time over the interval $]0,t[$, and find the identity

$$\frac{1}{2}\sum_{k=1}^{N} \lambda_k c_k^2(t) + \int\limits_0^t \sum_{k=1}^{N} |c_k'(\tau)|^2 d\tau$$

$$= \frac{1}{2}\sum_{k=1}^{N} \lambda_k a_k^2 + \int\limits_0^t \sum_{k=1}^{N} f_k(\tau)c_k'(\tau)d\tau,$$

which yields the bound

$$\sum_{k=1}^{N} \lambda_k c_k^2(t) + \int\limits_0^t \sum_{k=1}^{N} |c_k'(\tau)|^2 d\tau \leq \sum_{k=1}^{N} \lambda_k a_k^2 + \int\limits_0^t \sum_{k=1}^{N} f_k^2(\tau)d\tau$$

$$\leq \|\nabla a\|^2_{2,\Omega} + \|f\|^2_{2,Q_T}.$$

Passing to the limit as $N \to \infty$, we derive the following important estimate

$$\|\nabla u(\cdot,t)\|^2_{2,\Omega} + \int\limits_0^t \|\partial_t u(\cdot,\tau)\|^2_{2,\Omega} d\tau \leq \|\nabla a\|^2_{2,\Omega} + \|f\|^2_{2,Q_T} \qquad (4.4.5)$$

that is valid for any $t \in [0,T]$.

Now, our aim is to recover the pressure field. To this end, we proceed as follows. Consider the linear functional

$$l_t(v) = \int_\Omega (\nabla u(x,t) : \nabla v(x) + \partial_t u(x,t) \cdot v(x) - f(x,t) \cdot v(x))dx$$

for any $\overset{\circ}{L}{}^1_2(\Omega)$. It obeys the estimate

$$|l_t(v)| \leq \|\nabla u(\cdot,t)\|_{2,\Omega}\|\nabla v\|_{2,\Omega} + (\|\partial_t u(\cdot,t)\|_{2,\Omega} + \|f(\cdot,t)\|_{2,\Omega})\|v\|_{2,\Omega}.$$

According to Poincare's inequality, $\|v\|_{2,\Omega} \leq c(\Omega)\|\nabla v\|_{2,\Omega}$. So, the functional $v \mapsto l_t(v)$ is bounded for a.a. $t \in [0,T]$ and a bound of its norm is:

$$\|l_t\| \leq c(\Omega)(\|\nabla u(\cdot,t)\|_{2,\Omega} + \|\partial_t u(\cdot,t)\|_{2,\Omega} + \|f(\cdot,t)\|_{2,\Omega}). \qquad (4.4.6)$$

Moreover, $l_t(v) = 0$ for any $v \in \overset{\circ}{J}{}^1_2(\Omega)$. For bounded domains with Lipschitz boundary, there exists a function $t \mapsto p(\cdot, t) \in L_2(\Omega)$, see Chapter I, such that

$$l_t(v) = \int\limits_\Omega p(x,t)\mathrm{div} v dx, \qquad \|p(\cdot,t)\|_{2,\Omega} \le \|l_t\|.$$

It follows from (4.4.6) that

$$\|p\|_{2,Q_T} \le c\Big[\|\nabla u\|_{2,Q_T} + \|\partial_t u\|_{2,Q_T} + \|f\|_{2,Q_T}\Big]. \tag{4.4.7}$$

So, we have

$$\int\limits_\Omega \nabla u(x,t) : \nabla v(x) dx - \int\limits_\Omega p(x,t)\mathrm{div} v(x) dx$$
$$= \int\limits_\Omega f(x,t)\cdot v(x)dx - \int\limits_\Omega \partial_t u(x,t)\cdot v(x)dx \tag{4.4.8}$$

for any $v \in \overset{\circ}{L}{}^1_2(\Omega)$ and for a.a. $t \in [0,T]$.

For those t, i.e., for which (4.4.8) holds, we may apply the regularity theory developed in the case of the linear stationary Stokes system. More precisely, one can estimate higher derivatives in spatial variables:

$$\|\nabla^2 u(\cdot,t)\|_{2,\Omega} + \|\nabla p(\cdot,t)\|_{2,\Omega} \le c(\|f(\cdot,t)\|_{2,\Omega} + \|\partial_t u(\cdot,t)\|_{2,\Omega}$$

and thus

$$\|\nabla^2 u\|_{2,Q_T} + \|\nabla p\|_{2,Q_T} \le c(\|f\|_{2,Q_T} + \|\partial_t u\|_{2,Q_T}).$$

Combining the latter estimate with (4.4.5), we get the final bound:

$$\|\partial_t u\|_{2,Q_T} + \|\nabla^2 u\|_{2,Q_T} + \|\nabla p\|_{2,Q_T} \le c(\|f\|_{2,Q_T} + \|\nabla a\|_{2,\Omega}). \tag{4.4.9}$$

Now, let us show that $\nabla u \in C([0,T]; L_2(\Omega))$. To this end, we are going to use Theorem 2.2, introducing $H = \overset{\circ}{J}{}^1_2(\Omega)$ with scalar product $(u,v)_H = (\nabla u, \nabla v)$ and $V = \overset{\circ}{J}{}^1_2(\Omega) \cap W^2_2(\Omega)$ with the norm $\|v\|_V = \|\widetilde{\Delta} v\|_{2,\Omega}$. Now, we are going to verify that $V^* = \overset{\circ}{J}(\Omega)$ is dual to V with respect H.

Indeed, let $l \in V^*$. So, we have $|l(v)| \le c\|\widetilde{\Delta} v\|_{2,\Omega}$ for any $v \in V$. Since $\widetilde{\Delta}(V) = \overset{\circ}{J}(\Omega)$, one can define $G(p) = l(v)$, where $p = -\widetilde{\Delta} v$, for any $p \in \overset{\circ}{J}(\Omega)$. Obviously, $|G(p)| \le \|p\|_{2,\Omega}\|l\|$. Moreover, it is not difficult to check that $\|G\| = \|l\|$. By the Riesz theorem, there exists a unique $v^* \in \overset{\circ}{J}(\Omega)$ such that

$$G(p) = \int\limits_\Omega v^* \cdot p dx,$$

$\|v^*\|_{2,\Omega} = \|l\|$ and

$$l(v) = -\int_\Omega v^* \cdot \widetilde{\Delta} v dx = \int_\Omega \nabla v^* : \nabla v dx.$$

Now, every element v^* of $\overset{\circ}{J}(\Omega)$ defines a linear functional on V by formula

$$-\int_\Omega v^* \cdot \widetilde{\Delta} v dx = \int_\Omega \nabla v^* : \nabla v dx \leq \|v^*\|_{2,\Omega} \|\widetilde{\Delta} v\|_{2,\Omega} \leq \|v^*\|_{2,\Omega} \|v\|_V.$$

So, $V^* \simeq \overset{\circ}{J}(\Omega)$, i.e., spaces are isometrically isomorphic. By Theorem 2.2, $\nabla u \in C([0,T]; L_2(\Omega))$.

Summarizing mentioned above, one can formulate the following result.

Theorem 4.4. *Assume that the boundary of a bounded domain Ω is smooth and conditions (4.4.3) holds. Then,*

$$u \in W^{2,1}_2(Q_T), \qquad p \in W^{1,0}_2(Q_T),$$

with estimate (4.4.9). In addition, $\nabla u \in C([0,T]; L_2(\Omega))$ and equations

$$\partial_t u - \Delta u = f - \nabla p, \qquad \mathrm{div} u = 0$$

are satisfied a.e. in Q_T.

In fact, a more general statement is known about solutions to initial boundary value problems for the Stokes system.

Theorem 4.5. *Let Ω be a bounded domain with smooth boundary. Consider the following initial boundary value problem*

$$\partial_t u - \Delta u = f - \nabla p \quad \text{and} \quad \mathrm{div}\, u = 0 \quad \text{in} \quad Q_T = \Omega \times]0,T[,$$

$$\frac{1}{|\Omega|}\int_\Omega p(x,t)dx \equiv [p(\cdot,t)]_\Omega = 0, \qquad t \in [0,T], \tag{4.4.10}$$

$$u|_{\partial' Q_T} = 0,$$

where $\partial' Q_T$ is the parabolic boundary of Q_T.

Let $f \in L_{s,l}(Q_T) := L_l(0,T; L_s(\Omega))$ for some finite numbers $s > 1$ and $l > 1$. Then problem (4.4.10) has a unique solution such that $u \in W^{2,1}_{s,l}(Q_T)$ and $p \in W^{1,0}_{s,l}(Q_T)$, satisfying the following coercive estimate

$$\|u\|_{W^{2,1}_{s,l}(Q_T)} + \|p\|_{W^{1,0}_{s,l}(Q_T)} \leq c(\Omega, s, l, n)\|f\|_{L_{s,l}(Q_T)}.$$

Here, we have used the following notion:

$$W^{2,1}_{s,l}(Q_T) = \{v \in L_l(0,T; W^2_s(Q_T)),\ \partial_t v \in L_l(0,T; L_s(\Omega))\},$$

$$W^{1,0}_{s,l}(Q_T) = \{v \in L_l(0,T; W^1_s(Q_T))\},$$

and

$$W^{2,1}_s(Q_T) = W^{2,1}_{s,s}(Q_T), \qquad W^{1,0}_s(Q_T) = W^{1,0}_{s,s}(Q_T).$$

4.5 Uniqueness Results

Lemma 4.3. *Let $v \in L_1(0,T;\overset{\circ}{J}_m(\Omega))$ with $m > 1$ and Ω be a bounded domain in $\mathbb{R}^n$ with sufficiently smooth boundary. If $1 < m < 2$, assume in addition that $n = 2$ or 3. Suppose, further, that*

$$\int\limits_{Q_T} v \cdot (\partial_t w + \Delta w)dxdt = 0$$

for $w(x,t) = \chi(t)W(x)$ with an arbitrary function $\chi \in C^1([0,T])$ and an arbitrary divergence free field $W \in C^2(\overline{\Omega})$ subject to the end condition $\chi(T) = 0$ and to the boundary condition $W = 0$ on $\partial\Omega$, respectively.

Then v is identically zero in Q_T. Here, $Q_T = \Omega\times]0,T[$.

PROOF Take as a test function $w = \chi(t)\varphi_k(x)$, where φ_k is the k-th eigenfunction of the Stokes operator. $\chi(t)$ is a smooth function, satisfying the end condition $\chi(T) = 0$. Since Ω is a domain with smooth boundary, the eigenfunction φ_k is a smooth function as well. Indeed, it follows from embedding theorems, regularity theory for the Stokes system, and bootstrap arguments. Then, we have

$$\int\limits_{Q_T} v(x,t) \cdot (\chi'(t)\varphi_k(x) - \chi(t)\lambda_k\varphi_k(x))dxdt = 0$$

and thus

$$\int\limits_0^T v_k(t)(\chi'(t) - \chi(t)\lambda_k) = 0 \tag{4.5.1}$$

for any $\chi \in C^2([0,T])$ with $\chi(T) = 0$, where

$$v_k(t) = \int\limits_\Omega v(x,t) \cdot \varphi_k(x)dx.$$

From (4.5.1), it follows that

$$v_k'(t) + \lambda_k v_k(t) = 0$$
$$v_k(0) = 0.$$

The latter immediately implies that $v_k(t) = 0$ for $t \in [0,T]$.

Now, we wish to show that $v(x,t) = 0$ for any $x \in \Omega$ and $t \in [0,T]$.

Let us start with the simplest case $m \geq 2$. Obviously, for bounded domains, $\overset{\circ}{J}_m(\Omega) \subseteq \overset{\circ}{J}_2(\Omega)$. Hence, $v(\cdot,t) \in \overset{\circ}{J}_2(\Omega)$ and

$$\|v(\cdot,t)\|^2_{2,\Omega} = \sum_{k=1}^{\infty} v_k^2(t) = 0.$$

Let us consider now the case, in which

$$1 < m < 2. \tag{4.5.2}$$

First, we shall show

$$\int\limits_{\Omega} v(x,t)\cdot u(x)dx = 0 \tag{4.5.3}$$

for any $u \in C^{\infty}_{0,0}(\Omega)$ and for a.a. $t \in [0,T]$. To this end, fix an arbitrary test function $u \in C^{\infty}_{0,0}(\Omega)$ and let

$$S_N = \sum_{k=1}^{N} c_k\varphi_k,$$

where

$$c_k = \int_{\Omega} u\cdot\varphi_k dx.$$

We know that $S_N \to u$ in $L_2(\Omega)$ as $N \to \infty$. But it is not sufficient to justify (4.5.3) by taking the limit below

$$0 = \int\limits_{\Omega} v(x,t)\cdot S_N(x)dx \to \int\limits_{\Omega} v(x,t)\cdot u(x)dx, \qquad N \to \infty,$$

for a.a. $t \in [0,T]$. However, assuming additionally that $n = 2$ or $n = 3$, we will be able to show that sequence S_N is bounded in $L_{m'}(\Omega)$ and this will imply (4.5.3) in the case $1 < m < 2$.

Indeed, let us consider first $n = 2$. Then, by embedding theorems, we have

$$\Big(\int\limits_{\Omega} |S_N|^{m'}dx\Big)^{\frac{1}{m'}} \leq c(m,\Omega)\Big(\int\limits_{\Omega} |\nabla S_N|^2dx\Big)^{\frac{1}{2}}$$

$$\leq c(m,\Omega)\Big(\sum_{k=1}^{N}\lambda_k c_k^2\Big)^{\frac{1}{2}} \leq c(m,\Omega)\|\nabla u\|_{2,\Omega}.$$

So, required boundedness follows.

In the case $n = 3$, we have

$$\Big(\int\limits_{\Omega} |S_N|^{m'}dx\Big)^{\frac{1}{m'}} \leq c(m,\Omega)\Big(\int\limits_{\Omega} (|\nabla S_N|^2 + |\nabla^2 S_N|^2)dx\Big)^{\frac{1}{2}}.$$

If we let $-\widetilde{\Delta}S_N = f_N \in L_2(\Omega)$, then simply, by definition of the Stokes operator, the partial sum S^N solves the following boundary value problem

$$-\Delta S_N + \nabla p^N = f_N, \qquad \mathrm{div} S_N = 0 \qquad \textit{in } \Omega$$

$$S_N|_{\partial\Omega} = 0.$$

Now, we are again in a position to apply the regularity theory, developed for the stationary Stokes system, that gives the estimate

$$\|\nabla^2 S_N\|_{2,\Omega} + \|\nabla p^N\|_{2,\Omega} \leq c(\Omega)\|f_N\|_{2,\Omega}.$$

So, we have

$$\|\nabla^2 S_N\|_{2,\Omega} \leq c(\Omega)\|\widetilde{\Delta} S_N\|_{2,\Omega} \leq c(\Omega)\sum_{k=1}^{N}\lambda_k^2 c_k^2$$

$$\leq c(\Omega)\|\widetilde{\Delta} u\|_{2,\Omega} \leq c(\Omega)\|\Delta u\|_{2,\Omega} \leq c(\Omega)\|\nabla^2 u\|_{2,\Omega}.$$

And thus boundedness of $\|S_N\|_{m',\Omega}$ has been proven in the case $n = 3$ as well.

Now, the aim is to show that v is identically zero in Q_T. Fix $t \in [0,T]$ and consider a linear functional

$$l(w) = \int\limits_{\Omega} v \cdot w dx, \qquad w \in \overset{\circ}{L}{}^1_{m'}(\Omega).$$

By Poincaré's inequality, it is bounded in $\overset{\circ}{L}{}^1_{m'}(\Omega)$ and, by (4.5.3), vanishes on $\overset{\circ}{J}{}^1_{m'}(\Omega)$, i.e.,

$$l(w) = 0 \qquad \forall w \in \overset{\circ}{J}{}^1_{m'}(\Omega).$$

As we know, every functional, possessing the above properties, can be presented in the form

$$l(w) = \int\limits_{\Omega} p \,\mathrm{div}\, w dx, \qquad \forall w \in \overset{\circ}{L}{}^1_{m'}(\Omega)$$

for some $p \in L_m(\Omega)$. The latter means that $v = -\nabla p$.

Our next step is to show that p is a solution to the Neumann problem: $\Delta p = 0$ in Ω and $\partial p/\partial \nu = 0$ on $\partial\Omega$, where ν is the unit outward normal to the surface $\partial\Omega$, in the following sense

$$\int\limits_{\Omega} \nabla p \cdot \nabla q dx = 0, \qquad \forall q \in W^1_{m'}(\Omega). \tag{4.5.4}$$

Indeed, $\nabla p = -v \in \overset{\circ}{J}_m(\Omega)$. Therefore, there exists a sequence $w^{(k)} \in C^\infty_{0,0}(\Omega)$ such that $w^{(k)} \to \nabla p$ in $L_m(\Omega)$. So,

$$\int\limits_{\Omega} w^{(k)} \cdot \nabla q dx = 0 \to \int\limits_{\Omega} \nabla p \cdot \nabla q dx = 0$$

for any $q \in W^1_{m'}(\Omega)$.

Now, our problem has been reduced to the following uniqueness question: *Let* $p \in W^1_m(\Omega)$, $1 < m < 2$, *and*

$$\int_\Omega \nabla p \cdot \nabla q dx = 0 \qquad \forall q \in W^1_{m'}(\Omega).$$

Then p *must be a constant in* Ω.

Assume that f is a smooth function on $\partial\Omega$ and satisfies compatibility condition

$$\int_{\partial\Omega} f(s)ds = 0.$$

Consider the classical Neumann problem: $\Delta q = 0$ in Ω and $\partial q/\partial\nu = f$ on $\partial\Omega$. There exists a smooth solution to this problem. For it, we have

$$0 = \int_\Omega \nabla p \cdot \nabla q dx = \int_{\partial\Omega} p\frac{\partial q}{\partial \nu} ds = \int_{\partial\Omega} pf ds.$$

Since f is a smooth function satisfying the compatibility condition only, we can claim that

$$p = c_1$$

on $\partial\Omega$ for some constant c_1.

We let further $p_1 = p - c_1$ and p_1 is a solution to the homogeneous Dirichlet boundary problem: $\Delta p_1 = 0$ in Ω and $p_1 = 0$ on $\partial\Omega$. To show that p_1 is in fact identically zero, we find for any $q \in W^2_{m'}(\Omega)$ with $q = 0$ on $\partial\Omega$

$$\int_\Omega \nabla p_1 \cdot \nabla q dx = -\int_\Omega p_1 \Delta q dx.$$

We may select a function q in a special way so that $\Delta q = |p_1|^{m-1} \operatorname{sign} p_1 = f \in L_{m'}(\Omega)$ with $q = 0$ on $\partial\Omega$. It is well known that such a function exists and belongs to $W^2_{m'}(\Omega)$. Hence,

$$0 = \int_\Omega \nabla p_1 \cdot \nabla q dx = -\int_\Omega |p_1|^m dx$$

and thus p is a constant in Ω and $v = 0$ in Q_T. Lemma 4.3 is proved.

We have another uniqueness result.

Theorem 5.6. *Let* $v \in L_1(0,T; \overset{\circ}{J}{}^1_m(\Omega))$ *with* $m > 1$ *and* Ω *be a bounded domain with sufficiently smooth boundary. Assume*

$$\int_{Q_T} (v \cdot \partial_t w - \nabla v : \nabla w) dz = 0 \qquad (4.5.5)$$

for any $w(x,t) = \chi(t)W(x)$, *where* $\chi \in C^1([0,T])$ *such that* $\chi(T) = 0$ *and* $W \in C^\infty_{0,0}(\Omega)$.

Then v *is identically zero in* Q_T.

PROOF By density arguments, (4.5.5), of course, holds for any $W \in \overset{\circ}{J}{}^1_{m'}(\Omega)$.

Take any function $W \in C^2(\overline{\Omega})$ with $W = 0$ on $\partial\Omega$ and $\operatorname{div} W = 0$ in Ω. We know that $W \in \overset{\circ}{J}{}^1_{m'}(\Omega)$, see Chapter I, Theorem 4.3. So, v satisfies all assumptions of Lemma 4.3 and therefore $v \equiv 0$ in Q_T.

The above proof works well under additional assumption on n if $1 < m < 2$. However, we can give an alternative proof that does not need extra assumptions on the spatial dimension. We assume that (4.5.3) has been already proved. Then we can take test function in (4.5.3) in the following way $W = \nabla \wedge w$ with arbitrary function $w \in C^\infty_0(\Omega)$. This implies $\nabla \wedge v(\cdot,t) = 0$ in Ω. Taking into account the fact that v is divergence free, we deduce that $v(\cdot,t)$ is a harmonic function in Ω belonging to $\overset{\circ}{J}{}^1_m(\Omega)$. The rest of the proof is more or less the same as the final part of the proof of Lemma 4.3, see arguments providing $p_1 = 0$ there. Theorem 5.6 is proved.

4.6 Local Interior Regularity

In this section, we shall restrict ourselves to the 3D case just in order to reduce a number of parameters. Although it is clear that the extension to other dimensions is straightforward.

The problem of local interior regularity can be formulated as follows. Consider the Stokes system in a canonical domain, say, in $Q = B\times]-1,0[$

$$\partial_t u - \Delta u = f - \nabla p, \qquad \operatorname{div} u = 0. \tag{4.6.1}$$

We always assume that functions u and p have some starting differentiability properties. Keeping in mind the 3D non-stationary non-linear problem, we supposed that

$$u \in W^{1,0}_{m,n}(Q), \qquad p \in L_{m,n}(Q) \tag{4.6.2}$$

for some finite m and n being greater than 1.

Assuming that some additional information about the right-hand side f is given, we shall try to make some conclusions about smoothness of u and p in smaller parabolic balls $Q(r) = B(r)\times]-r^2,0[$.

It is known that, for stationary Stokes system as well as for the heat equation, solutions are smooth locally as long as f is smooth. However, in the case of non-stationary Stokes system, we have smoothing in spatial variables but not in time. This can be seen easily from the following example, in which $f = 0$ and

$$u(x,t) = c(t)\nabla h(x), \qquad p(x,t) = -c'(t)h(x).$$

Here, h is a harmonic function in B and c is a given function, defined on $[0,T]$. This solution is infinitely differentiable inside B but, under assumptions (4.6.2), it is just Hölder continuous in time. There is no smoothing in time despite the smoothness of f.

In general, we have the following statement.

Proposition 6.7. *Assume that u and p satisfy (4.6.1), conditions (4.6.2), and let*

$$f \in L_{s,n}(Q) \tag{4.6.3}$$

with $s \geq m$.

Then $u \in W^{2,1}_{s,n}(Q(1/2))$ and $p \in W^{1,0}_{s,n}(Q(1/2))$ and the estimate

$$\|\partial_t u\|_{s,n,Q(1/2)} + \|\nabla^2 u\|_{s,n,Q(1/2)} + \|\nabla p\|_{s,n,Q(1/2)}$$

$$\leq c(\|f\|_{s,n,Q} + \|u\|_{m,n,Q} + \|\nabla u\|_{m,n,Q} + \|p\|_{m,n,Q}) \tag{4.6.4}$$

holds.

PROOF It is sufficient to prove this proposition for case $s = m$. General case can be deduced from it by embedding theorems and bootstrap arguments.

Fix a non-negative cut-off function $\varphi \in C^\infty_0(B\times]-1,1[)$ so that $\varphi = 1$ in $B(1/2)\times]-(1/2)^2,(1/2)^2[$. For any $t \in]-1,0[$, we determine a function $w(\cdot,t)$ as a unique solution to the boundary value problem

$$\Delta w(\cdot,t) - \nabla q(\cdot,t) = 0, \qquad \operatorname{div} w(\cdot,t) = v(\cdot,t)\cdot\nabla\varphi(\cdot,t)$$

in B and

$$\int\limits_B q(x,t)dx = 0, \qquad w(\cdot,t) = 0$$

on ∂B. It satisfies the estimate

$$\|\nabla^2 w(\cdot,t)\|_{s,B} + \|q(\cdot,t)\|_{s,B} + \|\nabla q(\cdot,t)\|_{s,B} \leq$$

$$\leq c\|\nabla(v(\cdot,t)\cdot\nabla\varphi(\cdot,t))\|_{s,B}. \tag{4.6.5}$$

Letting

$$V = \varphi v - w, \qquad P = \varphi p - q,$$

$$F = \varphi f + v\partial_t\varphi - 2\nabla v\nabla\varphi - v\Delta\varphi + p\nabla\varphi - \partial_t w,$$

we observe that new functions V and P are a unique solution to the following initial boundary value problem

$$\partial_t V - \Delta V = F - \nabla P, \qquad \operatorname{div} V = 0$$

in Q,

$$V = 0$$

on $\partial' Q$. The statement of Theorem 4.5 and the estimate there yield

$$\|\partial_t v\|_{s,n,Q(1/2)} + \|\nabla^2 v\|_{s,n,Q(1/2)} + \|\nabla p\|_{s,n,Q(1/2)} \leq cA + c\|\partial_t w\|_{s,n,Q}, \tag{4.6.6}$$

where

$$A = \|f\|_{s,n,Q} + \|v\|_{s,n,Q} + \|\nabla v\|_{s,n,Q} + \|p\|_{s,n,Q}.$$

So, our task is to evaluate the last term on the right-hand side of (4.6.6). The main tool here is duality arguments developed by V. A. Solonnikov.

Introducing new notation $u = \partial_t w$ and $r = \partial_t q$, we can derive from the equations for w and q

$$\Delta u(\cdot,t) - \nabla r(\cdot,t) = 0,$$
$$\operatorname{div} u(\cdot,t) = \partial_t v(\cdot,t)\cdot\nabla\varphi(\cdot,t) + v(\cdot,t)\cdot\nabla\,\partial_t\varphi(\cdot,t) \tag{4.6.7}$$

in B,

$$\int\limits_B r(x,t)dx = 0, \qquad u(\cdot,t) = 0 \tag{4.6.8}$$

on ∂B.

Given $g \in L_{s'}(B)$ with $s' = s/(s-1)$, let us define a function $\widetilde{u}$ as a unique solution to the boundary value problem

$$\Delta\widetilde{u} - \nabla\widetilde{r} = g, \qquad \operatorname{div}\widetilde{u} = 0 \tag{4.6.9}$$

in B,

$$\int\limits_B \widetilde{r}(x)dx = 0, \qquad \widetilde{u} = 0 \tag{4.6.10}$$

on ∂B. Moreover, function $\widetilde{r}$ satisfies the estimate

$$\|\widetilde{r}\|_{s',B} + \|\nabla\widetilde{r}\|_{s',B} \leq c\|g\|_{s',B}. \tag{4.6.11}$$

Now, from (4.6.7)–(4.6.11), it follows that

$$\int\limits_B u(x,t)\cdot g(x)dx = \int\limits_B u(x,t)\cdot(\Delta\widetilde{u}(x) - \nabla\widetilde{r}(x))dx$$

$$= \int_B \widetilde{r}(x) \mathrm{div}\, u(x,t)dx = \int_B \widetilde{r}(x)(\partial_t v(x,t) \cdot \nabla \varphi(x,t) + v(x,t) \cdot \nabla\, \partial_t \varphi(x,t))dx.$$

The derivative $\partial_t v$ can be expressed from the Navier-Stokes equations. So, we derive from the previous identity the following:

$$\int_B u(x,t) \cdot g(x)dx = \int_B \widetilde{r}(x)(\Delta\, v(x,t) - \nabla\, p(x,t) + f(x,t)) \cdot \nabla \varphi(x,t)dx$$

$$+ \int_B \widetilde{r}(x) v(x,t) \cdot \nabla\, \partial_t \varphi(x,t)dx.$$

Integration by parts and estimate (4.6.11) imply

$$\int_B u(x,t) \cdot g(x)dx \leq c\|g\|_{s',B}(\|v(\cdot,t)\|_{s,B} + \|\nabla\, v(\cdot,t)\|_{s,B} + \|p(\cdot,t)\|_{s,B})$$

and thus

$$\|\partial_t w\|_{s,n,Q} \leq cA.$$

Proposition 6.7 is proved.

Keeping in mind the 3D non-stationary non-linear problem, one cannot expect that the number n is big. In such cases, the following embedding result can be useful.

Proposition 6.8. *Assume that* $v \in W^{2,1}_{s,n}(Q)$ *with*

$$1 < n \leq 2, \qquad \mu = 2 - \frac{2}{n} - \frac{3}{s} > 0.$$

Then

$$|v(z) - v(z')| \leq c(m,n,s)(|x - x'| + |t - t'|^{\frac{1}{2}})^{\mu}(\|v\|_{s,n,Q}$$

$$+ \|\nabla v\|_{s,n,Q} + \|\nabla^2 v\|_{s,n,Q} + \|\partial_t v\|_{s,n,Q})$$

for all $z = (x,t) \in Q(1/2)$ *and for all* $z' = (x',t') \in Q(1/2)$. *In other words,* v *is Hölder continuous with exponent* μ *relative to parabolic metric in the closure of* $Q(1/2)$.

Finally, using bootstrap arguments, we can prove the following statement which in a good accordance with the aforesaid example.

Proposition 6.9. *Assume that conditions (4.6.2) hold with* $1 < n < 2$ *and* $f = 0$. *Let* u *and* p *be an arbitrary solution to system (4.6.1). Then for any* $0 < \tau < 1$ *and for any* $k = 0, 1, ...,$ *the function* $(x,t) \mapsto \nabla^k u(x,t)$ *is Hölder continuous with any exponent less than* $2 - 2/n$ *in the closure of the set* $Q(\tau)$ *relative to the parabolic metric.*

4.7 Local Boundary Regularity

To describe the results of this section, we are going to exploit the following notation:

$$x = (x', x_3), \qquad x' = (x_1, x_2),$$

$$\mathcal{Q}_+(r) = \mathcal{C}_+(r) \times]-r^2, 0[\subset \mathbb{R}^3 \times \mathbb{R}, \qquad \mathcal{C}_+(r) = b(r) \times]0, r[\in \mathbb{R}^3,$$

$$b(r) = \{x' \in \mathbb{R}^2 : \ |x'| < r\}.$$

The complete analogue of Proposition 6.7 is as follows, see [Seregin (2002)], [Seregin (2002)], and [Seregin (2009)].

Proposition 7.10. *Assume that we are given three functions*

$$u \in W^{1,0}_{m,n}(\mathcal{Q}_+(2)), \qquad p \in L_{m,n}(\mathcal{Q}_+(2)), \qquad f \in L_{m_1,n}(\mathcal{Q}_+(2))$$

with $m_1 \geq m$ satisfying the system

$$\partial_t u - \Delta\, u = f - \nabla\, p, \qquad \operatorname{div} v = 0 \qquad in\ \ \mathcal{Q}_+(2),$$

and the homogeneous Dirichlet boundary condition

$$u(x', 0, t) = 0.$$

Then $u \in W^{2,1}_{m_1,n}(\mathcal{Q}_+(1))$ and $p \in W^{1,0}_{m_1,n}(\mathcal{Q}_+(1))$ with the estimate

$$\|\partial_t u\|_{L_{m_1,n}(\mathcal{Q}_+(1))} + \|\nabla^2 u\|_{L_{m_1,n}(\mathcal{Q}_+(1))} + \|\nabla p\|_{L_{m_1,n}(\mathcal{Q}_+(1))} \leq$$

$$\leq c(\|u\|_{L_{m,n}(\mathcal{Q}_+(2))} + \|\nabla u\|_{L_{m,n}(\mathcal{Q}_+(2))} + \|p\|_{L_{m,n}(\mathcal{Q}_+(2))} + \|f\|_{L_{m_1,n}(\mathcal{Q}_+(2))}).$$

If we assume $f = 0$ and $1 < n < 2$, then, by an embedding theorem similar to Proposition 4.6.2, u is Hölder continuous in the closure of the space-time cylinder $\mathcal{Q}_+(1)$. Hölder continuity is defined with respect to the parabolic metrics and the corresponding exponent does not exceed $2 - 2/n$. However, in general, the analogue of Proposition 6.9 is not true in the boundary regularity theory, i.e., in general there is no further smoothing even in spatial variables. Let us describe the corresponding counter-example.

We are looking for non-trivial solutions to the following homogeneous initial boundary value problem

$$\partial_t v - \Delta\, v = -\nabla\, q, \qquad \operatorname{div} v = 0 \tag{4.7.1}$$

in $\mathbb{R}^3_+\times]-4,0[$ under the homogeneous Dirichlet boundary condition

$$v(x',0,t)=0 \qquad x'\in\mathbb{R}^2, \quad -4<t<0, \tag{4.7.2}$$

and under homogeneous initial data

$$v(x,-4)=0 \qquad x\in\mathbb{R}^3_+. \tag{4.7.3}$$

Here $R^3_+=\{x=(x',x_3)\,:\,x_3>0\}$.

Taking an arbitrary function $f(t)$, we seek a non-trivial solution to (4.7.1)–(4.7.3) in the form of shear flow, say, along x_1-axis:

$$v(x,t)=(w(x_3,t),0,0), \qquad q(x,t)=-f(t)x_1.$$

Here, a scalar function u solves the following initial boundary value problem

$$\partial_t w(y,t)-w_{yy}(y,t)=f(t), \tag{4.7.4}$$

$$w(0,t)=0, \tag{4.7.5}$$

$$w(y,-4)=0, \tag{4.7.6}$$

where $0<y<+\infty$ and $-4<t<0$ and $w_{yy}=\partial^2 w/\partial y^2$.

It is not so difficult to solve (4.7.4)–(4.7.6) explicitly:

$$w(y,t)=\frac{2}{\sqrt{\pi}}\int\limits_{-4}^{t}f(t-\tau-4)d\tau\int\limits_{0}^{\frac{y}{\sqrt{4(\tau+4)}}}e^{-\xi^2}d\xi. \tag{4.7.7}$$

Keeping in mind that our aim is to construct irregular but integrable solution, we choose the function f as follows

$$f(t)=\frac{1}{|t|^{1-\alpha}}, \qquad 0<\alpha<1/2. \tag{4.7.8}$$

Then, direct calculations show us:
(i) w is a bounded smooth function in the strip $]0,+\infty[\times]-4,0[$ satisfying boundary and initial conditions;
(ii) $w_y(y,t)\geq c(\alpha)\frac{1}{y^{1-2\alpha}}$ for y and t subject to the inequalities $y^2\geq -4t$, $0<y\leq 3$, and $-9/8\leq t<0$.
(iii) Let s, s_1, l, and l_1 be numbers greater than 1 and satisfy the condition

$$K=\max\left\{\frac{1}{2}\Big(1-\frac{1}{s}\Big),1-\frac{1}{l_1}\right\}<\alpha<\frac{1}{2}. \tag{4.7.9}$$

Then

$$v\in W^{1,0}_{s,l}(\mathcal{C}_+(3)\times]-9/4,0[), \qquad q\in L_{s_1,l_1}(\mathcal{C}_+(3)\times]-9/4,0[).$$

Assume we are given numbers $1<m<+\infty$ and $1<n<2$. Letting $s=s_1=m$ and $l=l_1=n$ and choosing α so that inequality (4.7.9) holds. The functions v and q constructed above for the chosen α meet all the conditions of Proposition 7.10 with $f=0$. However, ∇v is unbounded in any neighborhood of the space-time point $z=(x,t)=0$. This is a counter-example taken from [Seregin and Šverák (2009)], which is an essential simplification of the corresponding counter-example in [Kang (2005)].

4.8 Comments

Chapter 4 contains standard material about existence, uniqueness, and regularity of solutions to the non-stationary Stokes system. A bit unusual facts for an introductory course are in the last three sections. In particular, fine uniqueness theorems and local regularity issues are discussed in Sections 5–7. They are needed for the local regularity analysis in Chapter 6.

Chapter 5

Non-linear Non-Stationary Problem

5.1 Compactness Results for Non-Stationary Problems

Our standing assumptions are as follows. We are given a triple of Banach spaces V_0, V, and V_1, having the following properties:

(i) $V_0 \subset V \subset V_1$, V_0 is a reflexive space;

(ii) imbedding $V_0 \subset V$ is compact;

(iii) imbedding $V \subset V_1$ is continuous;

(iv) $v \in V_0$ and $\|v\|_{V_1} = 0$ imply $\|v\|_V = 0$.

Lemma 5.1. *Given $\eta > 0$, there exists $C(\eta) > 0$ such that*

$$\|v\|_V \leq \eta \|v\|_{V_0} + C(\eta)\|v\|_{V_1} \tag{5.1.1}$$

for any $v \in V_0$.

PROOF Usual compactness arguments work. Assume that the statement is wrong. Then for any $n \in \mathbb{N}$ there exists $v_n \in V_0$ such that

$$\|v_n\|_V > \eta \|v_n\|_{V_0} + n\|v_n\|_{V_1}.$$

Then after normalization, we have

$$\|v_n'\|_V = 1 > \eta \|v_n'\|_{V_0} + n\|v_n'\|_{V_1},$$

where $v_n' = v_n/\|v_n\|_V$. The sequence v_n' is bounded in a reflexive space, and thus without loss of generality we may assume that

$$v_n' \rightharpoonup v_0$$

in V_0 and thus

$$v_n' \to v_0$$

in V and V_1. Since $n\|v'_n\|_{V_1}$ is bounded and therefore $\|v'_n\|_{V_1} \to 0 = \|v_0\|_{V_1}$. Hence, by assumption (iv), $\|v_0\|_V = 0$. However, $1 = \|v_n\|_V \to \|v_0\|_V$. This leads to contradiction. Lemma 5.1 is proved.

Proposition 1.1. *(Aubin-Lions lemma) Let $1 < p_0, p_1 < \infty$, V_1 is reflexive, and define*

$$W \equiv \Big\{ \|v\|_W = \|v\|_{L_{p_0}(0,T;V_0)} + \|\partial_t v\|_{L_{p_1}(0,T;V_1)} < \infty \Big\}.$$

Then W is compactly imbedded into $L_{p_0}(0,T;V)$.

PROOF Suppose that sequence $u^{(j)}$ is bounded in W. Then, without loss of generality, we may assume that

$$u^{(j)} \rightharpoonup u$$

in $L_{p_0}(0,T;V_0)$ and

$$\partial_t u^{(j)} \rightharpoonup \partial_t u$$

in $L_{p_1}(0,T;V_1)$. Setting $v^{(j)} = u^{(j)} - u$, we need to show that

$$v^{(j)} \to 0$$

in $L_{p_0}(0,T;V)$. By Lemma 5.1, we have for arbitrary number $\eta > 0$

$$\|v^{(j)}(\cdot,t)\|_V \leq \eta \|v^{(j)}(\cdot,t)\|_{V_0} + C(\eta)\|v^{(j)}(\cdot,t)\|_{V_1}$$

and thus

$$\|v^{(j)}\|_{L_{p_0}(0,T;V)} \leq \eta \|v^{(j)}\|_{L_{p_0}(0,T;V_0)} + C(\eta)\|v^{(j)}\|_{L_{p_0}(0,T;V_1)}$$

$$\leq c\eta + C(\eta)\|v^{(j)}\|_{L_{p_0}(0,T;V_1)}.$$

So, it is enough to show

$$v^{(j)} \to 0 \tag{5.1.2}$$

in $L_{p_0}(0,T;V_1)$. To this end, we are going first to prove that

$$\sup_j \sup_{0<t<T} \|v^{(j)}(\cdot,t)\|_{V_1} < \infty. \tag{5.1.3}$$

Indeed, if (5.1.3) would hold, then (5.1.2) would follow from

$$\|v^{(j)}(\cdot,t)\|_{V_1} \to 0 \tag{5.1.4}$$

for a.a. $t \in [0,T]$ and Lebesgue's theorem. So, our goal is to prove (5.1.3) and (5.1.4).

To prove (5.1.3), we exploit the following formula (it is a simple consequence of the definition of the derivative in time)

$$v^{(j)}(\cdot,t) = \int\limits_s^t \partial_t v^{(j)}(\cdot,\tau)d\tau + v^{(j)}(\cdot,s) \tag{5.1.5}$$

for any $0 \leq s, t \leq T$, which implies

$$\|v^{(j)}(\cdot,t)\|_{V_1} \leq \|v^{(j)}(\cdot,s)\|_{V_1} + \int\limits_s^t \|\partial_t v^{(j)}(\cdot,\tau)\|_{V_1} d\tau$$

$$\leq \|v^{(j)}(\cdot,s)\|_{V_1} + T^{\frac{1}{p_1'}} \|\partial_t v^{(j)}\|_{L_{p_1}(0,T;V_1)}.$$

The latter inequality can be integrated in s. As a result, we get (5.1.3).

Now, we wish to explain validity of (5.1.4). To this end, let us integrate (5.1.5) in s over the interval $]t, s_1[$

$$(s_1 - t)v^{(j)}(\cdot,t) = \int\limits_t^{s_1} ds \int\limits_s^t \partial_t v^{(j)}(\cdot,\tau)d\tau + \int\limits_t^{s_1} v^{(j)}(\cdot,s)ds.$$

After integration by parts in s in the first term of the right-hand side, we find

$$v^{(j)}(\cdot,t) = a^{(j)}(\cdot,t) + b^{(j)}(\cdot,t),$$

where

$$a^{(j)}(\cdot,t) = \frac{1}{s_1 - t} \int\limits_t^{s_1} v^{(j)}(\cdot,s)ds$$

and

$$b^{(j)}(\cdot,t) = \frac{1}{s_1 - t} \int\limits_t^{s_1} (s - s_1)\partial_t v^{(j)}(\cdot,s)ds.$$

Now, take any $\varepsilon > 0$ and fix it. Then

$$\|b^{(j)}(\cdot,t)\|_{V_1} \leq \frac{1}{|s_1 - t|} \Big(\int\limits_t^{s_1} |s_1 - s|^{p_1'} ds \Big)^{\frac{1}{p_1'}}$$

$$\times \Big(\int\limits_0^T \|\partial_t v^{(j)}(\cdot,s)\|_{V_1}^{p_1} ds \Big)^{\frac{1}{p_1}} \leq c|s_1 - t|^{\frac{1}{p_1'}} < \varepsilon$$

for any j and for s_1 sufficiently closed to t.

Next, we wish to show that for each fixed s_1 (for given t)

$$\|a^{(j)}(\cdot,t)\|_{V_1} \to 0. \tag{5.1.6}$$

To this end, we first notice that

$$a^{(j)}(\cdot,t) \rightharpoonup 0$$

in V_1. Then, if we would show boundedness of $a^{(j)}$ in V_0, (5.1.6) would follows from compactness of imbedding V_0 into V.

We have

$$\|a^{(j)}(\cdot,t)\|_{V_0} \leq \frac{1}{|s_1-t|}\int\limits_t^{s_1} \|v^{(j)}(\cdot,s)\|_{V_0} ds$$

$$\leq \frac{1}{|s_1-t|}|s_1-t|^{\frac{1}{p_0'}}\|v^{(j)}\|_{L_{p_0}(0,T;V_0)} \leq c|s_1-t|^{\frac{1}{p_0'}-1}.$$

So, given s_1, (5.1.6) holds and we may find $N(s_1,t)$ such that

$$\|a^{(j)}(\cdot,t)\|_{V_1} \leq \varepsilon$$

for any $j \geq N_1(s_1,t)$. This proves (5.1.4) and completes the proof of Proposition 1.1.

5.2 Auxiliary Problem

Assume that Ω is a bounded domain with sufficiently smooth boundary and that

$$a \in \overset{\circ}{J}(\Omega) \tag{5.2.1}$$

and

$$f \in L_2(0,T;(\overset{\circ}{J}{}^1_2(\Omega))'). \tag{5.2.2}$$

Proposition 2.2. *Let $Q_T = \Omega \times]0,T[$ and*

$$w \in L_\infty(Q_T), \qquad \operatorname{div} w = 0 \quad in\ Q_T. \tag{5.2.3}$$

There exists a unique solution v to the initial boundary value problem

$$\partial_t v - \Delta v + \operatorname{div} v \otimes w + \nabla q = f, \quad \operatorname{div} v = 0 \quad in\ Q_T,$$
$$v|_{\partial\Omega\times[0,T]} = 0, \tag{5.2.4}$$
$$v|_{t=0} = a$$

in the following sense:

$$v \in C([0,T]; L_2(\Omega)) \cap L_2(0,T; \overset{\circ}{J}{}^1_2(\Omega)), \quad \partial_t v \in L_2(0,T; (\overset{\circ}{J}{}^1_2(\Omega))');$$

for a.a. $t \in [0,T]$

$$\int_\Omega (\partial_t v(x,t) \cdot \widetilde{v}(x) + \nabla v(x,t) : \nabla \widetilde{v}(x))dx$$
$$= \int_\Omega (v(x,t) \otimes w(x,t) : \nabla \widetilde{v}(x) + f(x,t) \cdot \widetilde{v}(x))dx \tag{5.2.5}$$

for all $\widetilde{v} \in \overset{\circ}{J}{}^1_2(\Omega)$;

$$\|v(\cdot,t) - a(\cdot)\|_{2,\Omega} \to 0 \tag{5.2.6}$$

as $t \to +0$.

PROOF We are going to apply the Leray-Schauder principle, see Theorem 1.2 of Chapter III. We let

$$X = L_2(0,T; \overset{\circ}{J}(\Omega)).$$

Given $u \in X$, define $v = A(u)$ as a solution to the following problem:

$$v \in C([0,T]; L_2(\Omega)) \cap L_2(0,T; \overset{\circ}{J}{}^1_2(\Omega)), \quad \partial_t v \in L_2(0,T; (\overset{\circ}{J}{}^1_2(\Omega))'); \tag{5.2.7}$$

for a.a. $t \in [0,T]$

$$\int_\Omega (\partial_t v(x,t) \cdot \widetilde{v}(x) + \nabla v(x,t) : \nabla \widetilde{v}(x))dx$$
$$= \int_\Omega \widetilde{f}(x,t) \cdot \widetilde{v}(x)dx \tag{5.2.8}$$

for all $\widetilde{v} \in \overset{\circ}{J}{}^1_2(\Omega)$;

$$\|v(\cdot,t) - a(\cdot)\|_{2,\Omega} \to 0 \tag{5.2.9}$$

as $t \to +0$. Here, $\widetilde{f} = f - \operatorname{div} u \otimes w$.

Such a function v exists and is unique (for given u) according to Theorem 2.3 of Chapter 4 since

$$\widetilde{f} \in L_2(0,T; (\overset{\circ}{J}{}^1_2(\Omega))').$$

So, operator A is well defined. Let us check that it satisfies all the assumptions of Theorem 1.2 of Chapter 3.

Continuity: Let $v^1 = A(u^1)$ and $v^2 = A(u^2)$. Then

$$\int\limits_\Omega (\partial_t(v^1 - v^2)\cdot\widetilde{v} + \nabla(v^1 - v^2) : \nabla\widetilde{v})dx = \int\limits_\Omega (u^1 - u^2)\otimes w : \nabla\widetilde{v}dx$$

and letting $\widetilde{v} = v^1 - v^2$, we find

$$\frac{1}{2}\partial_t\|v^1 - v^2\|^2_{2,\Omega} + \|\nabla v^1 - \nabla v^2\|^2_{2,\Omega} \le c(w)\|u^1 - u^2\|_{2,\Omega}\|\nabla v^1 - \nabla v^2\|_{2,\Omega}$$

and thus

$$\sup_{0<t<T}\|v^1 - v^2\|_{2,\Omega} \le c(w)\|u^1 - u^2\|_{2,Q_T}.$$

The latter implies continuity.

Compactness: As in the previous case, we use the energy estimate

$$\sup_{0<t<T}\|v\|^2_{2,\Omega} + \|\nabla v\|^2_{2,\Omega} \le c(w)\|u\|^2_{2,Q_T} + c(\|f\|^2_{L_2(0,T;(\overset{\circ}{J}{}^1_2(\Omega))')} + \|a\|^2_\Omega).$$

The second estimate comes from (5.2.8) and has the form

$$\|\partial_t v\|^2_{L_2(0,T;(\overset{\circ}{J}{}^1_2(\Omega))')} \le \|\nabla v\|^2_{Q_T} + c(w)\|u\|^2_{2,Q_T} + c\|f\|^2_{L_2(0,T;(\overset{\circ}{J}{}^1_2(\Omega))')}.$$

Combining the above estimates, we observe that sets which are bounded in X remain to be bounded in

$$W = \{w \in L_2(0,T;\overset{\circ}{J}{}^1_2(\Omega)),\quad \partial_t w \in L_2(0,T;(\overset{\circ}{J}{}^1_2(\Omega))')\}.$$

By Proposition 1.1 for $V_0 = \overset{\circ}{J}{}^1_2(\Omega)$, $V = \overset{\circ}{J}(\Omega)$, and $V_1 = (\overset{\circ}{J}{}^1_2(\Omega))'$, such a set is precompact.

Now, we wish to verify the second condition in Theorem 1.2 of Chapter 3. For $v = \lambda A(v)$, after integration by parts, we find that, for a.a. $t \in [0,T]$,

$$\int\limits_\Omega (\partial_t v\cdot\widetilde{v} + \nabla v : \nabla\widetilde{v})dx = \lambda\int\limits_\Omega (f\cdot\widetilde{v} - (w\cdot\nabla v)\cdot\widetilde{v})dx$$

for any $\widetilde{v} \in \overset{\circ}{J}{}^1_2(\Omega)$. If we insert $\widetilde{v}(\cdot) = v(\cdot,t)$ into the latter relation, then the identity

$$\int\limits_\Omega (w\cdot\nabla v)\cdot v dx = 0,$$

ensures the following estimate:

$$\frac{1}{2}\partial_t\int\limits_\Omega |v|^2dx + \int\limits_\Omega |\nabla v|^2dx \le \|f\|_{(\overset{\circ}{J}{}^1_2(\Omega))'}\|\nabla v\|_{2,\Omega}$$

and thus

$$\|v\|^2_{Q_T} \le T \sup_{0<t<T} \int_\Omega |v(x,t)|^2 dx \le cT(\|f\|^2_{L_2(0,T;(\overset{\circ}{J}{}^1_2(\Omega))')} + \|a\|^2_{2,\Omega}) = R^2.$$

Now, all the statements of Proposition 2.2 follow from the Leray-Schauder principle. Proposition 2.2 is proved. □

We need a slightly stronger statement.

Proposition 2.3. *Assume that*

$$w \in L_{\infty,2}(Q_T), \qquad \operatorname{div} w = 0 \quad in\ Q_T. \tag{5.2.10}$$

Then all statements of Proposition 2.2 remain to be true.

PROOF Let

$$< h >_\varepsilon (t) := \int_0^T \nu_\varepsilon(t-s)h(s)ds$$

be a standard mollification of a function h with respect t. By assumption (5.2.10), the function $< w >_\varepsilon (x,t)$ belongs to $L_\infty(Q_T)$. Thanks to Proposition 2.2, for each fixed $\varepsilon > 0$, there exists a unique function $v = v^\varepsilon$ such that

$$v^\varepsilon \in C([0,T]; L_2(\Omega)) \cap L_2(0,T; \overset{\circ}{J}{}^1_2(\Omega)), \quad \partial_t v^\varepsilon \in L_2(0,T; (\overset{\circ}{J}{}^1_2(\Omega))');$$

for a.a. $t \in [0,T]$

$$\int_\Omega (\partial_t v^\varepsilon(x,t) \cdot \widetilde{v}(x) + \nabla v^\varepsilon(x,t) : \nabla \widetilde{v}(x))dx$$

$$= \int_\Omega (v^\varepsilon(x,t) \otimes < w >_\varepsilon (x,t) : \nabla \widetilde{v}(x) + f(x,t) \cdot \widetilde{v}(x))dx \tag{5.2.11}$$

for all $\widetilde{v} \in \overset{\circ}{J}{}^1_2(\Omega)$;

$$\|v^\varepsilon(\cdot,t) - a(\cdot)\|_{2,\Omega} \to 0 \tag{5.2.12}$$

as $t \to +0$.

Since div $< w >_\varepsilon = 0$, we can get the energy estimate

$$\sup_{0<t<T} \|v^\varepsilon(\cdot,t)\|^2_{2,\Omega} + \|\nabla v^\varepsilon\|^2_{2,Q_T} \le c(\|f\|^2_{L_2(0,T;(\overset{\circ}{J}{}^1_2(\Omega))')} + \|a\|^2_{2,\Omega}).$$

Moreover, the derivative in time has the upper bound

$$\|\partial_t v^\varepsilon\|^2_{L_2(0,T;(\overset{\circ}{J}{}^1_2(\Omega))')} \le c\|\nabla v^\varepsilon\|^2_{Q_T} + c\||v^\varepsilon| < w >_\varepsilon \||^2_{2,Q_T} +$$

$$+c\|f\|^2_{L_2(0,T;(\overset{\circ}{J}{}^1_2(\Omega))')}.$$

On the other hand, the second term on the right-hand side of the latter inequality can be estimated with the help of properties of mollifications. As a result, we have

$$\||v^\varepsilon|| < w >_\varepsilon \|_{2,Q_T} \leq \|v^\varepsilon\|_{2,\infty,Q_T} \| < w >_\varepsilon \|_{\infty,2,Q_T} \leq$$

$$\leq \|v^\varepsilon\|_{2,\infty,Q_T} \|w\|_{\infty,2,Q_T}.$$

Hence, the derivative in time is uniformly bounded (with respect to ε) in the above norm. Thus, without loss of generality, we may assume that, as $\varepsilon \to 0$,

$$v^\varepsilon \to v$$

in $L_2(Q_T)$,

$$\nabla v^\varepsilon \rightharpoonup \nabla v$$

in $L_2(Q_T)$,

$$\partial_t v^\varepsilon \rightharpoonup \partial_t v$$

in $L_2(0,T;(\overset{\circ}{J}{}^1_2(\Omega))')$. So, since

$$< w >_\varepsilon \rightharpoonup w$$

in $L_2(Q_T)$ at least, we deduce that

$$v^\varepsilon \otimes < w >_\varepsilon \rightharpoonup v \otimes w$$

in $L_1(Q_T)$. Taking a test function $\widetilde{v} \in \overset{\circ}{J}{}^1_2(\Omega))$ in (5.2.11), multiplying the corresponding idenity by a test function $\chi \in C_0^\infty(0,T)$, and integrating the product in t, we can pass to the limit as $\varepsilon \to 0$ and easily demonstrate that the function v satisfies (5.2.5). To show that initial condition (5.2.6) holds, let us notice that, for any $\varphi \in \overset{\circ}{J}{}^1_2(\Omega)$,

$$\int_\Omega (v^\varepsilon(x,t) - a(x)) \cdot \varphi(x)dx = \int_0^t \int_\Omega \partial_t v^\varepsilon(x,s) \cdot \varphi(x)dxds.$$

Since $v^\varepsilon(\cdot,t) \to v(\cdot,t)$ in $L_2(\Omega)$ for a.a. t and since $v \in C([0,T], L_2(\Omega))$, we conclude that the latter identity holds for the limit function v as well. Proposition 2.3 is proven.

Let ω_ϱ be a usual mollifier and let

$$(v)_\varrho(x,t) = \int_\Omega \omega_\varrho(x-x')v(x',t)dx'.$$

It is easy to check that $\operatorname{div}(v)_\varrho(\cdot,t) = 0$ if $t \mapsto v(\cdot,t) \in \overset{\circ}{J}(\Omega)$ (Exercise).

Now, we wish to show that there exists at least one function v^ϱ such that:

$$v^\varrho \in C([0,T]; L_2(\Omega)) \cap L_2(0,T; \overset{\circ}{J}{}^1_2(\Omega)), \quad \partial_t v^\varrho \in L_2(0,T; (\overset{\circ}{J}{}^1_2(\Omega))'); \tag{5.2.13}$$

for a.a. $t \in [0,T]$

$$\int_\Omega (\partial_t v^\varrho(x,t) \cdot \widetilde{v}(x) + \nabla v^\varrho(x,t) : \nabla \widetilde{v}(x))dx$$
$$= \int_\Omega (v^\varrho(x,t) \otimes (v^\varrho)_\varrho(x,t) : \nabla \widetilde{v}(x) + f(x,t) \cdot \widetilde{v}(x))dx \tag{5.2.14}$$

for all $\widetilde{v} \in \overset{\circ}{J}{}^1_2(\Omega)$;

$$\|v^\varrho(\cdot,t) - a(\cdot)\|_{2,\Omega} \to 0 \tag{5.2.15}$$

as $t \to +0$.

We note that (5.2.13)-(5.2.15) can be regarded as a weak form of the following initial boundary value problem

$$\partial_t v^\varrho - \Delta v^\varrho + (v^\varrho)_\varrho \cdot \nabla v^\varrho + \nabla q^\varrho = f, \quad \operatorname{div} v^\rho = 0 \quad in\ Q_T,$$
$$v^\varrho|_{\partial\Omega \times [0,T]} = 0, \tag{5.2.16}$$
$$v^\varrho|_{t=0} = a.$$

Proposition 2.4. *There exists at least one function v^ϱ satisfying (5.2.13)-(5.2.15). In addition, it satisfies the energy estimate*

$$|v^\varrho|^2_{2,Q_T} \equiv \sup_{0<t<T} \|v^\varrho(\cdot,t)\|^2_{2,\Omega} + \|\nabla v^\varrho\|^2_{2,Q_T}$$
$$\leq c(\|f\|^2_{L_2(0,T;(\overset{\circ}{J}{}^1_2(\Omega))')} + \|a\|^2_{2,\Omega}) \tag{5.2.17}$$

with a constant c independent of ϱ.

PROOF To simplify our notation, let us drop upper index ϱ. The idea is the same as in Proposition 2.2: to use the Leray-Schauder principle. The space X is the same as in Proposition 2.2. But the operator A will be defined in the different way: given $u \in X$, we are looking for $v = A(u)$ so that

$$v \in C([0,T]; L_2(\Omega)) \cap L_2(0,T; \overset{\circ}{J}{}^1_2(\Omega)), \quad \partial_t v \in L_2(0,T; (\overset{\circ}{J}{}^1_2(\Omega))'); \tag{5.2.18}$$

for a.a. $t \in [0,T]$

$$\int_\Omega (\partial_t v(x,t)\cdot \widetilde{v}(x) + \nabla v(x,t) : \nabla \widetilde{v}(x))dx$$

$$= \int_\Omega (v(x,t)\otimes (u)_\varrho(x,t) : \nabla\widetilde{v}(x) + f(x,t)\cdot\widetilde{v}(x))dx \qquad (5.2.19)$$

for all $\widetilde{v} \in \overset{\circ}{J}{}^1_2(\Omega)$;

$$\|v(\cdot,t) - a(\cdot)\|_{2,\Omega} \to 0 \qquad (5.2.20)$$

as $t \to +0$. By Proposition 2.3, such a function exists and is unique. We need to check that all the assumptions of Theorem 1.2 of Chapter 3 hold for our operator A.

Continuity: Do the same as in Proposition 2.2:

$$\frac{1}{2}\partial_t\|v^2 - v^1\|^2_{2,\Omega} + \|\nabla(v^2-v^1)\|^2_{2,\Omega}$$

$$= \int_\Omega \left(v^2\otimes (u^2)_\varrho - v^1\otimes(u^1)_\varrho\right) : \nabla(v^2-v^1)dx$$

$$= \int_\Omega (v^2-v^1)\otimes (u^2)_\varrho : \nabla(v^2-v^1)dx$$

$$+ \int_\Omega v^1\otimes(u^2-u^1)_\varrho : \nabla(v^2-v^1)dx.$$

The first integral in the right-hand side of the above identity is zero whereas the second one I can be bounded as follows

$$I \le \sup_{x\in\Omega}|(u^2-u^1)_\varrho(x,t)|\|v^1\|_{2,\Omega}\|\nabla(v^2-v^1)\|_{2,\Omega}.$$

So, by Hölder inequality, we have

$$\partial_t\|v^2-v^1\|^2_{2,\Omega} + \|\nabla(v^2-v^1)\|^2_{2,\Omega} \le c(\varrho)\|v^1\|^2_{2,\Omega}\|u^2-u^1\|^2_{2,\Omega}$$

and thus

$$|v^2-v^1|^2_{2,Q_T} \le c(\varrho)\sup_{0<t<T}\|v^1(\cdot,t)\|^2_{2,\Omega}\|u^2-u^1\|^2_{2,Q_T}.$$

The latter gives us continuity.

Compactness: Now, we wish to write down the energy estimate for v

$$\frac{1}{2}\partial_t\|v\|^2_{2,\Omega} + \|\nabla v\|^2_{2,\Omega} = \int_\Omega f\cdot v dx - \int_\Omega v\otimes(u)_\varrho : \nabla v dx$$

$$= \int_\Omega f\cdot v dx. \qquad (5.2.21)$$

As in the proof of the previous proposition, (5.2.21) implies the required energy estimate

$$|v|^2_{2,Q_T} \leq c(\|f\|^2_{L_2(0,T;(\overset{\circ}{J}{}^1_2(\Omega))')} + \|a\|^2_{2,\Omega}) = cS, \tag{5.2.22}$$

where a constant c is independent of ϱ.

Now, we need to evaluate the derivative in time. We have

$$\|\partial_t v\|^2_{(\overset{\circ}{J}{}^1_2(\Omega))'} \leq c\|\nabla v\|^2_{2,\Omega} + c\|f\|^2_{(\overset{\circ}{J}{}^1_2(\Omega))'} + c\int\limits_\Omega |v|^2|(u)_\varrho|^2 dx$$
$$\leq c\|\nabla v\|^2_{2,\Omega} + c\|f\|^2_{(\overset{\circ}{J}{}^1_2(\Omega))'} + c(\varrho)\Big(\int\limits_\Omega |u| dx\Big)^2 \|v\|^2_{2,\Omega}.$$

After integration in time, the following estimate comes out:

$$\|\partial_t v\|^2_{L_2(0,T;(\overset{\circ}{J}{}^1_2(\Omega))')} \leq c(\varrho)\Big(S + S\int\limits_{Q_T} |u|^2 dz\Big). \tag{5.2.23}$$

Making use of similar arguments as in the proof of Proposition 2.2, we conclude that for each fixed $\varrho > 0$ the operator A is compact.

Let us check the second condition of Theorem 1.2 of Chapter 3. The same idea used to prove (5.2.21) and (5.2.22) gives us:

$$|v|^2_{2,Q_T} \leq c(\lambda^2\|f\|^2_{L_2(0,T;(\overset{\circ}{J}{}^1_2(\Omega))')} + \|a\|^2_{2,\Omega}) < 2cS = R^2.$$

So, the existence is established.

The energy estimate can be proved along the lines of the proof of (5.2.22). Proposition 2.4 is proved.

5.3 Weak Leray-Hopf Solutions

Now, we consider the full non-stationary Navier-Stokes system in a bounded domain $\Omega \subset \mathbb{R}^n$ (n=2,3):

$$\begin{aligned} \partial_t v - \Delta v + \operatorname{div} v\otimes v + \nabla q = f, \quad \operatorname{div} v = 0 \quad in\ Q_T, \\ v|_{\partial\Omega\times[0,T]} = 0, \\ v|_{t=0} = a. \end{aligned} \tag{5.3.1}$$

We always assume that

$$f \in L_2(0,T;V') \tag{5.3.2}$$

and

$$a \in H, \tag{5.3.3}$$

where $V = \overset{\circ}{J}{}^1_2(\Omega)$ and $H = \overset{\circ}{J}(\Omega)$.

Definition 5.1. A function v is called a weak Leray-Hopf solution to initial boundary value problem (5.3.1)-(5.3.3) if it has the following properties:

(i) $v \in L_\infty(0,T;H) \cap L_2(0,T;V)$;

(ii) function $t \mapsto \int\limits_\Omega v(x,t)\cdot w(x)dx$ is continuous on $[0,T]$ for each $w \in L_2(\Omega)$;

(iii) $\int\limits_\Omega (-v\cdot\partial_t w - v\otimes v : \nabla w + \nabla v : \nabla w - f\cdot w)dz = 0$ for any test function w belonging to $C^\infty_{0,0}(Q_T) = \{w \in C^\infty_0(Q_T) : \operatorname{div} w = 0 \text{ in } Q_T\}$;

(iv) $\int\limits_\Omega |v(x,t) - a(x)|^2dx \to 0$ as $t \to +0$;

(v) $\frac{1}{2}\int\limits_\Omega |v(x,t)|^2dx + \int\limits_0^t\int\limits_\Omega |\nabla v|^2dxdt' \le \frac{1}{2}\int\limits_\Omega |a(x)|^2dx + \int\limits_0^t\int\limits_\Omega f\cdot vdxdt'$ for all $t \in [0,T]$.

Theorem 3.5. *Under assumptions (5.3.2) and (5.3.3), there exists at least one weak Leray-Hopf solution to (5.3.1).*

PROOF By Proposition 2.4, for any positive ϱ, there exists a function v^ϱ such that

$$v^\varrho \in C([0,T];H)\cap L_2(0,T;V), \quad \partial_t v^\varrho \in L_2(0,T;V'); \tag{5.3.4}$$

for a.a. $t \in [0,T]$

$$\int\limits_\Omega (\partial_t v^\varrho(x,t)\cdot\widetilde{v}(x) + \nabla v^\varrho(x,t) : \nabla\widetilde{v}(x))dx$$

$$= \int\limits_\Omega (v^\varrho(x,t)\otimes(v^\varrho)_\varrho(x,t) : \nabla\widetilde{v}(x) + f(x,t)\cdot\widetilde{v}(x))dx \tag{5.3.5}$$

for all $\widetilde{v} \in V$;

$$\|v^\varrho(\cdot,t) - a(\cdot)\|_{2,\Omega} \to 0 \tag{5.3.6}$$

as $t \to +0$. Moreover, v^ϱ has uniformly bounded energy

$$|v^\varrho|^2_{2,Q_T} \equiv \sup_{0<t<T}\|v^\varrho(\cdot,t)\|^2_{2,\Omega} + \|\nabla v^\varrho\|^2_{2,Q_T} \le A, \tag{5.3.7}$$

where a constant A depends only on T, $\|f\|_{L_2(0,T;V')}$, and $\|a\|_{2,\Omega}$.

Now, let us see what happens if $\varrho \to 0$. To apply Proposition 1.1 of this section on compactness, we need to estimate the derivative of v in t. To this end, we are going to use the following imbedding theorem

$$\overset{\circ}{J}{}^3_2(\Omega) \subset C^1(\overline{\Omega}),$$

which is true provided $n = 2, 3$. Then from (5.3.5) it follows

$$\int\limits_{\Omega} \partial_t v^\varrho \cdot \widetilde{v} dx \leq \|\nabla \widetilde{v}\|_{\infty,\Omega} \int\limits_{\Omega} |v^\varrho||(v^\varrho)_\varrho| dx + \|\nabla \widetilde{v}\|_{2,\Omega} \|\nabla v^\varrho\|_{2,\Omega}$$
$$+\|f\|_{V'} \|\nabla \widetilde{v}\|_{2,\Omega}$$

for any $v \in \overset{\circ}{J}{}^3_2(\Omega)$ and thus

$$\|\partial_t v^\varrho\|_{(\overset{\circ}{J}{}^3_2(\Omega))'} \leq c(\Omega)\Big(\|v^\varrho\|_{2,\Omega}\|(v^\varrho)_\varrho\|_{2,\Omega} + \|\nabla v^\varrho\|_{2,\Omega} + \|f\|_{V'}\Big)$$
$$\leq c(\Omega)\Big(|v^\varrho|_{2,Q_T}\|v^\varrho\|_{2,\Omega} + \|\nabla v^\varrho\|_{2,\Omega} + \|f\|_{V'}\Big).$$

Therefore

$$\|\partial_t v^\varrho\|_{L_2(0,T;(\overset{\circ}{J}{}^3_2(\Omega))')} \leq c(\Omega)\Big(|v^\varrho|_{2,Q_T}\|v^\varrho\|_{2,Q_T} + |v^\varrho|_{2,Q_T} + \|f\|_{L_2(0,T;V')}\Big).$$

Since $\|v^\varrho\|_{2,Q_T} \leq T^{\frac{1}{2}}|v^\varrho|_{2,Q_T}$, we get

$$\|\partial_t v^\varrho\|_{L_2(0,T;(\overset{\circ}{J}{}^3_2(\Omega))')} \leq A_1, \tag{5.3.8}$$

where a positive constant A_1 depends only on T, $\|f\|_{L_2(0,T;V')}$, and $\|a\|_{2,\Omega}$.

Now, we can apply Proposition 1.1 with the following choice of spaces

$$V \subset H \subset (\overset{\circ}{J}{}^3_2(\Omega))',$$

where the space $(\overset{\circ}{J}{}^3_2(\Omega))'$ is the space dual to $\overset{\circ}{J}{}^3_2(\Omega)$ relative to H, and state that after selecting a subsequence

$$v^\varrho \overset{*}{\rightharpoonup} v \qquad \text{in } L_\infty(0,T;H), \tag{5.3.9}$$

$$v^\varrho \rightharpoonup v \qquad \text{in } L_2(0,T;V), \tag{5.3.10}$$

$$v^\varrho \to v \qquad \text{in } L_2(0,T;H). \tag{5.3.11}$$

To see that

$$D_\varrho = \int\limits_{Q_T} |v^\varrho \otimes (v^\varrho)_\varrho - v \otimes v| dz \to 0$$

as $\varrho \to 0$, let us argue as follows:

$$D_\varrho \leq \int\limits_{Q_T} |(v^\varrho - v) \otimes (v^\varrho)_\varrho| dz + \int\limits_{Q_T} |v \otimes ((v^\varrho)_\varrho - v)| dz$$
$$\leq \|v^\varrho - v\|_{2,Q_T}\|(v^\varrho)_\varrho\|_{2,Q_T} + \int\limits_{Q_T} |v \otimes (v^\varrho - v)_\varrho| dz$$
$$+ \int\limits_{Q_T} |v \otimes ((v)_\varrho - v)| dz \leq \|v^\varrho - v\|_{2,Q_T} T^{\frac{1}{2}} |v^\varrho|_{2,Q_T}$$
$$+T^{\frac{1}{2}}|v|_{2,Q_T}\|v^\varrho - v\|_{2,Q_T} + T^{\frac{1}{2}}|v|_{2,Q_T}\|(v)_\varrho - v\|_{2,Q_T},$$

where the right-hand side tends to zero as $\varrho \to 0$ by (5.3.11) and (5.3.7).

Setting $\widetilde{w}(x) = w(x,t)$ in (5.3.5), with $w \in C^\infty_{0,0}(Q_T)$, integrating in t by parts in (5.3.5), and passing to the limit, we deduce that v satisfies (iii) of Definition 5.1. So, (i) and (iii) of Definition 5.1 have been verified.

Now, let us take and fix an arbitrary function $\widetilde{v} \in \overset{\circ}{J}{}^3_2(\Omega)$ and consider functions

$$t \mapsto f^\varrho_{\widetilde{v}}(t) = \int\limits_\Omega v^\varrho(x,t) \cdot \widetilde{v}(x)dx.$$

Now, our goal is to show that for every fixed $\widetilde{v}$, the set of functions $f^\varrho_{\widetilde{v}}$ is precompact in $C([0,T])$. Indeed, it is uniformly bounded since

$$\sup_{0<t<T} |f^\varrho_{\widetilde{v}}(t)| \leq |v^\varrho|_{2,Q_T}\|\widetilde{v}\|_{2,\Omega} \leq c|v^\varrho|_{2,Q_T}\|\widetilde{v}\|_{\overset{\circ}{J}{}^3_2(\Omega)} \leq cA\|\widetilde{v}\|_{\overset{\circ}{J}{}^3_2(\Omega)}.$$

Its equicontinuity follows from (5.3.8):

$$|f^\varrho_{\widetilde{v}}(t+\Delta t) - f^\varrho_{\widetilde{v}}(t)| = \Big|\int\limits_t^{t+\Delta t}\int\limits_\Omega \partial_t v^\varrho(x,\tau)\cdot\widetilde{v}(x)dxd\tau\Big|$$

$$\leq \int\limits_t^{t+\Delta t} \|\partial_t v^\varrho(\cdot,\tau)\|_{(\overset{\circ}{J}{}^3_2(\Omega))'}\|\widetilde{v}(\cdot)\|_{\overset{\circ}{J}{}^3_2(\Omega)}d\tau$$

$$\leq \sqrt{|\Delta t|}\|\partial_t v^\varrho\|_{L_2(0,T;(\overset{\circ}{J}{}^3_2(\Omega))')}\|\widetilde{v}\|_{\overset{\circ}{J}{}^3_2(\Omega)} \leq c\sqrt{|\Delta t|}A_1\|\widetilde{v}\|_{\overset{\circ}{J}{}^3_2(\Omega)}.$$

Now, let $\widetilde{v}^{(k)}$ be a countable set that is dense in $\overset{\circ}{J}{}^3_2(\Omega)$. Applying the diagonal Cantor procedure, we can select a subsequence such that

$$\int\limits_\Omega v^\varrho(x,t)\cdot\widetilde{v}^{(k)}(x)dx \to \int\limits_\Omega v(x,t)\cdot\widetilde{v}^{(k)}(x)dx$$

in $C([0,T])$. By boundedness of

$$\sup_{\varrho>0}\sup_{0<t<T}\|v^\varrho(\cdot,t)\|_{2,\Omega},$$

one can show (by density arguments)

$$\int\limits_\Omega v^\varrho(x,t)\cdot\widetilde{v}(x)dx \to \int\limits_\Omega v(x,t)\cdot\widetilde{v}(x)dx$$

in $C([0,T])$ for any $v \in \overset{\circ}{J}{}^3_2(\Omega)$ and then for any $v \in \overset{\circ}{J}(\Omega)$.

Now, a given w from the space $L_2(\Omega)$ can be decomposed as $w = u + \nabla p$, where $u \in \overset{\circ}{J}(\Omega)$ and thus

$$\int_\Omega v^\varrho(x,t)\cdot w(x)dx = \int_\Omega v^\varrho(x,t)\cdot u(x)dx$$
$$\to \int_\Omega v(x,t)\cdot u(x)dx = \int_\Omega v(x,t)\cdot w(x)dx \qquad (5.3.12)$$

in $C([0,T])$ as $\varrho \to 0$. So, (ii) of Definition 5.1 has been proved as well.

Next, we would like to justify (v) of Definition 5.1. To achieve this goal, let us pick up $\widetilde{v}(x)$ as $v^\varrho(x,t)$ in (5.3.5) and integrate the corresponding equality in t. Since

$$\int_\Omega v^\varrho \otimes (v^\varrho)_\varrho : \nabla v^\varrho dx = \int_\Omega v_i^\varrho (v_j^\varrho)_\varrho v_{i,j}^\varrho dx = \frac{1}{2}\int_\Omega (v_j^\varrho)_\varrho |v^\varrho|^2_{,j} = 0,$$

we have

$$\frac{1}{2}\int_\Omega |v^\varrho(x,t)|^2 dx + \int_0^t\int_\Omega |\nabla v^\varrho|^2 dxdt' = \frac{1}{2}\int_\Omega |a(x)|^2 dx$$
$$+ \int_0^t\int_\Omega f\cdot v^\varrho dxdt' \qquad (5.3.13)$$

for all $t \in [0,T]$ and for all $\varrho > 0$.

By (5.3.12),

$$\liminf_{\varrho\to 0} \int_\Omega |v^\varrho(x,t)|^2 dx \geq \int_\Omega |v(x,t)|^2 dx \qquad (5.3.14)$$

for any $t \in [0,T]$ and by (5.3.10)

$$\liminf_{\varrho\to 0} \int_0^t\int_\Omega |\nabla v^\varrho|^2 dxdt' \geq \int_0^t\int_\Omega |\nabla v|^2 dxdt' \qquad (5.3.15)$$

and

$$\liminf_{\varrho\to 0} \int_0^t\int_\Omega f\cdot v^\varrho dxdt' = \int_0^t\int_\Omega f\cdot v dxdt' \qquad (5.3.16)$$

for all $t \in [0,T]$. So, (v) of Definition 5.1 follows from (5.3.13)-(5.3.16).

It remains to prove validity of (iv) of Definition 5.1. To this end, notice that by (5.3.12)

$$a(\cdot) = v^{\varrho}(\cdot, 0) \rightharpoonup v(\cdot, 0)$$

in $L_2(\Omega)$. So, $v(\cdot,0) = a(\cdot)$. Moreover, according to (ii) of Definition 5.1

$$v(\cdot, t) \rightharpoonup v(\cdot, 0) = a(\cdot)$$

in $L_2(\Omega)$ as $t \to +0$. So,

$$\liminf_{t\to+0} \|v(\cdot,t)\|_{2,\Omega} \geq \|a\|_{2,\Omega}.$$

However, from the energy inequality it follows that

$$\limsup_{t\to+0} \|v(\cdot,t)\|_{2,\Omega} \leq \|a\|_{2,\Omega}.$$

The latter implies

$$\lim_{t\to+0} \|v(\cdot,t)\|_{2,\Omega} = \|a\|_{2,\Omega}$$

which together with week convergence gives (iv) of Definition 5.1. Theorem 3.5 is proved.

5.4 Multiplicative Inequalities and Related Questions

Case 1: n=2

Lemma 5.2. *(Ladyzhenskaya's inequality)*

$$\|u\|^4_{4,\Omega} \leq 2\|u\|^2_{2,\Omega}\|\nabla u\|^2_{2,\Omega}$$

for any $u \in C^\infty_0(\Omega)$.

PROOF. Obviously, it is enough to prove this inequality for $\Omega = \mathbb{R}^2$. We have

$$|u(x_1,x_2)|^2 = 2\int\limits_{-\infty}^{x_1} u(t,x_2)u_{,1}(t,x_2)dt$$

$$\leq 2\Big(\int\limits_{-\infty}^{\infty} |u(t,x_2)|^2 dt\Big)^{\frac12}\Big(\int\limits_{-\infty}^{\infty} |u_{,1}(t,x_2)|^2 dt\Big)^{\frac12}.$$

And then

$$\int\limits_{-\infty}^{\infty}\int\limits_{-\infty}^{\infty} |u(x_1,x_2)|^4 dx_1 dx_2 = \int\limits_{-\infty}^{\infty}\int\limits_{-\infty}^{\infty} |u(x_1,x_2)|^2 |u(x_1,x_2)|^2 dx_1 dx_2$$

$$\leq 4\int_{-\infty}^{\infty} dx_1dx_2\Big\{\Big(\int_{-\infty}^{\infty}|u(t,x_2)|^2dt\Big)^{\frac{1}{2}}\Big(\int_{-\infty}^{\infty}|u_{,1}(t,x_2)|^2dt\Big)^{\frac{1}{2}}$$

$$\times\Big(\int_{-\infty}^{\infty}|u(x_1,s)|^2ds\Big)^{\frac{1}{2}}\Big(\int_{-\infty}^{\infty}|u_{,2}(x_1,s)|^2ds\Big)^{\frac{1}{2}}\Big\}$$

$$=\int_{-\infty}^{\infty}\Big(\int_{-\infty}^{\infty}|u(t,x_2)|^2dt\Big)^{\frac{1}{2}}\Big(\int_{-\infty}^{\infty}|u_{,1}(t,x_2)|^2dt\Big)^{\frac{1}{2}}dx_2$$

$$\times\int_{-\infty}^{\infty}\Big(\int_{-\infty}^{\infty}|u(x_1,s)|^2ds\Big)^{\frac{1}{2}}\Big(\int_{-\infty}^{\infty}|u_{,2}(x_1,s)|^2ds\Big)^{\frac{1}{2}}dx_1$$

$$\leq 4\|u\|_2\|u_{,1}\|_2\|u\|_2\|u_{,2}\|_2\leq 2\|u\|_2^2(\|u_{,1}\|_2^2+\|u_{,2}\|_2^2)$$

$$=2\|u\|_2^2\|\nabla u\|_2^2. \qquad \square$$

Corollary 4.6. *Let $u\in L_\infty(0,T;H)\cap L_2(0,T;V)$. Then*

$$\|u\|_{4,Q_T}\leq 2^{\frac{1}{4}}|u|_{2,Q_T}.$$

PROOF We have

$$\|u(\cdot,t)\|_{4,\Omega}^4\leq 2\|u(\cdot,t)\|_{2,\Omega}^2\|\nabla u(\cdot,t)\|_{2,\Omega}^2\leq 2|u|_{2,Q_T}^2\|\nabla u(\cdot,t)\|_{2,\Omega}^2.$$

After integration in t, we get the required inequality. □

Case 2: n=3

Lemma 5.3. *Let $2\leq s\leq 6$ and $\alpha=\frac{3(s-2)}{2s}$. Then, for any $u\in C_0^\infty(\Omega)$,*

$$\|u\|_{s,\Omega}\leq c(s)\|u\|_{2,\Omega}^{1-\alpha}\|\nabla u\|_{2,\Omega}^{\alpha}.$$

PROOF The Gagliardo-Nirenberg inequality in dimension three reads

$$\|u\|_{6,\Omega}\leq c\|\nabla u\|_{2,\Omega}, \qquad u\in C_0^\infty(\Omega),$$

with a constant c independent of Ω. It remains to use interpolation in L_s

$$\|u\|_{s,\Omega}\leq\|u\|_{2,\Omega}^{1-\alpha}\|u\|_{6,\Omega}^{\alpha}$$

with $\alpha=\frac{3(s-2)}{2s}$. □

Corollary 4.7. *Let $u\in L_\infty(0,T;H)\cap L_2(0,T;V)$. Then*

$$\|u\|_{s,l,Q_T}\leq c(s)|u|_{2,Q_T}$$

for $2\leq s\leq 6$ and l satisfying

$$\frac{3}{s}+\frac{2}{l}=\frac{3}{2}.$$

Here, $\|u\|_{s,l,Q_T}=\|u\|_{L_s(0,T;L_l(\Omega))}$.

PROOF By Lemma 5.3,

$$\|u(\cdot,t)\|_{s,\Omega} \leq c(s)|u|_{2,Q_T}^{1-\alpha}\|\nabla u(\cdot,t)\|_{2,\Omega}^{\alpha}$$

and

$$\Big(\int_0^T \|u(\cdot,t)\|_{s,\Omega}^l dt\Big)^{\frac{1}{l}} \leq c(s)|u|_{2,Q_T}^{1-\alpha}\Big(\int_0^T \|\nabla u(\cdot,t)\|_{2,\Omega}^{\alpha l} dt\Big)^{\frac{1}{l}}.$$

If $\alpha l = 2$, then the required inequality follows and

$$\frac{3}{s}+\frac{2}{l}=\frac{3}{s}+\alpha=\frac{3}{s}+\frac{3(2-s)}{2s}=\frac{3}{2}. \qquad \Box$$

Corollary 4.8. *Let $u \in L_\infty(0,T;H)\cap L_2(0,T;V)$. Then*

$$\|u\cdot\nabla u\|_{s,l,Q_T} \leq c(s)|u|_{2,Q_T}^2$$

with s and l greater than one and subject to the identity

$$\frac{3}{s}+\frac{2}{l}=4.$$

PROOF By Hölder inequality,

$$\int_\Omega |u\cdot\nabla u|^s dx \leq \Big(\int_\Omega |\nabla u|^2 dx\Big)^{\frac{s}{2}}\Big(\int_\Omega |u|^{s_1} dx\Big)^{\frac{s}{s_1}}$$

with $s_1 = \frac{2s}{2-s}$ and, hence, after integration in t and application of Hölder inequality, one can derive a bound:

$$\|u\cdot\nabla u\|_{s,l,Q_T} \leq \|\nabla u\|_{2,Q_T}\Big(\int_0^T\Big(\int_\Omega |u|^{s_1}dx\Big)^{\frac{2l}{s_1(2-l)}}dt\Big)^{\frac{2-l}{2l}}.$$

It is easy to verify that

$$\frac{3}{s_1}+\frac{2}{l_1}=\frac{3}{2}, \qquad l_1=\frac{2l}{l-2}. \tag{5.4.1}$$

The required inequality follows now from Corollary 4.7.

Let us discuss some consequences of (5.4.1):

$$\frac{3}{s_1}+\frac{2}{l_1}=\frac{3}{\frac{2s}{2-s}}+\frac{2-l}{l}=\frac{3(2-s)}{2s}+\frac{2}{l}-1=\frac{3}{2},$$

which implies

$$\frac{3}{s}+\frac{2}{l}=4. \qquad \Box$$

5.5 Uniqueness of Weak Leray-Hopf Solutions. 2D Case

Theorem 5.9. *(O. Ladyzhenskaya) Let $n = 2$. Then, under assumptions (5.3.2), (5.3.3), a weak Leray-Hopf solution to initial boundary value problem (5.3.1) is unique.*

PROOF We let $\widetilde{f} = -\operatorname{div} v \otimes v + f$, where v is a weak Leray-Hopf solution to initial boundary value problem (5.3.1). By Corollary 4.6, $\widetilde{f}$ belongs to $L_2(0,T;V')$ (since $v \in L_4(Q_T)$). By Theorem 2.3 of Chapter 4, we know that there exists a unique function u having the following properties:

$$u \in C([0,T];H) \cap L_2(0,T;V), \qquad \partial_t u \in L_2(0,T;V'); \tag{5.5.1}$$

for a.a. $t \in [0,T]$,

$$\int\limits_{\Omega} \Big[\partial_t u(x,t) \cdot w(x) + \nabla u(x,t) : \nabla w(x) - \widetilde{f}(x,t) \cdot w(x)\Big] dx = 0 \tag{5.5.2}$$

for all $w \in V$;

$$u(x,0) = a(x), \qquad x \in \Omega. \tag{5.5.3}$$

Recalling Definition 5.1, part (iii), we get from (5.5.2) that $\overline{v} = v - u$ satisfies the identity

$$\int\limits_{Q_T} \Big[-\overline{v}(x,t) \cdot W(x)\partial_t \chi(t) + \nabla\overline{v}(x,t) : \nabla W(x)\chi(t)\Big] dz = 0 \tag{5.5.4}$$

for all $W \in C^\infty_{0,0}(\Omega)$ and for any $\chi \in C^1_0(0,T)$. It is not difficult to show that (5.5.4) can be extended to all functions χ that are Lipschitz continuous in $[0,T]$ and satisfy the end conditions $\chi(0) = \chi(T) = 0$.

Now, our aim is to get rid of the assumption that χ vanishes at $t = 0$. To this end, we are going to use the following fact:

$$\|\overline{v}(\cdot,t)\|_{2,\Omega} \to 0 \tag{5.5.5}$$

as $t \to +0$. Take any function $\chi \in C^1([0,T])$ so that $\chi(T) = 0$ and a function φ_ε having the properties: $\varphi_\varepsilon(t) = 1$ if $t \geq \varepsilon$, $\varphi_\varepsilon(t) = 0$ if $0 < t \leq \varepsilon/2$, and $\varphi_\varepsilon(t) = (2t-\varepsilon)/\varepsilon$ if $\varepsilon/2 < t < \varepsilon$. Then, by (5.5.4), with $\varphi_\varepsilon\chi$ as χ,

$$\int\limits_{Q_T} \varphi_\varepsilon(t)\Big[-\overline{v}(x,t) \cdot W(x)\partial_t \chi(t) + \nabla\overline{v}(x,t) : \nabla W(x)\chi(t)\Big] dz$$

$$= \frac{2}{\varepsilon}\int\limits_{\frac{\varepsilon}{2}}^{\varepsilon}\int\limits_{\Omega} \overline{v}(x,t) \cdot W(x)\chi(t) dz = I_\varepsilon.$$

For the right-hand side, we have

$$|I_\varepsilon| \leq \sup_{x\in\Omega}|W(x)| \sup_{\tau\in[0,T]} \chi(\tau)\sqrt{|\Omega|} \sup_{0<t\leq\varepsilon} \|\overline{v}(\cdot,t)\|_{2,\Omega} \to 0$$

as $\varepsilon \to 0$.

So,

$$\int_{Q_T} \Big[-\overline{v}(x,t)\cdot W(x)\partial_t\chi(t) + \nabla\overline{v}(x,t) : \nabla W(x)\chi(t)\Big]dz = 0$$

for all $W \in C^\infty_{0,0}(\Omega)$ and any $\chi \in C^1([0,T])$ with $\chi(T) = 0$. This means that $\overline{v} \equiv 0$ in Q_T, see Theorem 5.6 of Chapter 4, and thus any weak Leray-Hopf solution v has the following properties:

$$v \in C([0,T];H) \cap L_2(0,T;V), \qquad \partial_t v \in L_2(0,T;V'); \tag{5.5.6}$$

for a.a. $t \in [0,T]$,

$$\int_\Omega \Big[\partial_t v(x,t)\cdot w(x) + \nabla v(x,t) : \nabla w(x)\Big]dx$$

$$= \int_\Omega \Big[v(x,t)\otimes v(x,t) : \nabla w(x) + \widetilde{f}(x,t)\cdot w(x)\Big]dx \tag{5.5.7}$$

for all $w \in V$;

$$v(x,0) = a(x), \qquad x \in \Omega. \tag{5.5.8}$$

Now, assume that we have two different solutions v^1 and v^2. Letting $u = v^2 - v^1$, one deduce from (5.5.7) that:

$$\int_\Omega \Big[\partial_t u(x,t)\cdot w(x) + \nabla u(x,t) : \nabla w(x)\Big]dx =$$

$$= \int_\Omega (v^2(x,t)\otimes v^2(x,t) - v^1(x,t)\otimes v^1(x,t)) : \nabla w(x)dx$$

for any $w \in V$. Taking $w(x) = u(x,t)$ in the above identity,

$$\frac{1}{2}\partial_t\|u(\cdot,t)\|^2_{2,\Omega} + \|\nabla u(\cdot,t)\|^2_{2,\Omega} =$$

$$= \int_\Omega (u(x,t)\otimes v^2(x,t) : \nabla u(x,t) + v^1(x,t)\otimes u(x,t) : \nabla u(x,t))dx \leq$$

$$\leq \|\nabla u(\cdot,t)\|_{2,\Omega}\|u(\cdot,t)\|_{4,\Omega}\|v^1(\cdot,t)\|_{4,\Omega}.$$

By using Ladyzhenskaya's inequality twice,

$$\|v^1(\cdot,t)\|^4_{4,\Omega} \leq 2\|v^1(\cdot,t)\|^2_{2,\Omega}\|\nabla v^1(\cdot,t)\|^2_{2,\Omega} \leq 2|v^1|^2_{2,Q_T}\|\nabla v^1(\cdot,t)\|^2_{2,\Omega} \leq$$
$$\leq c(a,f)\|\nabla v^1(\cdot,t)\|^2_{2,\Omega}$$

and

$$\|u(\cdot,t)\|^4_{4,\Omega} \leq 2\|u(\cdot,t)\|^2_{2,\Omega}\|\nabla u(\cdot,t)\|^2_{2,\Omega}.$$

And thus

$$\frac{1}{2}\partial_t\|u(\cdot,t)\|^2_{2,\Omega} + \|\nabla u(\cdot,t)\|^2_{2,\Omega} \leq$$
$$\leq c(a,f)\|\nabla u(\cdot,t)\|^{\frac{3}{2}}_{2,\Omega}\|u(\cdot,t)\|^{\frac{1}{2}}_{2,\Omega}\|\nabla v^1(\cdot,t)\|^{\frac{1}{2}}_{2,\Omega}.$$

Applying Young's inequality, we find

$$\partial_t\|u(\cdot,t)\|^2_{2,\Omega} + \|\nabla u(\cdot,t)\|^2_{2,\Omega} \leq c_0\|u(\cdot,t)\|^2_{2,\Omega}\|\nabla v^1(\cdot,t)\|^2_{2,\Omega}$$

and thus

$$\partial_t\|u(\cdot,t)\|^2_{2,\Omega} \leq c_0 y(t)\|u(\cdot,t)\|^2_{2,\Omega},$$

where $y(t) := \|\nabla v^1(\cdot,t)\|^2_{2,\Omega}$. From this differential inequality, it is not difficult to derive

$$\partial_t\Big(e^{-c_0\int\limits_0^t y(\tau)d\tau}\|u(\cdot,t)\|^2_{2,\Omega}\Big) \leq 0$$

and

$$e^{-c_0\int\limits_0^t y(\tau)d\tau}\|u(\cdot,t)\|^2_{2,\Omega} \leq \|u(\cdot,0)\|^2_{2,\Omega} = 0.$$

Therefore, $\|u(\cdot,t)\|^2_{2,\Omega} \equiv 0$. $\square$

Let us discuss further regularity of 2D weak Leray-Hopf solutions.

Theorem 5.10. *Assume that $a \in V$, and $f \in L_2(Q_T)$. Let v be a unique solution to initial boundary value problem (5.3.1). Then*

$$v \in W^{2,1}_2(Q_T), \qquad \nabla v \in C([0,T];L_2(\Omega)).$$

Moreover, there exists $q \in W^{1,0}_2(Q_T)$ such that

$$\partial_t v + v\cdot\nabla v - \Delta v = f - \nabla q, \qquad \operatorname{div} v = 0$$

a.e. in Q_T. It is supposed that Ω is a bounded domain with smooth boundary.

PROOF Let us go back to problem (5.3.4)-(5.3.6), where a function v^ϱ defined by the following relations:

$$v^\varrho \in C([0,T];H)\cap L_2(0,T;V), \quad \partial_t v^\varrho \in L_2(0,T;V'); \tag{5.5.9}$$

for a.a. $t\in[0,T]$

$$\int_\Omega (\partial_t v^\varrho(x,t)\cdot\widetilde{v}(x)+\nabla v^\varrho(x,t):\nabla\widetilde{v}(x))dx = \int_\Omega \widetilde{f}(x,t)\cdot\widetilde{v}(x)dx \tag{5.5.10}$$

for all $\widetilde{v}\in V$, where

$$\widetilde{f}=f-(v^\varrho)_\varrho\cdot\nabla v^\varrho\in L_2(Q_T);$$

$$\|v^\varrho(\cdot,t)-a(\cdot)\|_{2,\Omega}\to 0 \tag{5.5.11}$$

as $t\to+0$.

According to Theorem 4.4 of Chapter 4, there exists a functions $u^\varrho\in W^{2,1}_2(Q_T)$ with $\nabla u^\varrho\in C([0,T];L_2(\Omega))$ and $p^\varrho\in W^{1,0}_2(Q_T)$ such that

$$\partial_t u^\varrho-\Delta u^\varrho=\widetilde{f}-\nabla p^\varrho, \qquad \operatorname{div} u^\varrho=0$$

a.e. in Q_T and $u^\varrho(\cdot,0)=a(\cdot)$. Using the same arguments as in the proof of the previous statement, we can claim that $v^\varrho=u^\varrho$. Multiplying then the equation

$$\partial_t v^\varrho+(v^\varrho_k)_\varrho v^\varrho_{,k}-\Delta v^\varrho=f-\nabla p^\varrho$$

by $\widetilde{\Delta} v^\varrho$ and integrating each term in the product with respect to x, we find

$$\int_\Omega \Delta v^\varrho\cdot\widetilde{\Delta} v^\varrho dx=\int_\Omega|\widetilde{\Delta} v^\varrho|^2dx,$$

$$\int_\Omega \nabla p^\varrho\cdot\widetilde{\Delta} v^\varrho dx=0,$$

$$\int_\Omega \partial_t v^\varrho\cdot\widetilde{\Delta} v^\varrho dx=\int_\Omega \partial_t v^\varrho\cdot\Delta v^\varrho dx=-\frac{1}{2}\partial_t\|\nabla v^\varrho\|^2_{2,\Omega}$$

since $\widetilde{\Delta} v^\varrho\in H$ and $\partial_t v^\varrho\in H$ for each fixed t. So, we derive the inequality

$$\frac{1}{2}\partial_t\|\nabla v^\varrho\|^2_{2,\Omega}+\|\widetilde{\Delta} v^\varrho\|^2_{2,\Omega}\leq\|f\|_{2,\Omega}\|\widetilde{\Delta} v^\varrho\|_{2,\Omega}+$$
$$+\|(v^\varrho)_\varrho\|_{4,\Omega}\|\nabla v^\varrho\|_{4,\Omega}\|\widetilde{\Delta} v^\varrho\|_{2,\Omega}\leq$$
$$\leq\|f\|_{2,\Omega}\|\widetilde{\Delta} v^\varrho\|_{2,\Omega}+\|v^\varrho\|_{4,\Omega}\|\nabla v^\varrho\|_{4,\Omega}\|\widetilde{\Delta} v^\varrho\|_{2,\Omega}.$$

To estimate the first and the second factors in the last term of the right-hand side of the above inequality, we are going to exploit Ladyzhenskaya's inequality one more time

$$\|v^\varrho\|^4_{4,\Omega} \leq 2\|v^\varrho\|^2_{2,\Omega}\|\nabla v^\varrho\|^2_{2,\Omega} \leq c(a,f)\|\nabla v^\varrho\|^2_{2,\Omega}$$

and

$$\|\nabla v^\varrho\|^4_{4,\Omega} \leq c(\Omega)\|\nabla v^\varrho\|^2_{2,\Omega}(\|\nabla^2 v^\varrho\|^2_{2,\Omega} + \|\nabla v^\varrho\|^2_{2,\Omega}).$$

We also need the Cattabriga-Solonnikov inequality

$$\|\nabla^2 v^\varrho\|_{2,\Omega} \leq c(\Omega)\|\widetilde{\Delta} v^\varrho\|_{2,\Omega}.$$

So, combining the latter results, we find the basic estimate

$$\frac{1}{2}\partial_t\|\nabla v^\varrho\|^2_{2,\Omega} + \|\widetilde{\Delta} v^\varrho\|^2_{2,\Omega} \leq \|f\|_{2,\Omega}\|\widetilde{\Delta} v^\varrho\|_{2,\Omega}$$

$$+c(a,f,\Omega)\|\nabla v^\varrho\|^{\frac{1}{2}}_{2,\Omega}\|\nabla v^\varrho\|^{\frac{1}{2}}_{2,\Omega}(\|\widetilde{\Delta} v^\varrho\|^{\frac{1}{2}}_{2,\Omega} + \|\nabla v^\varrho\|^{\frac{1}{2}}_{2,\Omega})\|\widetilde{\Delta} v^\varrho\|_{2,\Omega}$$

Now, if we apply Young's inequalities in an appropriate way, we can derive the following differential inequality

$$\partial_t y + \|\widetilde{\Delta} v^\varrho\|^2_{2,\Omega} \leq c(a,f,\Omega)(\|f\|^2_{2,\Omega} + y + \|\nabla v^\varrho\|^2_{2,\Omega} y),$$

where function $y(t) = \|\nabla v^\varrho(\cdot,t)\|^2_{2,\Omega}$ obeys the initial condition $y(0) = \|\nabla a\|^2_{2,\Omega}$.

The latter, together with energy estimate (5.3.7), gives two estimates

$$\sup_{0<t<T}\|\nabla v^\varrho(\cdot,t)\|^2_{2,\Omega} + \|\nabla^2 v^\varrho\|^2_{2,Q_T} \leq c(a,f,\Omega) < \infty. \tag{5.5.12}$$

To get all remaining estimates, we make use of Theorem 4.4 of Section 4, which reads

$$\|\partial_t v^\varrho\|_{2,Q_T} + \|\nabla^2 v^\varrho\|_{2,Q_T} + \|\nabla p^\varrho\|_{2,Q_T} \leq c\Big[\|\widetilde{f}\|_{2,Q_T} + \|\nabla a\|_{2,\Omega}\Big].$$

But

$$\|\widetilde{f}\|_{2,Q_T} \leq \|f\|_{2,Q_T} + \|(v^\varrho)_\varrho\cdot\nabla v^\varrho\|_{2,Q_T} \leq$$

$$\leq \|f\|_{2,Q_T} + \|(v^\varrho)_\varrho\|_{4,Q_T}\|\nabla v^\varrho\|_{4,Q_T} \leq$$

$$\leq c(a,f)\Big[1 + \|v^\varrho\|_{4,Q_T}\|\nabla v^\varrho\|_{4,Q_T}\Big].$$

The right-hand side of the above inequality can be evaluated with the help of Ladyzhenskaya's inequality and (5.5.12).

Finally, we can pass to the limit as $\varrho \to 0$ and get all the statements of Theorem 5.10. Theorem 5.10 is proved.

5.6 Further Properties of Weak Leray-Hopf Solutions

Theorem 6.11. *Let Ω be a bounded domain with smooth boundary. Assume*

$$f \in L_2(Q_T), \qquad a \in H. \tag{5.6.1}$$

I. Let v be an arbitrary weak Leray-Hopf solution in Q_T for the right-hand side f and initial data a satisfying (5.6.1). Then, for each $\delta > 0$ and for any numbers $s, l > 1$ subject to the condition

$$\frac{3}{s} + \frac{2}{l} = 4,$$

we have

$$v \in W^{2,1}_{s,l}(Q_{\delta,T}),$$

where $Q_{\delta,T} = \Omega \times]\delta, T[$. Moreover, there exists a function q (pressure) which belongs to the following spaces

$$q \in W^{1,0}_{s,l}(Q_{\delta,T}) \cap L_{s',l'}(Q_{\delta,T}) \tag{5.6.2}$$

with the same δ, s and l as above and

$$s' = \frac{3s}{3-s}, \qquad l' = l$$

so that

$$\frac{3}{s'} + \frac{2}{l'} = 3.$$

The Navier-Stokes equations

$$\partial_t v + \operatorname{div} v \otimes v - \Delta v = f - \nabla q, \qquad \operatorname{div} v = 0$$

hold in the sense of distributions and a.e. in Q_T.

II. Given right-hand side f and initial data a satisfying (5.6.1), there exists at least one weak Leray-Hopf solution v and a pressure q with the properties mentioned in Part I such that, for any $t_0 \in]0, T]$, the local energy inequality

$$\int_\Omega |v(x,t_0)|^2 \varphi(x,t_0)dx + \int_0^{t_0}\int_\Omega \varphi |\nabla v|^2 dxdt \leq \int_0^{t_0}\int_\Omega \Big(|v|^2(\partial_t \varphi + \Delta \varphi) +$$

$$+ v \cdot \nabla \varphi (|v|^2 + 2q) + 2\varphi f \cdot v\Big) dxdt \tag{5.6.3}$$

holds for any non-negative function $\varphi \in C^\infty_0(\mathbb{R}^3 \times]0, \infty[)$.

PROOF I. Take $\chi \in C_0^\infty(0,\infty)$ $(0 \leq \chi \leq 1)$ and let $u = \chi v$. Obviously,

$$u \in L_\infty(0,T;H) \cap L_2(0,T;V).$$

Next, insert χw with $w \in C_{0,0}^\infty(Q_T)$ into identity (iii) of Definition 5.1 and get for u:

$$\int\limits_{Q_T} (-u \cdot \partial_t w + \nabla u : \nabla w) dz = \int\limits_{Q_T} \widetilde{f} \cdot w dz,$$

where

$$\widetilde{f} = \chi f - \chi \operatorname{div} v \otimes v - \partial_t \chi v = \chi f - \chi v \cdot \nabla v - \partial_t \chi v.$$

By Corollary 4.8,

$$\widetilde{f} \in L_{s,l}(Q_T)$$

for any $s, l > 1$ satisfying $3/s + 2/l = 4$. Moreover, $u = 0$ for sufficiently small t. On the other hand, the linear theory ensures that, for such $\widetilde{f}$, there exist functions $\widetilde{v}$ and $\widetilde{q}$ such that

$$\widetilde{v} \in W_{s,l}^{2,1}(Q_T), \qquad \widetilde{q} \in W_{s,l}^{1,0}(Q_T)$$

with finite numbers $s, l > 1$ satisfying $3/s + 2/l = 4$ and

$$\partial_t \widetilde{v} - \Delta \widetilde{v} + \nabla \widetilde{q} = \widetilde{f}, \qquad \operatorname{div} \widetilde{v} = 0$$

in Q_T,

$$\widetilde{v}(x,t) = 0 \quad x \in \partial\Omega, \qquad \int\limits_{\Omega} \widetilde{q}(x,t) dx = 0$$

for $t \in [0,T]$,

$$\widetilde{v}(\cdot,0) = 0$$

in Ω.

Using essentially the same arguments as in 2D-case, we can show that for $\widehat{v} = u - \widetilde{v}$

$$\int\limits_{Q_T} (\widehat{v} \cdot \partial_t w - \nabla \widehat{v} : \nabla w) dz = 0$$

for $w = \widetilde{\chi} W$ with $W \in C_{0,0}^\infty(\Omega)$ and with $\widetilde{\chi} \in C^2([0,T])$ and $\widetilde{\chi}(T) = 0$.

Now, the uniqueness results for the linear theory imply that $\widehat{u} = 0$. Hence, $\chi v \in W_{s,l}^{2,1}(Q_T)$ and

$$\chi(\partial_t v + v \cdot \nabla v - \Delta v - f) = -\nabla \widetilde{q}.$$

Next, take any $\delta > 0$ and assume that $\chi(t) = \chi_\delta(t) = 1$ if $t > \delta$. A pressure $\widetilde{q}$ corresponding to chosen δ is denoted by q_δ. Obviously, $q_\delta \in W^{1,0}_{s,l}(Q_{\delta,T})$ with required s and l. Assuming that $\delta_1 > \delta_2$ then $q_{\delta_1} = q_{\delta_2}$ in $Q_{\delta_1,T}$. This allows us to introduce the function q so that

$$q(\cdot,t) = q_\delta(\cdot,t)$$

if $t > \delta > 0$. It is well defined and satisfies the required properties. So, the first part of the theorem is proved.

Part II. Now, let us go back to the proof of Theorem 3.5 on the existence of weak Leray-Hopf solutions and try to apply the procedure, described in the proof of Part I, to regularized problem. Letting $u^\varrho = \chi v^\varrho$, where χ is a function of t from $C^2_0(0,\infty)$, we state that u^ϱ is a solution to the problem:

$$u^\varrho \in C([0,T];H) \cap L_2(0,T;V), \quad \partial_t u^\varrho \in L_2(0,T;V'); \tag{5.6.4}$$

for a.a. $t \in [0,T]$

$$\int\limits_\Omega (\partial_t u^\varrho(x,t) \cdot \widetilde{v}(x) + \nabla u^\varrho(x,t) : \nabla \widetilde{v}(x))dx$$

$$= \int\limits_\Omega \widetilde{f}^\varrho(x,t) \cdot \widetilde{v}(x)dx \tag{5.6.5}$$

for all $\widetilde{v} \in V$;

$$\|u^\varrho(\cdot,t)\|_{2,\Omega} \to 0 \tag{5.6.6}$$

as $t \to +0$.

Here,

$$\widetilde{f}^\varrho = \widetilde{f}^\varrho_1 + \widetilde{f}^\varrho_2$$

with

$$\widetilde{f}^\varrho_1(x,t) = \chi(t)f(x,t) + \partial_t\chi(t)v^\varrho(x,t),$$

$$\widetilde{f}^\varrho_2(x,t) = -\chi(t)(v^\varrho)_\varrho(x,t) \cdot \nabla v^\varrho(x,t).$$

Simply repeating the proof of Corollary 4.7 and Corollary 4.8, we estimate the second part

$$\|\widetilde{f}^\varrho_2\|_{s',l',Q_T} \leq c(s')|v^\varrho|^2_{2,Q_T} \leq c(s',a,f)$$

with $s', l' > 1$ and $3/s' + 2/l' = 4$. Moreover, we can claim that the whole right-hand side $\widetilde{f}^\varrho$ is estimated similarly:

$$\|\widetilde{f}^\varrho\|_{s',l',Q_T} \leq c(s',a,f,\chi,\Omega).$$

Now, according to Theorem 4.5 of Chapter 4, there exists functions $\widetilde{u}^\varrho$ and q^ϱ_χ satisfy the relations:

$$\widetilde{u}^\varrho \in W^{2,1}_{s',l'}(Q_T), \qquad q^\varrho_\chi \in W^{1,0}_{s',l'}(Q_T);$$

$$\partial_t \widetilde{u}^\varrho - \Delta\, \widetilde{u}^\varrho = \widetilde{f}^\varrho - \nabla\, q^\varrho_\chi, \qquad \operatorname{div} \widetilde{u}^\varrho = 0$$

in Q_T;

$$\widetilde{u}^\varrho|_{\partial' Q_T} = 0, \qquad [q^\varrho_\chi(\cdot, t)]_\Omega = 0.$$

By the same theorem, these functions have the bound

$$\begin{aligned} &\|\nabla^2 \widetilde{u}^\varrho\|_{s',l',Q_T} + \|\partial_t \widetilde{u}^\varrho\|_{s',l',Q_T} + \|\nabla q^\varrho_\chi\|_{s',l',Q_T} \\ &\qquad \leq c(s', \Omega)\|\widetilde{f}^\varrho\|_{s',l',Q_T} \leq c(s', \Omega, a, f, \chi) \end{aligned}$$

Applying Theorem 5.6 of Chapter 4 on uniqueness and similar arguments to those that are used in 2D case, we can state

$$u^\varrho = \widetilde{u}^\varrho.$$

Next, we consider the sequence of functions χ_δ with $\delta = T/k$, $k \in \mathbb{N}$. And thus

$$\begin{aligned} &\|\nabla^2 v^\varrho\|_{s',l',Q_{\delta,T}} + \|\partial_t v^\varrho\|_{s',l',Q_{\delta,T}} + \|\nabla q^\varrho_\delta\|_{s',l',Q_{\delta,T}} \\ &\qquad \leq c(s', \Omega, a, f, \delta). \end{aligned} \tag{5.6.7}$$

Let v^ϱ be a sequence constructed in the proof of Theorem 3.5, i.e.,

$$v^\varrho \overset{*}{\rightharpoonup} v$$

in $L_\infty(0, T; H)$,

$$\nabla v^\varrho \rightharpoonup \nabla v$$

in $L_2(0, T; V)$,

$$v^\varrho \to v$$

in $L_2(Q_T)$,

$$\int\limits_\Omega v^\varrho(x,t) \cdot w(x)dx \to \int\limits_\Omega v(x,t) \cdot w(x)dx$$

in $C([0, T])$ for each $w \in L_2(\Omega)$.

In addition, we know that

$$\|v^\varrho\|_{s'',l'',Q_T} \leq c|v^\varrho|_{2,Q_T} \leq c(a, f)$$

with $3/s''+2/l''=3/2$, see Corollary 4.7. In particular, we may assume that (take $s''=l''=10/3$)

$$v^\varrho \rightharpoonup v$$

in $L_{\frac{10}{3}}(Q_T)$ and therefore

$$v^\varrho \to v \tag{5.6.8}$$

in $L_3(Q_T)$ and

$$(v^\varrho)_\varrho \cdot \nabla v^\varrho \rightharpoonup v \cdot \nabla v$$

in $L_1(Q_T)$.

Given $s,l>1$ with 3/s+2/l=3, we find $l'=l$ and $s'=3s/(s+3)$. It is easy to check, $3/s'+2/l'=4$. Using the diagonal Cantor procedure and bounds (5.6.7), one can ensure that

$$\partial_t v^\varrho \rightharpoonup \partial_t v, \qquad \nabla^2 v^\varrho \rightharpoonup \nabla^2 v$$

in $L_{s',l'}(Q_{\delta,T})$ for each $\delta>0$. As to the pressure, the diagonal Cantor procedure can be used one more time to show that

$$\nabla q^\varrho_\delta \rightharpoonup \nabla q_\delta$$

in $L_{s',l'}(Q_{\delta,T})$ and

$$q^\varrho_\delta \rightharpoonup q_\delta$$

in $L_{s,l}(Q_{\delta,T})$ for each δ. Here $[q_\delta(\cdot,t)]_\Omega=0$.

It is worthy to note that $q_{\delta_1}=q_{\delta_2}$ in $Q_{\varrho_2,T}$ if $\delta_1<\delta_2$. Indeed, it follows from two identities for $\delta=\delta_1$ and $\delta=\delta_2$

$$\nabla q_\delta = f-\partial_t v - v\cdot\nabla v+\Delta v$$

in $Q_{\delta,T}$ and $[q_\delta(\cdot,t)]_\Omega=0$.

So, the function, defined as

$$q(z)=q_\delta(z), \qquad z\in Q_{\delta,T},$$

belongs $W^{1,0}_{s',l'}(Q_T)$ and $L_{s,l}(Q_T)$ for each $\delta>0$.

The last thing is to check validity of the local energy inequality. It is known that the regularised solution is smooth enough and, therefore, obeys the local energy identity. So, a given non-negative function $\varphi\in C^\infty_0(\mathbb{R}^3\times]0,\infty[)$, we choose δ so small that $\operatorname{spt}\varphi\in\mathbb{R}^3\times]\delta,\infty[$. After multiplication of the equation

$$\partial_t v^\varrho+(v^\varrho)_\varrho\cdot\nabla v^\varrho-\Delta v^\varrho=f-\nabla q^\varrho_\delta$$

by φv^ϱ and integration of the product by parts (which is legal for the regularized solution)

$$\int\limits_\Omega |v^\varrho(x,t_0)|^2\varphi(x,t_0)dx + \int\limits_0^{t_0}\int\limits_\Omega \varphi|\nabla v^\varrho|^2 dxdt = \int\limits_0^{t_0}\int\limits_\Omega \Big(|v^\varrho|^2(\partial_t\varphi + \Delta\varphi)+$$

$$+(v^\varrho)_\varrho\cdot\nabla\varphi(|v^\varrho|^2 + 2q^\varrho_\delta) + 2f\cdot v\Big)dxdt \tag{5.6.9}$$

for any $t \in [\delta, T]$.

We also know that

$$q^\varrho_\delta \rightharpoonup q$$

in $L_{\frac{3}{2}}(Q_{\delta,T})$. (Indeed, $q^\varrho_\delta \rightharpoonup q$ in $L_{\frac{5}{3}}(Q_{\delta,T})$ since $3/(5/3) + 2/(5/3) = 3$). Taking into account (5.6.8) and using the same arguments as in the proof of Theorem 3.5, one can pass to the limit in (5.6.9) as $\varrho \to 0$ and get required local energy inequality (5.6.3). □

5.7 Strong Solutions

Definition 5.2. A weak Leray-Hopf solution is called a strong solution, if

$$\nabla v \in L_\infty(0,T;L_2(\Omega)). \tag{5.7.1}$$

Theorem 7.12. *(Global existence of strong solutions for "small" data). There exists a constant $c_0(\Omega)$ such that if*

$$\arctan(\|\nabla a\|^2_{2,\Omega}) + c_0(\Omega)(\|a\|^2_{2,\Omega} + \|f\|^2_{2,Q_T}) < \frac{\pi}{2}, \tag{5.7.2}$$

then there exists a strong solution to initial boundary value problem (5.3.1).

PROOF Let us go back to problem (5.3.4)-(5.3.6), see the proof of Theorem 3.5,

$$v^\varrho \in C([0,T];H)\cap L_2(0,T;V), \quad \partial_t v^\varrho \in L_2(0,T;V'); \tag{5.7.3}$$

for a.a. $t \in [0,T]$

$$\int\limits_\Omega (\partial_t v^\varrho(x,t)\cdot\widetilde{v}(x) + \nabla v^\varrho(x,t) : \nabla\widetilde{v}(x))dx =$$

$$= \int\limits_\Omega f^\varrho(x,t)\cdot\widetilde{v}(x)dx \tag{5.7.4}$$

for all $\widetilde{v} \in V$;

$$\|v^\varrho(\cdot,t) - a(\cdot)\|_{2,\Omega} \to 0 \qquad (5.7.5)$$

as $t \to +0$. Here,

$$f^\varrho = f - (v^\varrho)_\varrho \cdot \nabla v^\varrho \in L_2(Q_T).$$

Using the similar arguments as in the proof of Theorem 5.9 of this section and Theorems 5.6, 4.5 of Section 4 on uniqueness and regularity for non-stationary Stokes problem, we can conclude that

$$v^\varrho \in W^{2,1}_2(Q_T), \qquad \nabla v^\varrho \in C([0,T]; L_2(\Omega)).$$

Moreover, there exists a pressure field $p^\varrho \in W^{1,0}_2(Q_T)$ such that the regularized Navier-Stokes equations

$$\partial_t v^\varrho - \Delta v^\varrho = f^\varrho - \nabla p^\varrho, \qquad \operatorname{div} v^\varrho = 0$$

hold a.e. in Q_T. This is the starting point for the proof of our theorem. We know that sequence v^ϱ converges to a weak Leray-Hopf solution to corresponding initial boundary value problem (5.3.1). So, what we need is to get uniform estimates of ∇v^ϱ. Let

$$y(t) := \int\limits_\Omega |\nabla v^\varrho(x,t)|^2 dx.$$

We proceed as in the proof of Theorem 5.10, multiplying the equation by $\widetilde{\Delta} v^\varrho$ and arguing exactly as it has been done there. As a result, after obvious applications of Cauchy and Hölder inequalities, we find

$$y' + 3/2 \int\limits_\Omega |\widetilde{\Delta} v^\varrho|^2 dx \leq c \int\limits_\Omega |(v^\varrho)_\varrho|^2 |\nabla v^\varrho|^2 dx + c\|f\|^2_{2,\Omega} \leq$$

$$\leq c\|v^\varrho\|^2_{6,\Omega} \|\nabla v^\varrho\|^2_{3,\Omega} + c\|f\|^2_{2,\Omega}. \qquad (5.7.6)$$

The first term on the right-hand side of the above inequality can be evaluated with the help of the Gagliardo-Nirenberg inequality in dimension three

$$\|v^\varrho\|^2_{6,\Omega} \leq cy,$$

and the multiplicative inequality

$$\|\nabla v^\varrho\|^2_{3,\Omega} \leq c(\Omega) y^{\frac{1}{2}} \Big(\int\limits_\Omega |\nabla^2 v^\varrho|^2 dx + y \Big)^{\frac{1}{2}}.$$

In addition, the Cattabriga-Solonnikov inequality of the form

$$\int_\Omega |\nabla^2 v^\varrho|^2 dx \le c(\Omega) \int_\Omega |\widetilde{\Delta} v^\varrho|^2 dx \tag{5.7.7}$$

is needed. So, from (5.7.6) and (5.7.7), it follows that:

$$y' + 5/4 \int_\Omega |\widetilde{\Delta} v^\varrho|^2 dx \le c(\Omega)(y^3 + y^2) + c\|f\|^2_{2,\Omega}. \tag{5.7.8}$$

Recalling the properties of eigenvalues λ_k and eigenfunctions φ_k of the Stokes operator, we observe that:

$$\int_\Omega |\widetilde{\Delta} v^\varrho|^2 = \sum_{k=1}^{\infty} d_k^2 \lambda_k^2 \ge \lambda_1 \sum_{k=1}^{\infty} d_k^2 \lambda_k = \lambda_1 y,$$

where

$$d_k(t) = \int_\Omega v^\varrho(x,t) \cdot \varphi_k(x) dx.$$

So, (5.7.8) yields the final differential inequality

$$y'(t) + \lambda_1 y(t) \le c_1(\Omega)(y^3(t) + g(t)), \tag{5.7.9}$$

with

$$y(0) = \|\nabla a\|^2_{2,\Omega}, \qquad g(t) = \|f(\cdot,t)\|^2_{2,\Omega}.$$

A weaker versions of (5.7.9) is

$$\frac{y'(t)}{1 + y^2(t)} \le c_1(\Omega)\Big[y(t) + g(t)\Big]$$

so that after integration of it and application of the energy inequality, we derive the bound

$$\arctan(y(t)) \le \arctan(\|\nabla a\|^2_{2,\Omega}) + c_1(\Omega)\Big[\int_0^T y(t)dt + \int_0^T g(t)dt\Big]$$
$$\le \arctan(\|\nabla a\|^2_{2,\Omega}) + c_0(\Omega)\Big[\|a\|^2_{2,\Omega} + \|f\|^2_{2,Q_T}\Big] < \frac{\pi}{2}$$

for any $t \in [0,T]$, which implies

$$y(t) \le C, \qquad t \in [0,T],$$

with a constant c independent of ϱ and t. □

Theorem 7.13. *Assume that*

$$a \in V, \qquad f \in L_2(Q_T).$$

Then there exists $T' \in]0,T]$ such that initial boundary value problem (5.3.1) has a strong solution in $Q_{T'}$.

PROOF Arguing as in the proof of Theorem 7.12, let us go back to inequality (5.7.9). We need to show that there exists $T' \leq T$, where $y(t)$ has an upper bound independent of ϱ and $t \in [0, T']$. To achieve this goal, let us make a substitution $z(t) = y(t) - y(0)$ and, after application of Young's inequality, get the following modification of estimate (5.7.9)

$$z'(t) + \lambda_1 z(t) \leq c_1(\Omega)(z^3(t) + y^3(0) + g(t)).$$

An equivalent form of it is:

$$z' + \lambda_1(1 - \frac{c_1}{\lambda_1} z^2) z \leq c_1(y^3(0) + g(t))$$

with $z(0) = 0$. By continuity,

$$\frac{c_1}{\lambda_1} z^2(t) < 1 \tag{5.7.10}$$

for small positive t. Without loss of generality, we may assume that there exists $\varrho_0 > 0$ such that, for all $0 < \varrho < \varrho_0$, there is $0 < t_\varrho \leq T$ with the following properties: inequality (5.7.10) holds for $0 < t < t_\varrho$ and

$$\frac{c_1}{\lambda_1} z^2(t_\varrho) = 1.$$

Next, we take the largest value $T' \in]0, T]$ so that

$$\int_0^{T'} \Big[c_1(y^3(0) + g(t)) + \lambda_1 y(0)\Big] dt \leq \frac{1}{2}\sqrt{\frac{\lambda_1}{c_1}}.$$

From inequality (5.7.10) and the definition of t_ϱ, it follows that:

$$z'(t) - \lambda_1(1 - \frac{c_1}{\lambda_1} z^2) y(0) \leq c_1(y^3(0) + g(t))$$

for all $0 \leq t \leq t_\varrho$. Therefore,

$$z(t) \leq \int_0^t (c_1(y^3(0) + g(s)) + \lambda_1 y(0)) ds$$

for all $0 \leq t \leq t_\varrho$. And thus for $t = t_\varrho$, we have

$$z(t_\varrho) = \sqrt{\frac{\lambda_1}{c_1}} \leq \int_0^{t_\varrho} (c_1(y^3(0) + g(s)) + \lambda_1 y(0)) ds$$

and, in a view of the definition of T',

$$\int_0^{T'} \Big[c_1(y^3(0) + g(t)) + \lambda_1 y(0)\Big] dt < \int_0^{t_\varrho} \Big[c_1(y^3(0) + g(s)) + \lambda_1 y(0)\Big] ds.$$

The latter implies $t_\varrho \geq T'$ for all $\varrho > 0$ and the required estimate

$$y(t) \leq \|\nabla a\|^2_{2,\Omega} + \int\limits_0^{T'} \Big[c_1(y^3(0) + g(t)) + \lambda_1 y(0)\Big] dt \leq \|\nabla a\|^2_{2,\Omega} + \frac{1}{2}\sqrt{\frac{\lambda_1}{c_1}}$$

for all $0 < t \leq T'$ and for all $\varrho > 0$. □

Remark 5.1. If $f = 0$, the lower bound for T' can be improved

$$T' \geq \frac{c_4(\Omega)}{\|\nabla a\|^4_{2,\Omega}}$$

and this is the celebrated Leray estimate.

PROOF In this case, one can deduce from inequality (5.7.9) the following:

$$\frac{y'}{y^3} \leq c_1$$

for $0 \leq t \leq T$. The integration gives us:

$$\frac{1}{y^2(0)} - \frac{1}{y^2(t)} \leq 2c_1 t$$

and thus

$$y^2(t)(1 - 2c_1 t y^2(0)) \leq y^2(0).$$

We let $T'_0 = \frac{1}{4c_1 y^2(0)}$. Then $1 - 2c_1 t y^2(0) \geq 1/2$ and $y(t) \leq \sqrt{2}y(0)$ for $0 < t \leq T'_0$ which implies $T' \geq T'_0$ and we get the required estimate with an appropriated constant.

Remark 5.2. Solutions, constructed in Theorems 7.12 and 7.13, have the following regularity properties: $v \in W^{2,1}_2(Q_T)$ and there exists a function q such that $\nabla q \in L_2(Q_T)$ and

$$\partial_t v + v \cdot \nabla v - \Delta v = f - \nabla q, \qquad \operatorname{div} v = 0$$

a.e. in Q_T. (One should replace Q_T with $Q_{T'}$ in the case of Theorem 7.13).

PROOF In fact, from (5.7.8) and from the Cattabriga-Solonnikov inequality (5.7.7), it follows that:

$$\|\nabla^2 v^\varrho\|^2_{2,Q_T} \leq c(\|\nabla a\|_{2,\Omega}, \|f\|_{2,Q_T})$$

and, in a view of derivation (5.7.8), we get

$$\int\limits_{Q_T} |(v^\varrho)_\varrho|^2 |\nabla v^\varrho|^2 dz \leq c(\|\nabla a\|_{2,\Omega}, \|f\|_{2,Q_T}).$$

The linear theory, applied to the initial boundary value problem

$$\partial_t v^\varrho - \Delta v^\varrho + \nabla q^\varrho = f - (v^\varrho)_\varrho \cdot \nabla v^\varrho, \qquad \operatorname{div} v^\varrho = 0 \qquad \text{in } Q_T$$
$$v^\varrho|_{\partial\Omega\times[0,T]} = 0, \qquad v^\varrho|_{t=0} = a,$$

leads to all other statements of Remark 5.7.5.

The main result of this section is:

Theorem 7.14. *(Uniqueness of strong solutions in the class of weak Leray-Hopf solutions) Assume that u^1 and u^2 are weak Leray-Hopf solutions to the initial boundary value problem*

$$\partial_t v + v\cdot\nabla v - \Delta v = f - \nabla q, \qquad \operatorname{div} v = 0 \qquad \text{in } Q_T$$
$$v|_{\partial\Omega\times[0,T]} = 0, \qquad v|_{t=0} = a$$

with $a \in V$ and $f \in L_2(Q_T)$. Let u^2 be a strong solution then $u^1 = u^2$,

We start with several auxiliary propositions.

Proposition 7.15. *(Uniqueness of strong solutions in the class of strong solutions) Assume that u^1 and u^2 are strong solutions to the initial boundary value problem*

$$\partial_t v + v\cdot\nabla v - \Delta v = f - \nabla q, \qquad \operatorname{div} v = 0 \qquad \text{in } Q_T$$
$$v|_{\partial\Omega\times[0,T]} = 0, \qquad v|_{t=0} = a,$$

with $a \in V$ and $f \in L_2(Q_T)$. Then $u^1 = u^2$.

PROOF First we notice that if u is a strong solution to the above initial boundary value problem then

$$\partial_t u \in L_2(0,T;V').$$

Indeed,

$$\Big|\int\limits_\Omega (u\cdot\nabla u)\cdot w dx\Big| = \Big|\int\limits_\Omega u\otimes u : \nabla w dx\Big| \le \|u\|^2_{4,\Omega}\|\nabla w\|_{2,\Omega}$$
$$\le c(\Omega)\|\nabla u\|^2_{2,\Omega}\|\nabla w\|_{2,\Omega} \le C(\Omega,u)\|\nabla w\|_{2,\Omega}$$

for any $w \in V$. So, $\widetilde{f} = f - u\cdot\nabla u \in L_2(0,T;V')$ and thus $\partial_t u \in L_2(0,T;V')$. Then, by the definition of weak solution, we have

$$\int\limits_{Q_T} \chi(t)\Big[\partial_t u(x,t)\cdot w(x) - (u\otimes u)(x,t) : \nabla w(x)$$
$$+\nabla u(x,t) : \nabla w(x)\Big] dz = \int\limits_{Q_T} \chi(t) f(x,t)\cdot w(x) dz$$

for any $w \in C^\infty_{0,0}(\Omega)$ and for any $\chi \in C^\infty_0(0,T)$. It is easy to see

$$\int\limits_\Omega \Big[\partial_t u(x,t)\cdot w(x) - (u\otimes u)(x,t) : \nabla w(x)$$

$$+\nabla u(x,t) : \nabla w(x)\Big]dx = \int\limits_\Omega f(x,t)\cdot w(x)dx$$

for any $w \in V$ and for a.a. $t \in [0,T]$.

So, assume that u^1 and u^2 are two different strong solutions and let $v = u^1 - u^2$. Then, we have

$$\frac{1}{2}\partial_t\|v\|^2_{2,\Omega} + \|\nabla v\|^2_{2,\Omega} = \int\limits_\Omega (u^1\otimes u^1 - u^2\otimes u^2) : \nabla v dx$$

$$= \int\limits_\Omega (v\otimes u^1 + u^2\otimes v) : \nabla v dx = \int\limits_\Omega u^2\otimes v : \nabla v dx$$

$$= -\int\limits_\Omega v\otimes v : \nabla u^2 dx \leq \|\nabla u^2\|_{2,\Omega}\|v\|^2_{4,\Omega}. \qquad (5.7.11)$$

Let us recall the following 3D multiplicative inequality

$$\|v\|_{4,\Omega} \leq c\|v\|^{\frac{1}{4}}_{2,\Omega}\|\nabla v\|^{\frac{3}{4}}_{2,\Omega}.$$

So, after application of Young's inequality

$$\partial_t\|v\|^2_{2,\Omega} + \|\nabla v\|^2_{2,\Omega} \leq c\|\nabla u^2\|^4_{2,\Omega}\|v\|^2_{2,\Omega}.$$

Since $\|\nabla u^2\|_{2,\infty,Q_T} \leq C_1$,

$$\partial_t\|v\|^2_{2,\Omega} \leq C_1^4\|v\|^2_{2,\Omega},$$

which implies

$$e^{-C_1 t}\int\limits_\Omega |v(x,t)|^2 dx \leq \int\limits_\Omega |v(x,0)|^2 dx = 0.$$

Proposition 7.15 is proved.

Proposition 7.16. *(Smoothness of strong solutions) Let a vector field u be a strong solution to the following initial boundary value problem:*

$$\partial_t u + u\cdot\nabla u - \Delta u = f - \nabla p, \qquad \operatorname{div} u = 0 \qquad in\ Q_T$$

$$u|_{\partial\Omega\times[0,T]} = 0, \qquad u|_{t=0} = a,$$

with $a \in V$ and $f \in L_2(Q_T)$. Then $u \in W^{2,1}_2(Q_T)$ and $p \in W^{1,0}_2(Q_T)$.

PROOF Let us denote $\|\nabla u\|_{2,\infty,Q_T}$ by A. Coming back to our proof of Theorem 7.13, define a positive number T_A so that

$$c_1\Big[T_A A^3 + \int\limits_{]0,T[\cap]t,t+T_A[} g(s)ds\Big] \leq \frac{1}{2}\sqrt{\frac{\lambda_1}{c_1}} \qquad (5.7.12)$$

for any $t \in [0,T]$.

By (5.7.12) and by Theorem 7.13, there exists a strong solution u^1 in Q_{T_A} with initial data $u^1|_{t=0} = a$. This solution belongs to $W^{2,1}_2(Q_{T_A})$ and the corresponding pressure q^1 belongs to $W^{1,0}_2(Q_{T_A})$, see Remark 5.2. By Proposition 7.15, $u^1 = u$ in Q_{T_A}. We know also $\nabla u \in C(0,T;L_2(\Omega))$. So, we can apply Theorem 7.13 one more time in $Q_{T_A/2,3T_A/2}$ and find a strong solution u^2 there with initial data $u(\cdot,T_A/2)$. By Proposition 7.15, $u^2 = u$ in $Q_{T_A/2,3T_A/2}$ and $\nabla q^1 = \nabla q^2$ in $Q_{T_A/2,T_A}$. After a finite number of steps, we find that $u \in W^{2,1}_2(Q_T)$ and can easily recover a function $p \in W^{1,0}_2(Q_T)$ such that $\nabla p = \nabla q^k$ on $Q_T \cap Q_{T_A k/2, T_A(k+1)/2}$, where $k = 1,2,....$ □

PROOF OF THEOREM 7.14 Since u^2 is a strong solution, it satisfies the identity

$$\int\limits_\Omega \Big[\partial_t u^2(x,t)\cdot w(x) + (u^2(x,t)\cdot\nabla u^2(x,t))\cdot w(x)$$

$$+\nabla u^2(x,t) : \nabla w(x)\Big]dx = \int\limits_\Omega f(x,t)\cdot w(x)dx \qquad (5.7.13)$$

for any $w \in V$ and for a.a. $t \in [0,T]$. Regarding to u^1, we have a weaker identity

$$\int\limits_{Q_T} (-u^1\cdot\partial_t w - u^1\otimes u^1 : \nabla w + \nabla u^1 : \nabla w - f\cdot w)dz = 0 \qquad (5.7.14)$$

for any $w \in C^\infty_{0,0}(Q_T)$.

We would like to test (5.7.14) with u^2 but it should be justified. Indeed, we know that

$$u^1 \in L_{\frac{10}{3}}(Q_T)$$

and, by density arguments, (5.7.14) must be true for $w(x,t) = \chi(t)v(x)$ with $v \in \overset{\circ}{J}{}^1_{\frac{5}{2}}(\Omega)$ and $\chi \in \overset{\circ}{W}{}^1_1(0,T) = \{\chi \in W^1_1(0,T) :\ \chi(0) = 0,\ \chi(T) = 0\}$.

Since $u^2 \in W^{2,1}_2(Q_T)$, the series $\sum_{k=1}^\infty c_k(t)\varphi_k(x)$ converges to u^2 in $W^{2,1}_2(Q_T)$. Here, φ_k is the kth eigenfunction of the Stokes operator. This, in turn, implies that the series $\sum_{k=1}^\infty c_k(t)\nabla\varphi_k(x)$ converges to ∇u^2 in

$L_{\frac{5}{2}}(Q_T)$. To justify that, we need two inequalities. The first of them is multiplicative one:

$$\|\nabla v\|_{\frac{5}{2},Q_T} \leq c(\Omega,T) \sup_{0<t<T} \|\nabla v\|^{\frac{7}{10}}_{2,\Omega}(\|\nabla^2 v\|_{2,Q_T} + \|\nabla v\|_{2,Q_T})^{\frac{3}{10}}.$$

The second inequality is

$$\|\nabla v\|^2_{2,\infty,Q_T} \leq \frac{1}{T}\|\nabla v\|^2_{2,Q_T} + 2\|\partial_t v\|_{2,Q_T}\|\Delta v\|_{2,Q_T}.$$

This inequality can be easily derived from the identity

$$\partial_t \int\limits_\Omega |\nabla v^N(x,t)|^2 dx = -2\int\limits_\Omega \partial_t v^N(x,t)\cdot \Delta v^N(x,t)dx,$$

where $v^N(x,t) = \sum_{k=1}^N c_k(t)\varphi_k(x)$.

So, if χ is a Lipschitz function on $[0,T]$ with $\chi(0) = \chi(T) = 0$, then identity (5.7.14) holds for $\chi(t)c_k(t)\varphi_k(x)$ with any number k. Taking into account what is mentioned above, it is not so difficult to show that (5.7.14) can be tested with $\chi_{\alpha,\beta}u^2$, where $\chi_{\alpha,\beta}(t) = t/\alpha$ if $0 \leq t \leq \alpha$, $\chi_{\alpha,\beta}(t) = 1$ if $\alpha < t < t_0$, $\chi_{\alpha,\beta}(t) = (t_0+\beta-t)/\beta$ if $t_0 \leq t \leq t_0+\beta$, and $\chi_{\alpha,\beta}(t) = 0$ if $t_0+\beta < t \leq T$. Inserting $w = \chi_{\alpha,\beta}u^2$ into (5.7.14), we find

$$\int\limits_{Q_T} \chi_{\alpha,\beta}(-u^1\cdot\partial_t u^2 - u^1\otimes u^1 : \nabla u^2 + \nabla u^1 : \nabla u^2 - f\cdot u^2)dz$$

$$= \int\limits_{Q_T} u^1\cdot u^2 \chi'_{\alpha,\beta} dz = I_\alpha + I_\beta,$$

where

$$I_\alpha = \frac{1}{\alpha}\int\limits_0^\alpha\int\limits_\Omega u^1\cdot u^2 dxdt = \frac{1}{\alpha}\int\limits_0^\alpha\int\limits_\Omega (u^1-a)\cdot(u^2-a)dxdt$$

$$+\frac{1}{\alpha}\int\limits_0^\alpha\int\limits_\Omega a\cdot(u^2-a)dxdt + \frac{1}{\alpha}\int\limits_0^\alpha\int\limits_\Omega a\cdot(u^1-a)dxdt$$

$$+\frac{1}{\alpha}\int\limits_0^\alpha\int\limits_\Omega |a|^2 dxdt.$$

Since $\|u^1(\cdot,t)-a(\cdot)\|_{2,\Omega}$ and $\|u^2(\cdot,t)-a(\cdot)\|_{2,\Omega}$ go to zero as $t\to 0$, we can observe that

$$I_\alpha \to \int\limits_\Omega |a|^2 dx$$

as $\alpha \to 0$.

The analogous result takes place at the right end point:

$$-I_\beta = \frac{1}{\beta}\int\limits_{t_0}^{t_0+\beta}\int\limits_\Omega u^1 \cdot u^2 dxdt$$

$$= \frac{1}{\beta}\int\limits_{t_0}^{t_0+\beta}\int\limits_\Omega u^1(x,t)\cdot(u^2(x,t) - u^2(x,t_0))dxdt$$

$$+\frac{1}{\beta}\int\limits_{t_0}^{t_0+\beta}\int\limits_\Omega (u^1(x,t) - u^1(x,t_0))\cdot u^2(x,t_0)dxdt$$

$$+\int\limits_\Omega u^1(x,t_0)\cdot u^2(x,t_0)dx$$

By strong continuity in $L_2(\Omega)$ of the strong solution u^2, the first term on the right-hand side goes to zero and by weak continuity of weak solution u^1 the second term there goes to zero as well. So,

$$I_\beta \to -\int\limits_\Omega u^1(x,t_0)\cdot u^2(x,t_0)dx$$

as $\beta \to 0$. Finally, we have

$$\int\limits_0^{t_0}\int\limits_\Omega (-u^1\cdot\partial_t u^2 - u^1\otimes u^1 : \nabla u^2 + \nabla u^1 : \nabla u^2 - f\cdot u^2)dz$$

$$+\int\limits_\Omega u^1(x,t_0)\cdot u^2(x,t_0)dx - \int\limits_\Omega |a|^2 dx = 0 \quad (5.7.15)$$

for any $t_0 \in [0,T]$.

Now, we are going to test (5.7.13) with $w(x) = u^1(x,t) - u^2(x,t)$, which, after integration over $]0,t_0[$, gives us:

$$\int\limits_0^{t_0}\int\limits_\Omega \partial_t u^2(x,t)\cdot u^1(x,t)dxdt - \frac{1}{2}\int\limits_\Omega |u^2(x,t_0)|^2 dx + \frac{1}{2}\int\limits_\Omega |a|^2 dx$$

$$+\int\limits_0^{t_0}\int\limits_\Omega \Big(-u^2\otimes u^2 : \nabla(u^1 - u^2) + \nabla u^2 : \nabla(u^1 - u^2)$$

$$-f\cdot(u^1 - u^2)\Big)dxdt = 0.$$

So, adding the latter to (5.7.15), we find

$$\int\limits_{\Omega}\Big[-\frac{1}{2}|a(x)|^2+u^1(x,t_0)\cdot u^2(x,t_0)-\frac{1}{2}|u^2(x,t_0)|^2\Big]dx$$

$$\int\limits_0^{t_0}\int\limits_{\Omega}\Big[-u^1\otimes u^1:\nabla u^2-u^2\otimes u^2:\nabla(u^1-u^2)-|\nabla u^2|^2$$

$$+2\nabla u^1\cdot\nabla u^2-f\cdot u^1\Big]dxdt=0. \quad (5.7.16)$$

We also know that weak solution satisfies the energy inequality

$$\frac{1}{2}\int\limits_{\Omega}|u^1(x,t_0)|^2dx+\int\limits_0^{t_0}\int\limits_{\Omega}|\nabla u^1|^2dxdt\leq\frac{1}{2}\int\limits_{\Omega}|a|^2dx+\int\limits_0^{t_0}\int\limits_{\Omega}f\cdot u^1dxdt.$$

Subtracting (5.7.16) from the energy inequality, we show

$$\frac{1}{2}\int\limits_{\Omega}|u^1(x,t_0)-u^2(x,t_0)|^2dx+\int\limits_0^{t_0}\int\limits_{\Omega}|\nabla(u^1-u^2)|^2dxdt$$

$$\leq-\int\limits_0^{t_0}\int\limits_{\Omega}(u^1\otimes u^1:\nabla u^2+u^2\otimes u^2:\nabla(u^1-u^2))dxdt=I. \quad (5.7.17)$$

The rest of the proof is similar to the proof of Theorem 7.12. Indeed,

$$I=-\int\limits_0^{t_0}\int\limits_{\Omega}(u^1-u^2)\otimes(u^1-u^2):\nabla u^2dxdt$$

$$\leq\int\limits_0^{t_0}\|\nabla u^2\|_{2,\Omega}\|u^2-u^1\|^2_{4,\Omega}dt$$

$$\leq\sup_{0<t<T}\|\nabla u^2\|_{2,\Omega}\int\limits_0^{t_0}\|u^2-u^1\|^2_{4,\Omega}dt.$$

Since u^2 is a strong solution, the quantity $\sup_{0<t<T}\|\nabla u^2(\cdot,t)\|_{2,\Omega}=\|\nabla u^2\|_{2,\infty,Q_T}$ is finite. Applying multiplicative inequality, we have

$$I\leq c\|\nabla u^2\|_{2,\infty,Q_T}\int\limits_0^{t_0}\|u^2-u^1\|^{\frac{1}{2}}_{2,\Omega}\|\nabla(u^2-u^1)\|^{\frac{3}{2}}_{2,\Omega}dt$$

$$\leq c\|\nabla u^2\|_{2,\infty,Q_T}\Big(\int\limits_0^{t_0}\int\limits_{\Omega}|u^2-u^1|^2dxdt\Big)^{\frac{1}{4}}\Big(\int\limits_0^{t_0}\int\limits_{\Omega}|\nabla(u^2-u^1)|^2dxdt\Big)^{\frac{3}{4}}.$$

From here and from (5.7.17), it follows that:

$$y'(t_0) \leq c\|\nabla u^2\|^4_{2,\infty,Q_T} y(t_0), \qquad y(t_0) := \int\limits_0^{t_0}\int\limits_\Omega |u^1(x,t) - u^2(x,t)|^2 dxdt$$

for all $t_0 \in [0,T]$ with $y(0) = 0$. This immediately implies $u^1 = u^2$. □

Theorem 7.17. *(Ladyzhenskaya-Prodi-Serrin condition) Let $a \in V$ and $f \in L_2(Q_T)$. Assume that we have two weak Leray-Hopf solutions u^1 and u^2. Assume that u^2 obeys the Ladyzhenskaya-Prodi-Serrin condition, i.e.,*

$$u^2 \in L_{s,l}(Q_T)$$

with $s, l \geq 1$, satisfying

$$\frac{3}{s} + \frac{2}{l} = 1.$$

Then $u^1 = u^2$.

Proof Just for simplicity let us assume $f = 0$. We also suppose that $s > 3$. The case $s = 3$ and $l = \infty$ is much more complicated and will be discussed later. Our aim is to show that a weak Leray-Hopf solution, satisfying the Ladyzhenskaya-Prodi-Serrin condition, is in fact a strong one. Then, the statement of the theorem follows from Theorem 7.14.

We know that, by Theorem 7.13, there exists a strong solution to our initial-boundary value problem on a small time interval $[0,T]$, which, by Theorem 7.14, coincides with any weak solution and, in particular, with $u \equiv u^2$. Let us denote by $T_0 (\leq T)$ the first instance of time, for which u is not a strong solution on $[0, T_0]$. By Proposition 7.16, we have for any $T' < T_0$,

$$u \in W^{2,1}_2(Q_{T'}), \qquad \nabla u \in C([0,T']; L_2(Q_{T'}))$$

and there exists a pressure field $p \in W^{1,0}_2(Q_{T'})$ so that

$$\partial_t u + u \cdot \nabla u - \Delta u = -\nabla p, \qquad \operatorname{div} u = 0$$

a.e. in $Q_{T'}$.

By Remark 5.1,

$$T_0 - t \geq \frac{c_4}{\|\nabla u(t)\|^4_{2,\Omega}}$$

for all $t < T_0$. So, we have

$$\lim_{t \to T_0 - 0} \|\nabla u(t)\|_{2,\Omega} = \infty. \tag{5.7.18}$$

We proceed as in the proof of (5.7.6) simply replacing v^ϱ with u. As a result,

$$y' + \int_\Omega |\widetilde{\Delta} u|^2 dx \leq c \int_\Omega |u|^2 |\nabla u|^2 dx, \tag{5.7.19}$$

where $y(s) = \|\nabla u\|^2_{2,\Omega}$.

Applying consequently Hölder inequality, an appropriated multiplicative inequality, and the Cattabriga-Solonnikov inequality to the right-hand side of (5.7.19), we have

$$\int_\Omega |u|^2 |\nabla u|^2 dx \leq \|u\|^2_{s,\Omega} \|\nabla u\|^2_{\frac{2s}{s-2},\Omega}$$

$$\leq c(\Omega, s)\|u\|^2_s \|\nabla u\|_2^{2(1-\frac{3}{s})} (\|\nabla^2 u\|_2 + \|\nabla u\|_2)^{\frac{6}{s}}$$

$$\leq c(\Omega, s)\|u\|^2_{s,\Omega} \|\nabla u\|^{\frac{4}{l}}_{2,\Omega} \|\widetilde{\Delta} u\|^{\frac{6}{s}}_{2,\Omega}$$

Using Young's inequality, we arrive at the final inequality

$$y'(t) \leq c(\Omega, s)\|u(\cdot, t)\|^l_{s,\Omega} y(t)$$

for all $t < T_0$. Integrating it, we find

$$y(t) \leq y(0) e^{c(\Omega,s)\|u\|^l_{s,l,Q_T}}$$

for all $t < T_0$. This contradicts (5.7.18). □

Remark 5.3. Unfortunately, we do not know whether any weak Leray-Hopf solution u satisfies the Ladyzhenskaya-Prodi-Serrin condition. What we know is that

$$u \in L_{s',l'}(Q_T)$$

with $s', l' \leq 1$ and

$$\frac{3}{s'} + \frac{2}{l'} = \frac{3}{2}.$$

So, there is a finite gap.

The problem of uniqueness of weak solution is still open.

If we show that any weak (Leray-Hopf) solution is smooth (for example, it is strong), then we have uniqueness in the class of weak solutions.

The problem of smoothness of weak solutions is one of seven Millennium problems.

5.8 Comments

Chapter 5 contains an introduction to the theory of energy solutions developed in [Leray (1934)] and later on in [Hopf (1950-1951)]. Our proof of the global well-posedness of the 2D dimensional problem is due to [Ladyzhenskaya (1958)]. As to local in time well-posedness of the 3D problem, we follow [Leray (1934)] and [Kiselev and Ladyzhenskaya (1957)]. We also prove classical results related to the Ladyzhenskaya-Prodi-Serrin condition and the uniqueness of strong solutions in the class of weak solutions, see for example [Ladyzhenskaya (1967)], [Prodi (1959)], [Serrin (1962)].

Chapter 6

Local Regularity Theory for Non-Stationary Navier-Stokes Equations

6.1 ε-Regularity Theory

The aim of this section is so-called suitable weak solutions to the Navier-Stokes equations and their smoothness. Those solutions were introduced in [Caffarelli *et al.* (1982)], see also [Scheffer (1976)]-[Scheffer (1982)], [Lin (1998)], and [Ladyzhenskaya and Seregin (1999)]. Our version is due to [Lin (1998)].

Definition 6.1. Let ω be a domain in $\mathbb{R}^3$. We say that a pair u and p is a *suitable weak* solution to the Navier-Stokes equations in $\omega\times]T_1,T[$ if u and p obey the conditions:

$$u \in L_{2,\infty}(\omega\times]T_1,T[) \cap L_2(T_1,T;W^1_2(\omega)); \tag{6.1.1}$$

$$p \in L_{\frac{3}{2}}(\omega\times]T_1,T[); \tag{6.1.2}$$

$$\partial_t u + u\cdot\nabla u - \Delta u = -\nabla p, \qquad \operatorname{div} u = 0 \tag{6.1.3}$$

in the sense of distributions;

the local energy inequality

$$\left.\begin{aligned} &\int\limits_\omega \varphi(x,t)|u(x,t)|^2\,dx + 2\int\limits_{\omega\times]T_1,t[} \varphi|\nabla u|^2\,dxdt' \\ &\leq \int\limits_{\omega\times]T_1,t[} (|u|^2(\Delta\varphi+\partial_t\varphi) + u\cdot\nabla\varphi(|u|^2+2q))\,dxdt' \end{aligned}\right\} \tag{6.1.4}$$

holds for a.a. $t\in]T_1,T[$ and all nonnegative functions $\varphi\in C^\infty_0(\omega\times]T_1,\infty[)$.

One of the main results of the theory of suitable weak solutions reads:

Lemma 6.1. *There exist absolute positive constants ε_0 and c_{0k}, $k = 1, 2, ...$, with the following property. Assume that a pair U and P is a suitable weak solution to the Navier-Stokes equations in Q and satisfies the condition*

$$\int\limits_Q \left(|U|^3 + |P|^{\frac{3}{2}}\right) dz < \varepsilon_0. \tag{6.1.5}$$

Then, for any natural number k, $\nabla^{k-1}U$ is Hölder continuous in $\overline{Q}(\frac{1}{2})$ and the following bound is valid:

$$\max_{z \in Q(\frac{1}{2})} |\nabla^{k-1}U(z)| < c_{0k}. \tag{6.1.6}$$

To formulate Lemma 6.1, we exploit the following notation and abbreviations:

$$z = (x,t), \quad z_0 = (x_0, t_0); \qquad B(x_0, R) = \{|x - x_0| < R\};$$

$$Q(z_0, R) = B(x_0, R) \times]t_0 - R^2, t_0[;$$

$$B(r) = B(0,r), \; Q(r) = Q(0,r), \; B = B(1), \; Q = Q(1).$$

Remark 6.1. For $k = 1$, Lemma 6.1 has been proven essentially in [Caffarelli *et al.* (1982)], see Corollary 1. For alternative approach, we refer the reader to [Ladyzhenskaya and Seregin (1999)], see Lemma 3.1. The case $k > 1$ was treated in [Necas *et al.* (1996)], see Proposition 2.1, with the help of the case $k = 1$ and regularity results for linear Stokes type systems.

In turn, if $k = 1$, Lemma 6.1 is a consequence of Proposition 1.1 below. To state it, we need to introduce certain integral quantities that play an important role in the regularity theory:

$$Y(z_0, R; v, q) = Y^1(z_0, R; v) + Y^2(z_0, R; q),$$

$$Y^1(z_0, R; v) = \Big(\frac{1}{|Q(R)|} \int\limits_{Q(z_0,R)} |v - (v)_{z_0,R}|^3 \, dz\Big)^{\frac{1}{3}},$$

$$Y^2(z_0, R; q) = R\Big(\frac{1}{|Q(R)|} \int\limits_{Q(z_0,R)} |q - [q]_{z_0,R}|^{\frac{3}{2}} \, dz\Big)^{\frac{2}{3}},$$

$$(v)_{z_0,R} = \frac{1}{|Q(R)|} \int\limits_{Q(z_0,R)} v \, dz, \qquad [q]_{x_0,R} = \frac{1}{|B(R)|} \int\limits_{B(x_0,R)} q \, dx,$$

$$Y^1_\vartheta(v) = Y^1(0,\vartheta;v), \qquad Y^2_\vartheta(q) = Y^2(0,\vartheta;q),$$

$$Y_\vartheta(v,q) = Y(0,\vartheta;v,q), \qquad (v)_{,\vartheta} = (v)_{0,\vartheta}, \qquad [q]_{,\vartheta} = [q]_{0,\vartheta}.$$

Proposition 1.1. *Given numbers $\vartheta \in]0,1/2[$ and $M > 3$, there are two constants $\varepsilon_1(\vartheta, M) > 0$ and $c_1(M) > 0$ such that, for any suitable weak solution v and q to the Navier-Stokes equations in Q, satisfying the additional conditions*

$$|(v)_{,1}| < M, \qquad Y_1(v,q) < \varepsilon_1, \tag{6.1.7}$$

the following estimate is valid:

$$Y_\vartheta(v,q) \le c_1 \vartheta^{\frac{2}{3}} Y_1(v,q). \tag{6.1.8}$$

PROOF OF PROPOSITION 1.1 Assume that the statement is false. This means that a number $\vartheta \in]0,1/2[$ and a sequence of suitable weak solutions v^k and q^k (in Q) exist such that:

$$Y_1(v^k, q^k) = \varepsilon_{1k} \to 0 \tag{6.1.9}$$

as $k \to +\infty$,

$$Y_\vartheta(v^k, q^k) > c_1 \varepsilon_{1k} \vartheta^{\frac{3}{2}} \tag{6.1.10}$$

for all $k \in \mathbb{N}$. A constant c_1 will be specified later.

Let us introduce functions

$$u^k = (v^k - (v^k)_{,1})/\varepsilon_{1k}, \qquad p^k = (q^k - [q^k]_{,1})/\varepsilon_{1k}.$$

They obey the following relations

$$Y_1(u^k, p^k) = 1, \tag{6.1.11}$$

$$Y_\vartheta(u^k, p^k) > c_1 \vartheta^{\frac{2}{3}}, \tag{6.1.12}$$

and the system

$$\left.\begin{array}{c} \partial_t u^k + \frac{1}{\varepsilon_{1k}} \operatorname{div}\left((v^k)_{,1} + \varepsilon_{1k} u^k\right) \otimes \left((v^k)_{,1} + \varepsilon_{1k} u^k\right) \\ -\Delta\, u^k = -\nabla\, p^k, \quad \operatorname{div} u^k = 0 \end{array}\right\} \quad \text{in } Q \tag{6.1.13}$$

in the sense of distributions.

Without loss of generality, one may assume that:

$$\begin{cases} u^k \rightharpoonup u & \text{in} \quad L_3(Q) \\ p^k \rightharpoonup p & \text{in} \quad L_{\frac{3}{2}}(Q) \\ (v^k)_{,1} \to b & \text{in} \quad \mathbb{R}^3 \end{cases} \tag{6.1.14}$$

and

$$\left.\begin{array}{r}\partial_t u + \operatorname{div} u \otimes b - \Delta u = -\nabla p \\ \operatorname{div} u = 0\end{array}\right\} \quad \text{in } Q \tag{6.1.15}$$

in the sense of distributions. By (6.1.11) and (6.1.14), we have

$$|b| < M, \qquad Y_1(u,p) \leq 1, \qquad [p(\cdot,t)]_{,1} = 0 \quad \text{for all } t \in]-1,0[. \tag{6.1.16}$$

Choosing a cut-off function φ in an appropriate way in the local energy inequality, we find the energy estimate for u^k

$$\|u^k\|_{2,\infty,Q(3/4)} + \|\nabla u^k\|_{2,Q(3/4)} \leq c_2(M) \tag{6.1.17}$$

that remains to be true for the limit function u

$$\|u\|_{2,\infty,Q(3/4)} + \|\nabla u\|_{2,Q(3/4)} \leq c_2(M).$$

It is easy to check that p is a harmonic function depending on t as a parameter. After application of bootstrap arguments, we find

$$\sup_{z\in Q(2/3)} \Big(|\nabla u(z)| + |\nabla^2 u(z)|\Big)$$

$$+\Big(\sup_{x\in B(2/3)} \int\limits_{-(2/3)^2}^{0} |\partial_t u(x,t)|^{\frac{3}{2}} dt\Big)^{\frac{2}{3}} \leq c_3(M).$$

From the above estimate, a parabolic embedding theorem and scaling, it follows that

$$\Big(\frac{1}{|Q(\tau)|}\int\limits_{Q(\tau)} |u - (\nabla u)_{,\tau} x - (u)_{,\tau}|^3 dz\Big)^{\frac{1}{3}}$$

$$\leq c\tau^2\Big(\frac{1}{|Q(\tau)|}\int\limits_{Q(\tau)} (|\nabla^2 u|^{\frac{3}{2}} + |\partial_t u|^{\frac{3}{2}}) dz\Big)^{\frac{2}{3}}$$

$$\leq c\tau^2\Big(C(M) + \frac{1}{\tau^2}C(M)\Big)^{\frac{2}{3}} \leq C(M)\tau^{\frac{2}{3}}$$

for all $0 < \tau < 2/3$. The latter estimates gives us:

$$Y^1_\vartheta(u) \leq \widetilde{c}_1(M)\vartheta^{\frac{2}{3}}. \tag{6.1.18}$$

Using the known multiplicative inequality, see the previous chapter, we derive from (6.1.17) another estimate

$$\|u^k\|_{\frac{10}{3},Q(3/4)} \leq c_4(M). \tag{6.1.19}$$

Let us find a bound of the first derivative in time with the help of duality arguments. Indeed, we have from (6.1.13) and (6.1.16)

$$\|\partial_t u^k\|_{L_{\frac{3}{2}}(-(3/4)^2,0;(\overset{\circ}{W}{}^2_2(B(3/4)))')} \leq c_5(M). \tag{6.1.20}$$

Here, $\overset{\circ}{W}{}^2_2(B(3/4))$ is the completion of $C_0^\infty(B(2/3))$ in $W_2^2(B(2/3))$. By the compactness arguments used in the previous section, a subsequence can be selected so that

$$u^k \to u \qquad \text{in} \qquad L_3(Q(3/4)). \tag{6.1.21}$$

Now, taking into account (6.1.21) and (6.1.18), we pass to the limit in (6.1.12) and find

$$c_1\vartheta^{\frac{2}{3}} \leq \widetilde{c}_1\vartheta^{\frac{2}{3}} + \limsup_{k\to\infty} Y_\vartheta^2(p^k). \tag{6.1.22}$$

In order to pass to the limit in the last term of the right-hand side in (6.1.22), let us decompose the pressure p^k as follows (see [Seregin (1999, 2001, 2002)]):

$$p^k = p_1^k + p_2^k. \tag{6.1.23}$$

Here, the first function p_1^k is defined as a unique solution to the following boundary value problem: find $p_1^k(\cdot,t) \in L_{\frac{3}{2}}(B)$ such that

$$\int\limits_B p_1^k(x,t)\Delta\psi(x)\,dx = -\varepsilon_{1k}\int\limits_B u^k(x,t)\otimes u^k(x,t) : \nabla^2\psi(x)\,dx$$

for all smooth test functions ψ subjected to the boundary condition $\psi|_{\partial B} = 0$. It is easy to see that

$$\Delta p_2^k(\cdot,t) = 0 \qquad \text{in } B \tag{6.1.24}$$

and, by the coercive estimates for Laplace's operator with the homogeneous Dirichlet boundary condition, we get the bound for p_1^k:

$$\int\limits_B |p_1^k(x,t)|^{\frac{3}{2}}\,dx \leq c\varepsilon_{1k}^{\frac{3}{2}}\int\limits_B |u^k(x,t)|^3\,dx. \tag{6.1.25}$$

Passing to the limit in (6.1.22), we show with the help of (6.1.25) that:

$$c_1\vartheta^{\frac{2}{3}} \leq \widetilde{c}_1\vartheta^{\frac{2}{3}} + \limsup_{k\to\infty} Y_\vartheta^2(p_2^k). \tag{6.1.26}$$

By Poincare's inequality, (6.1.26) can be reduced to the form

$$c_1\vartheta^{\frac{2}{3}} \leq \widetilde{c}_1\vartheta^{\frac{2}{3}} + c\vartheta^2\limsup_{k\to\infty}\Big(\frac{1}{|Q(\vartheta)|}\int\limits_{Q(\vartheta)} |\nabla p_2^k|^{\frac{3}{2}}\,dz\Big)^{\frac{2}{3}}. \tag{6.1.27}$$

We know that the function $p_2^k(\cdot,t)$ is harmonic in B and, by the mean value theorem, estimate

$$\sup_{x\in B(3/4)} |\nabla p_2^k(x,t)|^{\frac{3}{2}} \leq c\int\limits_B |p_2^k(x,t)|^{\frac{3}{2}}\,dx$$

holds, which in turns implies

$$\frac{1}{|Q(\vartheta)|}\int\limits_{Q(\vartheta)} |\nabla p_2^k|^{\frac{3}{2}}\,dz \leq \frac{c}{\vartheta^2}\int\limits_Q |p_2^k|^{\frac{3}{2}}\,dz$$

$$\leq c\Big(\frac{1}{\vartheta^2}+\frac{1}{\vartheta^2}\int\limits_Q |p_1^k|^{\frac{3}{2}}\,dz\Big).$$

The latter inequality, together with (6.1.25), allows us to take the limit in (6.1.27). As a result, we show that

$$c_1\vartheta^{\frac{2}{3}} \leq \widetilde{c}_1\vartheta^{\frac{2}{3}} + c\vartheta^{\frac{2}{3}}. \tag{6.1.28}$$

If, from the very beginning, c_1 is chosen so that

$$c_1 = 2(\widetilde{c}_1 + c),$$

we arrive at the contradiction. Proposition 1.1 is proved.

Proposition 1.1 admits the following iterations.

Proposition 1.2. *Given numbers $M > 3$ and $\beta \in [0,2/3[$, we choose $\vartheta \in]0,1/2[$ so that*

$$c_1(M)\vartheta^{\frac{2-3\beta}{6}} < 1. \tag{6.1.29}$$

Let $\overline{\varepsilon}_1(\vartheta,M) = \min\{\varepsilon_1(\vartheta,M),\vartheta^5 M/2\}$. If

$$|(v)_{,1}| < M, \qquad Y_1(v,q) < \overline{\varepsilon}_1, \tag{6.1.30}$$

then, for any $k=1,2,...,$

$$\begin{aligned} \vartheta^{k-1}|(v)_{,\vartheta^{k-1}}| < M, \qquad Y_{\vartheta^{k-1}}(v,q) < \overline{\varepsilon}_1 \leq \varepsilon_1, \\ Y_{\vartheta^k}(v,q) \leq \vartheta^{\frac{2+3\beta}{6}} Y_{\vartheta^{k-1}}(v,q). \end{aligned} \tag{6.1.31}$$

PROOF We use induction on k. For $k=1$, this is nothing but Proposition 1.1.

Assume now that statements (6.1.31) are valid for $s=1,2,...,k\geq 2$. Our goal is to prove that they are valid for $s=k+1$ as well. Obviously, by induction,

$$Y_{\vartheta^k}(v,q) < \overline{\varepsilon}_1 \leq \varepsilon_1,$$

and

$$|(v^k)_{,1}| = \vartheta^k |(v)_{,\vartheta^k}| \leq \vartheta^k |(v)_{,\vartheta^k} - (v)_{,\vartheta^{k-1}}| + \vartheta^k |(v)_{,\vartheta^{k-1}}|$$

$$\leq \frac{1}{\vartheta^5} Y_{\vartheta^{k-1}}(v,q) + \frac{1}{2}\vartheta^{k-1}|(v)_{,\vartheta^{k-1}}| < \frac{1}{\vartheta^5}\overline{\varepsilon}_1 + M/2 \leq M.$$

Introducing scaled functions

$$v^k(y,s) = \vartheta^k v(\vartheta^k y, \vartheta^{2k} s), \qquad q^k(y,s) = \vartheta^{2k} q(\vartheta^k y, \vartheta^{2k} s)$$

for $(y,s) \in Q$, we observe that v^k and q^k are a suitable weak solution in Q. Since

$$Y_1(v^k, q^k) = \vartheta^k Y_{\vartheta^k}(v,q) < \overline{\varepsilon}_1 \leq \varepsilon_1$$

and

$$|(v^k)_{,1}| = \vartheta^k |(v)_{,\vartheta^k}| < M,$$

we conclude

$$Y_\vartheta(v^k, q^k) \leq c_1 \vartheta^{\frac{2}{3}} Y_1(v^k, q^k) < \vartheta^{\frac{2+3\beta}{6}} Y_1(v^k, q^k),$$

which is equivalent to the third relation in (6.1.31). Proposition 1.2 is proved.

A direct consequence of Proposition 1.2 and the Navier-Stokes scaling

$$v^R(y,s) = Rv(x_0 + Ry, t_0 + R^2 s), \qquad q^R(y,s) = R^2 q(x_0 + Ry, t_0 + R^2 s)$$

is the following statement.

Proposition 1.3. *Let M, β, ϑ, and $\overline{\varepsilon}_1$ be as in Proposition 1.2. Let a pair v and q be an arbitrary suitable weak solution to the Navier-Stokes equations in the parabolic cylinder $Q(z_0, R)$, satisfying the additional conditions*

$$R|(v)_{z_0,R}| < M, \qquad RY(z_0, R; v, q) < \overline{\varepsilon}_1. \tag{6.1.32}$$

Then, for any $k = 1, 2, \ldots$, the estimates

$$Y(z_0, \vartheta^k R; v, q) \leq \vartheta^{\frac{2+3\beta}{6}k} Y(z_0, R; v, q) \tag{6.1.33}$$

hold.

Proof of Lemma 6.1 We start with the case $k = 1$. Define

$$A = \int\limits_Q \left(|U|^3 + |P|^{\frac{3}{2}} \right) dz.$$

Then, let $M = 2002$, $\beta = 1/3$, and let ϑ be chosen according to (6.1.29) and fix.

First, we observe that

$$Q(z_0,1/4)\subset Q \qquad \text{if} \qquad z_0\in\overline{Q}(3/4)$$

and

$$\frac{1}{4}Y(z_0,1/4;U,P)\leq c(A^{\frac{1}{3}}+A^{\frac{2}{3}}),\qquad \frac{1}{4}|(U)_{z_0,\frac{1}{4}}|\leq cA^{\frac{1}{3}}.$$

Selecting ε_0 so that

$$c(\varepsilon_0^{\frac{1}{3}}+\varepsilon_0^{\frac{2}{3}})<\overline{\varepsilon}_1,\qquad c\varepsilon_0^{\frac{1}{3}}<2002.$$

Then, by (6.1.5), we have

$$\frac{1}{4}Y(z_0,1/4;U,P)<\overline{\varepsilon}_1,\qquad \frac{1}{4}|(U)_{z_0,\frac{1}{4}}|<M,$$

and thus, by Proposition 1.3,

$$Y(z_0,\vartheta^k/4;U,P)\leq\vartheta^{\frac{k}{2}}Y(z_0,1/4;U,P)\leq\vartheta^{\frac{k}{2}}\overline{\varepsilon}_1$$

for all $z_0\in\overline{Q}(3/4)$ and for all $k=1,2,\ldots$. Hölder continuity of v on the set $\overline{Q}(2/3)$ follows from Campanato's type condition. Moreover, the quantity

$$\sup_{z\in\overline{Q}(2/3)}|v(z)|$$

is bounded by an absolute constant.

The case $k>1$ is treated with the help of the regularity theory for the Stokes equations and bootstrap arguments, for details, see [Necas *et al.* (1996)], Proposition 2.1. Lemma 6.1 is proved.

In what follows, the scaled energy quantities, i.e., the energy quantities that are invariant with respect to the Navier-Stokes scaling,

$$A(v;z_0,r)\equiv\sup_{t_0-r^2\leq t\leq t_0}\frac{1}{r}\int\limits_{B(x_0,r)}|v(x,t)|^2\,dx,\quad E(v;z_0,r)\equiv\frac{1}{r}\int\limits_{Q(z_0,r)}|\nabla v|^2\,dz,$$

$$C(v;z_0,r)\equiv\frac{1}{r^2}\int\limits_{Q(z_0,r)}|v|^3\,dz,\quad D_0(q;z_0,r)\equiv\frac{1}{r^2}\int\limits_{Q(z_0,r)}|q-[q]_{x_0,r}|^{\frac{3}{2}}\,dz$$

will be exploited. We are also going to use abbreviations for them such as $A(r)=A(v;0,r)$, etc.

Our aim is to prove a version of the Caffarelli-Kohn-Nireberg theorem (Here, we follow F.-H. Lin's arguments, see [Lin (1998)]).

Theorem 1.4. *Let v and q be a suitable weak solution to the Navier-Stokes equations in Q. There exists a positive universal constant ε such that if*

$$\sup_{0<r<1}E(r)<\varepsilon,$$

then $z=0$ is regular point of v, i.e., v is Hölder continuous in the closure of the parabolic cylinder $Q(\varrho)$ with some positive $\varrho<r$.

Let us start with the proof of auxiliary lemmata. In fact, the first statement is a scaled version of a particular multiplicative inequality.

Lemma 6.2. *For all* $0 < r \leq \varrho \leq 1$,

$$C(r) \leq c\Big[\Big(\frac{r}{\rho}\Big)^3 A^{\frac{3}{2}}(\rho) + \Big(\frac{\varrho}{r}\Big)^3 A^{\frac{3}{4}}(\varrho)E^{\frac{3}{4}}(\varrho)\Big]. \tag{6.1.34}$$

PROOF We have

$$\int\limits_{B(r)} |v|^2\,dx = \int\limits_{B(r)} \Big(|v|^2 - [|v|^2]_{,\varrho}\Big)\,dx + \int\limits_{B(r)} [|v|^2]_{,\varrho}\,dx \leq$$

$$\leq \int\limits_{B(\varrho)} \Big||v|^2 - [|v|^2]_{,\varrho}\Big|\,dx + \Big(\frac{r}{\varrho}\Big)^3 \int\limits_{B(\varrho)} |v|^2\,dx.$$

By the Poincaré-Sobolev inequality,

$$\int\limits_{B(\varrho)} \Big||v|^2 - [|v|^2]_{,\varrho}\Big|\,dx \leq c\varrho \int\limits_{B(\varrho)} |\nabla v|\,|v|\,dx,$$

where c is an absolute positive constant. So, we get

$$\left.\begin{aligned} \int\limits_{B(r)} |v|^2\,dx &\leq c\varrho\Big(\int\limits_{B(\varrho)} |\nabla v|^2\,dx\Big)^{\frac{1}{2}}\Big(\int\limits_{B(\varrho)} |v|^2\,dx\Big)^{\frac{1}{2}} + \\ &\quad + \Big(\frac{r}{\varrho}\Big)^3 \int\limits_{B(\varrho)} |v|^2\,dx \leq \\ &\leq c\varrho^{\frac{3}{2}} A^{\frac{1}{2}}(\varrho)\Big(\int\limits_{B(\varrho)} |\nabla v|^2\,dx\Big)^{\frac{1}{2}} + \Big(\frac{r}{\varrho}\Big)^3 \varrho A(\varrho). \end{aligned}\right\} \tag{6.1.35}$$

Using the known multiplicative inequality, one can find

$$\int\limits_{B(r)} |v|^3\,dx \leq c\Big[\Big(\int\limits_{B(r)} |\nabla v|^2\,dx\Big)^{\frac{3}{4}}\Big(\int\limits_{B(r)} |v|^2\,dx\Big)^{\frac{3}{4}} +$$

$$+\frac{1}{r^{\frac{3}{2}}}\Big(\int\limits_{B(r)} |v|^2\,dx\Big)^{\frac{3}{2}}\Big] \leq \Big(\text{see } (6.1.35)\Big) \leq$$

$$\leq c\Big\{\varrho^{\frac{3}{4}} A^{\frac{3}{4}}(\varrho)\Big(\int\limits_{B(r)} |\nabla v|^2\,dx\Big)^{\frac{3}{4}} +$$

$$+\frac{1}{r^{\frac{3}{2}}}\Big[c\varrho^{\frac{3}{2}} A^{\frac{1}{2}}(\varrho)\Big(\int\limits_{B(\varrho)} |\nabla v|^2\,dx\Big)^{\frac{1}{2}} + \Big(\frac{r}{\varrho}\Big)^3 \varrho A(\varrho)\Big]^{\frac{3}{2}}\Big\} \leq$$

$$\leq c\Big\{\Big(\frac{r}{\varrho}\Big)^3 A^{\frac{3}{2}}(\varrho) + \Big(\int\limits_{B(\varrho)} |\nabla v|^2\,dx\Big)^{\frac{3}{4}}\Big[\varrho^{\frac{3}{4}} + \frac{\varrho^{\frac{9}{4}}}{r^{\frac{3}{2}}}\Big] A^{\frac{3}{4}}(\varrho)\Big\}.$$

Integrating the latter inequality in t on $]t_0-r^2,t_0[$, we establish

$$\int\limits_{Q(r)}|v|^3\,dz\leq c\Big\{r^2\Big(\frac{r}{\varrho}\Big)^3A^{\frac{3}{2}}(\varrho)+$$

$$+\Big[\varrho^{\frac{3}{4}}+\frac{\varrho^{\frac{9}{4}}}{r^{\frac{3}{2}}}\Big]A^{\frac{3}{4}}(\varrho)\int\limits_{t_0-r^2}^{t_0}dt\Big(\int\limits_{B(x_0,\varrho)}|\nabla v|^2\,dx\Big)^{\frac{3}{4}}\Big\}\leq$$

$$\leq c\Big\{r^2\Big(\frac{r}{\varrho}\Big)^3A^{\frac{3}{2}}(\varrho)+\Big[\varrho^{\frac{3}{4}}+\frac{\varrho^{\frac{9}{4}}}{r^{\frac{3}{2}}}\Big]A^{\frac{3}{4}}(\varrho)r^{\frac{1}{2}}\Big(\int\limits_{Q(\varrho)}|\nabla v|^2\,dz\Big)^{\frac{3}{4}}\Big\}\leq$$

$$\leq c\Big\{r^2\Big(\frac{r}{\varrho}\Big)^3A^{\frac{3}{2}}(\varrho)+\Big[\varrho^{\frac{3}{4}}+\frac{\varrho^{\frac{9}{4}}}{r^{\frac{3}{2}}}\Big]A^{\frac{3}{4}}(\varrho)r^{\frac{1}{2}}E^{\frac{3}{4}}(\varrho)\,\varrho^{\frac{3}{4}}\Big\}.$$

It remains to notice that

$$\Big[\varrho^{\frac{3}{4}}+\frac{\varrho^{\frac{9}{4}}}{r^{\frac{3}{2}}}\Big]r^{\frac{1}{2}}\varrho^{\frac{3}{4}}=\Big[\Big(\frac{\varrho}{r}\Big)^{\frac{3}{2}}+\Big(\frac{\varrho}{r}\Big)^3\Big]r^2\leq 2\Big(\frac{\varrho}{r}\Big)^3r^2$$

and then complete the proof of Lemma 6.2.

Lemma 6.3. *For any* $0<R\leq 1$,

$$A(R/2)+E(R/2)\leq c\Big[C^{\frac{2}{3}}(R)+C^{\frac{1}{3}}(R)D_0^{\frac{2}{3}}(R)$$
$$+A^{\frac{1}{2}}(R)C^{\frac{1}{3}}(R)E^{\frac{1}{2}}(R)\Big].\qquad(6.1.36)$$

PROOF Picking up a suitable cut-off function in energy inequality (6.1.4), we get the following estimates

$$A(R/2)+E(R/2)\leq c\Big\{\frac{1}{R^3}\int\limits_{Q(R)}|v|^2\,dz+$$

$$+\frac{1}{R^2}\int\limits_{Q(R)}\Big||v|^2-[|v|^2]_{,R}\Big|\,|v|\,dz+$$

$$+\frac{1}{R^2}\Big(\int\limits_{Q(R)}|q-[q]_{,R}|^{\frac{3}{2}}\,dz\Big)^{\frac{2}{3}}\Big(\int\limits_{Q(R)}|v|^3\,dz\Big)^{\frac{1}{3}}\Big\}.$$

Since

$$\frac{1}{R^3}\int\limits_{Q(z_0,R)}|v|^2\,dz\leq cC^{\frac{2}{3}}(R),$$

we find

$$\left.\begin{aligned} A(R/2)+E(R/2)\leq c\Big\{C^{\frac{2}{3}}(R)+C^{\frac{1}{3}}(R)\,D_0^{\frac{2}{3}}(R)+\\ +\tfrac{1}{R^2}\int\limits_{Q(z_0,R)}\Big||v|^2-[|v|^2]_{,R}\Big|\,|v|\,dz\Big\}. \end{aligned}\right\} \qquad (6.1.37)$$

Application of Hölder inequality to the last term on the right-hand side of (6.1.37) gives:

$$S\equiv\int\limits_{Q(R)}\Big||v|^2-[|v|^2]_{,R}\Big|\,|v|\,dz\leq$$

$$\leq\int\limits_{-R^2}^{0}dt\Big(\int\limits_{B(R)}\Big||v|^2-[|v|^2]_{,R}\Big|^{\frac{3}{2}}dx\Big)^{\frac{2}{3}}\Big(\int\limits_{B(R)}|v|^3\,dx\Big)^{\frac{1}{3}}.$$

By the Gagliardo-Nirenberg inequality

$$\Big(\int\limits_{B(R)}\Big||v|^2-[|v|^2]_{,R}\Big|^{\frac{3}{2}}dx\Big)^{\frac{2}{3}}\leq c\int\limits_{B(R)}|\nabla v|\,|v|\,dx,$$

we have

$$S\leq c\int\limits_{-R^2}^{0}dt\Big(\int\limits_{B(R)}|\nabla v|^2\,dx\Big)^{\frac{1}{2}}\Big(\int\limits_{B(R)}|v|^2\,dx\Big)^{\frac{1}{2}}\Big(\int\limits_{B(R)}|v|^3\,dx\Big)^{\frac{1}{3}}\leq$$

$$\leq c\,R^{\frac{1}{2}}A^{\frac{1}{2}}(R)\int\limits_{-R^2}^{0}dt\Big(\int\limits_{B(R)}|\nabla v|^2\,dx\Big)^{\frac{1}{2}}\Big(\int\limits_{B(R)}|v|^3\,dx\Big)^{\frac{1}{3}}\leq$$

$$\leq c\,R^{\frac{1}{2}}A^{\frac{1}{2}}(R)\Big(\int\limits_{Q(R)}|v|^3\,dz\Big)^{\frac{1}{3}}\Big(\int\limits_{-R^2}^{0}dt\Big(\int\limits_{B(R)}|\nabla v|^2\,dx\Big)^{\frac{3}{4}}\Big)^{\frac{2}{3}}\leq$$

$$\leq R^{\frac{1}{2}+\frac{2}{3}}A^{\frac{1}{2}}(R)C^{\frac{1}{3}}(R)R^{\frac{1}{3}}\Big(\int\limits_{Q(R)}|\nabla v|^2\,dz\Big)^{\frac{1}{2}}\leq$$

$$\leq c\,R^2A^{\frac{1}{2}}(R)C^{\frac{1}{3}}(R)E^{\frac{1}{2}}(R).$$

Now, (6.1.36) follows from the latter relation and from (6.1.37). Lemma 6.3 is proved.

Now, our goal is to work out an estimate for the pressure.

Lemma 6.4. *Let* $0<\varrho\leq 1$. *Then*

$$D_0(r)\leq c\Big[\Big(\frac{r}{\varrho}\Big)^{\frac{5}{2}}D_0(\varrho)+\Big(\frac{\varrho}{r}\Big)^{2}A^{\frac{1}{2}}(\varrho)E(\varrho)\Big] \qquad (6.1.38)$$

for all $r\in]0,\varrho]$.

PROOF We split the pressure in two parts

$$q = p_1 + p_2 \tag{6.1.39}$$

in $B(\varrho)$ so that p_1 is a unique solution to the variational identity

$$\int\limits_{B(\varrho)} p_1 \Delta\varphi dx = -\int\limits_{B(\varrho)} (\tau - \tau_\varrho) : \nabla^2 \varphi dx, \tag{6.1.40}$$

in which φ is an arbitrary test function of $W^2_3(B(\varrho))$ satisfying the boundary condition $\varphi|_{\partial B(\varrho)} = 0$ and

$$\tau := (v - c_\varrho) \otimes (v - c_\varrho), \quad \tau_\varrho := [(v - c_\varrho) \otimes (v - c_\varrho)]_{,\varrho}, \quad c_\varrho := [v]_{,\varrho}.$$

Here, time t is considered as a parameter. Clearly,

$$\Delta p_2 = 0 \tag{6.1.41}$$

in $B(\varrho)$.

We can easily find the bound for p_1 (by a suitable choice of the test function in (6.1.40))

$$\int\limits_{B(\varrho)} |p_1|^{\frac{3}{2}} dx \leq c \int\limits_{B(\varrho)} |\tau - \tau_\varrho|^{\frac{3}{2}} dx.$$

The Gagliardo-Nirenberg inequality

$$\int\limits_{B(\varrho)} |p_1|^{\frac{3}{2}} dx \leq c\Big(\int\limits_{B(\varrho)} |v - c_\varrho||\nabla v| dx\Big)^{\frac{3}{2}}$$

and Hölder inequality imply

$$\int\limits_{B(\varrho)} |p_1|^{\frac{3}{2}} dx \leq c\Big(\int\limits_{B(\varrho)} |v - c_\varrho|^2 dx\Big)^{\frac{3}{4}} \Big(\int\limits_{B(\varrho)} |\nabla v|^2 dx\Big)^{\frac{3}{4}}.$$

On the other hand, Poincaré's inequality

$$\int\limits_{B(\varrho)} |v - c_\varrho|^2 dx \leq c\varrho^2 \int\limits_{B(\varrho)} |\nabla v|^2 dx$$

and the minimality property of c_ϱ

$$\int\limits_{B(\varrho)} |v - c_\varrho|^2 dx \leq \int\limits_{B(\varrho)} |v|^2 dx$$

lead to the estimate

$$\frac{1}{\varrho^2} \int\limits_{-\varrho^2}^{0} \int\limits_{B(\varrho)} |p_1|^{\frac{3}{2}} dz \leq cE(\varrho) A^{\frac{1}{2}}(\varrho). \tag{6.1.42}$$

By the mean value theorem for harmonic function p_2, we have for $0 < r \leq \varrho/2$

$$\sup_{x\in B(r)} |p_2(x,t) - [p_2]_{,r}(t)|^{\frac{3}{2}} \leq c r^{\frac{3}{2}} \sup_{x\in B(\varrho/2)} |\nabla p_2(x,t)|^{\frac{3}{2}}$$

$$\leq c\Big(\frac{r}{\varrho^4} \int\limits_{B(\varrho)} |p_2(x,t) - [p_2]_{,\varrho}(t)|dx\Big)^{\frac{3}{2}} \tag{6.1.43}$$

$$\leq \frac{c}{\varrho^3}\Big(\frac{r}{\varrho}\Big)^{\frac{3}{2}} \int\limits_{B(\varrho)} |p_2(x,t) - [p_2]_{,\varrho}(t)|^{\frac{3}{2}} dx.$$

Next, by (6.1.39) and (6.1.43),

$$D_0(r) \leq \frac{c}{r^2} \int\limits_{Q(r)} |p_1 - [p_1]_{,r}|^{\frac{3}{2}} dz + \frac{c}{r^2} \int\limits_{Q(r)} |p_2 - [p_2]_{,r}|^{\frac{3}{2}} dz$$

$$\leq \frac{c}{r^2} \int\limits_{Q(r)} |p_1|^{\frac{3}{2}} dz + \frac{c}{r^2}\frac{1}{\varrho^3}\Big(\frac{r}{\varrho}\Big)^{\frac{3}{2}} \int\limits_{-r^2}^{0} r^3 \int\limits_{B(\varrho)} |p_2(x,t) - [p_2]_{,\varrho}(t)|^{\frac{3}{2}} dx$$

$$\leq c\Big(\frac{\varrho}{r}\Big)^2 E(\varrho) A^{\frac{1}{2}}(\varrho) + c\Big(\frac{r}{\varrho}\Big)^{\frac{5}{2}} \frac{1}{\varrho^2} \int\limits_{Q(\varrho)} |p_2 - [p_2]_{,\varrho}|^{\frac{3}{2}} dz$$

$$\leq c\Big(\frac{\varrho}{r}\Big)^2 E(\varrho) A^{\frac{1}{2}}(\varrho) + c\Big(\frac{r}{\varrho}\Big)^{\frac{5}{2}} \Big[\frac{1}{\varrho^2} \int\limits_{Q(\varrho)} |q - [q]_{,\varrho}|^{\frac{3}{2}} dz$$

$$+\frac{1}{\varrho^2} \int\limits_{Q(\varrho)} |p_1 - [p_1]_{,\varrho}|^{\frac{3}{2}} dz\Big]$$

$$\leq c\Big[\Big(\frac{r}{\varrho}\Big)^{\frac{5}{2}} D_0(\varrho) + \Big(\frac{\varrho}{r}\Big)^2 E(\varrho) A^{\frac{1}{2}}(\varrho)\Big].$$

So, inequality (6.1.38) is shown. Lemma 6.4 is proved.

Proof of Theorem 1.4 It follows from (6.1.34), (6.1.38), and the assumptions of Theorem 1.4 that:

$$C(r) \leq c\Big[\Big(\frac{\varrho}{r}\Big)^3 A^{\frac{3}{4}}(\varrho)\varepsilon^{\frac{3}{4}} + \Big(\frac{r}{\varrho}\Big)^3 A^{\frac{3}{2}}(\varrho)\Big] \tag{6.1.44}$$

and

$$D_0(r) \leq c\Big[\Big(\frac{r}{\varrho}\Big)^{\frac{5}{2}} D_0(\varrho) + \Big(\frac{\varrho}{r}\Big)^2 A^{\frac{1}{2}}(\varrho)\varepsilon\Big]. \tag{6.1.45}$$

Introducing the new quantity

$$\mathcal{E}(r) = A^{\frac{3}{2}}(r) + D_0^2(r),$$

we derive from local energy inequality (6.1.36) the following estimate

$$\mathcal{E}(r) \leq c\Big[C(2r) + C^{\frac{1}{2}}(2r)D_0(2r) + A^{\frac{3}{4}}(2r)C^{\frac{1}{2}}(2r)\varepsilon^{\frac{3}{4}}\Big] + D_0^2(r)$$

$$\leq c\Big[C(2r) + D_0^2(2r) + A^{\frac{3}{4}}(2r)C^{\frac{1}{2}}(2r)\varepsilon^{\frac{3}{4}}\Big]. \tag{6.1.46}$$

Now, let us assume that $0 < r \leq \varrho/2 < \varrho \leq 1$. Replacing r with $2r$ in (6.1.44) and (6.1.45), we can reduce (6.1.46) to the form

$$\mathcal{E}(r) \leq c\Big[\Big(\frac{\varrho}{r}\Big)^3 A^{\frac{3}{4}}(\varrho)\varepsilon^{\frac{3}{4}} + \Big(\frac{r}{\varrho}\Big)^3 A^{\frac{3}{2}}(\varrho)$$

$$+\Big(\frac{r}{\varrho}\Big)^5 D_0^2(\varrho) + \Big(\frac{\varrho}{r}\Big)^4 A(\varrho)\varepsilon^2$$

$$+A^{\frac{3}{4}}(2r)\Big(\Big(\frac{\varrho}{r}\Big)^3 A^{\frac{3}{4}}(\varrho)\varepsilon^{\frac{3}{4}} + \Big(\frac{r}{\varrho}\Big)^3 A^{\frac{3}{2}}(\varrho)\Big)^{\frac{1}{2}} E_0^{\frac{3}{4}}\Big]$$

$$\leq c\Big[\Big(\frac{r}{\varrho}\Big)^3 A^{\frac{3}{2}}(\varrho) + \Big(\frac{r}{\varrho}\Big)^5 D_0^2(\varrho) + \Big(\frac{r}{\varrho}\Big)^{\frac{3}{2}} A^{\frac{3}{4}}(\varrho)\varepsilon^{\frac{3}{4}} A^{\frac{3}{4}}(\varrho)\Big(\frac{\varrho}{r}\Big)^{\frac{3}{4}}$$

$$+\Big(\frac{\varrho}{r}\Big)^{\frac{3}{2}+\frac{3}{4}} A^{\frac{3}{4}+\frac{3}{8}}(\varrho)\varepsilon^{\frac{3}{4}+\frac{3}{8}} + \Big(\frac{\varrho}{r}\Big)^4 A(\varrho)\varepsilon^2 + \Big(\frac{\varrho}{r}\Big)^3 A^{\frac{3}{4}}(\varrho)\varepsilon^{\frac{3}{4}}\Big].$$

Here, the obvious inequality $A(2r) \leq c\varrho A(\varrho)/r$ has been used. Applying Young inequality with an arbitrary positive constant δ, we show that

$$\mathcal{E}(r) \leq c\Big(\frac{r}{\varrho}\Big)^{\frac{3}{4}}(\varepsilon^{\frac{3}{4}} + 1)\mathcal{E}(\varrho) + c\delta\mathcal{E}(\varrho)$$

$$+c(\delta)\Big(\Big(\frac{\varrho}{r}\Big)^6 \varepsilon^{\frac{3}{2}} + \Big(\frac{\varrho}{r}\Big)^{12}\varepsilon^6 + \Big(\frac{\varrho}{r}\Big)^9 \varepsilon^{\frac{9}{2}}\Big)\Big].$$

Therefore,

$$\mathcal{E}(r) \leq c\Big[\Big(\frac{r}{\varrho}\Big)^{\frac{3}{4}}(\varepsilon^{\frac{3}{4}} + 1) + \delta\Big]\mathcal{E}(\varrho) + c(\delta)\Big(\frac{\varrho}{r}\Big)^{12}(\varepsilon^6 + \varepsilon^{\frac{9}{2}} + \varepsilon^{\frac{3}{2}}). \tag{6.1.47}$$

Inequality (6.1.47) holds for $r \leq \varrho/2$ and can be rewritten as follows:

$$\mathcal{E}(\vartheta\varrho) \leq c\Big[\vartheta^{\frac{3}{4}}(\varepsilon^{\frac{3}{4}}+1)+\delta\Big]\mathcal{E}(\varrho)+c(\delta)\vartheta^{-12}(\varepsilon^{6}+\varepsilon^{\frac{9}{2}}+\varepsilon^{\frac{3}{2}}) \tag{6.1.48}$$

for any $0 < \vartheta \leq 1/2$ and for any $0 < \varrho \leq 1$.

Now, assuming that $\varepsilon \leq 1$, let us fix ϑ and δ to provide the conditions:

$$2c\vartheta^{\frac{1}{4}} < 1/2, \quad 0 < \vartheta \leq 1/2, \quad c\delta < \vartheta^{\frac{1}{2}}/2. \tag{6.1.49}$$

Obviously, ϑ and δ are independent of ε. So,

$$\mathcal{E}(\vartheta\varrho) \leq \vartheta^{\frac{1}{2}}\mathcal{E}(\varrho)+G \tag{6.1.50}$$

for any $0 < \varrho \leq 1$, where $G = G(\varepsilon) \to 0$ as $\varepsilon \to 0$.

Iterations of (6.1.50) give us

$$\mathcal{E}(\vartheta^k\varrho) \leq \vartheta^{\frac{k}{2}}\mathcal{E}(\varrho)+cG$$

for any natural numbers k and for any $0 < \varrho \leq 1$. Letting $\varrho = 1$, we find

$$\mathcal{E}(\vartheta^k) \leq \vartheta^{\frac{k}{2}}\mathcal{E}(1)+cG \tag{6.1.51}$$

for the same values of k. It can be easily deduced from (6.1.51) that

$$\mathcal{E}(r) \leq c(r^{\frac{1}{2}}\mathcal{E}(1)+G(\varepsilon)) \tag{6.1.52}$$

for all $0 < r \leq 1/2$. Now, (6.1.44) and (6.1.45) imply

$$C(r)+D_0(r) \leq c\Big[A^{\frac{3}{4}}(2r)\varepsilon^{\frac{3}{4}}+A^{\frac{3}{2}}(2r)\Big]+c(r^{\frac{1}{4}}\mathcal{E}^{\frac{1}{2}}(1)+G^{\frac{1}{2}}(\varepsilon))$$

$$\leq c\Big[A^{\frac{3}{2}}(2r)+\varepsilon^{\frac{3}{2}}\Big]+c(r^{\frac{1}{4}}\mathcal{E}^{\frac{1}{2}}(1)+G^{\frac{1}{2}}(\varepsilon))$$

$$\leq c\Big[c((2r)^{\frac{1}{2}}\mathcal{E}(1)+G(\varepsilon))+\varepsilon^{\frac{3}{2}}\Big]+c(r^{\frac{1}{4}}\mathcal{E}^{\frac{1}{2}}(1)+G^{\frac{1}{2}}(\varepsilon)).$$

Now we see that, for sufficiently small ε and sufficiently small r_0,

$$C(r_0)+D_0(r_0) < \varepsilon_0,$$

where ε_0 is a number of Lemma 6.1. Since v and $q-[q]_{,r_0}$ are a suitable weak solution in $Q(r_0)$, Lemma 6.1 and the Navier-Stokes scaling yields required statement. Theorem 1.4 is proved.

Now, we are in a position to speculate about ε-regularity theory. Quantities that are invariant with respect to the Navier-Stokes scaling

$$v^\lambda(y,s) = \lambda v(x_0+\lambda y, t_0+\lambda^2 s),$$
$$q^\lambda(y,s) = \lambda^2 q(x_0+\lambda y, t_0+\lambda^2 s) \tag{6.1.53}$$

play the crucial role in this theory. By the definition, such quantities are defined on parabolic balls $Q(r)$ and have the property

$$F(v,q;r) = F(v^\lambda, q^\lambda; r/\lambda).$$

There are two types of statements in the ε-regularity theory for suitable weak solutions to the Navier-Stokes equations and the first one reads:

Suppose that v and q are a suitable weak solution to the Navier-Stokes equations in Q. There exist universal positive constants ε and $\{c_k\}_{k=1}^\infty$ such that if $F(v,q;1) < \varepsilon$ then $|\nabla^k v(0)| < c_k$, $k = 0,1,2,\ldots$. Moreover, the function $z \mapsto \nabla^k v(z)$ is Hölder continuous (relative to the parabolic metric) with any exponent less $1/3$ in the closure of $Q(1/2)$.

An important example of such kind of quantities appears in Lemma 6.1 and is as follows:

$$F(v,q;r) = \frac{1}{r^2} \int\limits_{Q(r)} \left(|v|^3 + |q|^{\frac{3}{2}}\right) dz.$$

In the other type of statements, it is supposed that our quantity F is independent of the pressure q:

Let v and q be a suitable weak solution in Q. There exists a universal positive constant ε with the property: if $\sup_{0<r<1} F(v;r) < \varepsilon$ then $z = 0$ is a regular point. Moreover, for any $k = 0,1,2,\ldots$, the function $z \mapsto \nabla^k v(z)$ is Hölder continuous with any exponent less $1/3$ in the closure of $Q(r)$ for some positive r.

Dependence on the pressure in the above statement is hidden. In fact, the radius r is determined by the $L_{\frac{3}{2}}$-norm of the pressure over the whole parabolic cylinder Q.

To illustrate the second statement, let us consider several examples. In the first one, we deal with the Ladyzhenskaya-Prodi-Serrin type quantities

$$F(v;r) = M_{s,l}(v;r) = \|v\|^l_{s,l,Q(r)} = \int\limits_{-r^2}^{0} \Big(\int\limits_{B(r)} |v|^s dx \Big)^{\frac{l}{s}} dt$$

provided

$$\frac{3}{s} + \frac{2}{l} = 1$$

and $s \geq 3$. Local regularity results connected with those quantities have been proved partially by J. Serrin in [Serrin (1962)] and then by M. Struwe in [Sruwe (1988)] for the velocity field v having finite energy even with no assumption on the pressure. However, in such a case, we might loose Hölder continuity.

Energy scale-invariant quantities present an important example of the second kind of quantities. Some of them have been listed above. For more examples of scaled energy quantities, we refer to the paper [Gustafson *et al.* (2007)]. It is worthy to note that the second statement applied to the scaled dissipation E is the famous Caffarelli-Kohn-Nirenberg theorem, which is Theorem 1.4. It gives the best estimate for Hausdorff's dimension of the singular set for a class of weak Leray-Hopf solutions to the Cauchy problem. A certain generalization of the Caffarelli-Kohn-Nirenberg theorem itself has been proved in [Seregin (2007)] and is formulated as follows.

Proposition 1.5. *Let v and q be a suitable weak solution to the Navier-Stokes equations in Q. Given $M > 0$, there exists a positive number $\varepsilon(M)$ having the property: if two inequalities* $\limsup_{r\to 0} E(r) < M$ *and*

$$\liminf_{r\to 0} E(r) < \varepsilon(M)$$

hold, then $z = 0$ is a regular point of v.

Typical examples of the third group of quantities invariant to the Navier-Stokes scaling are:

$$G_1(v;r) = \sup_{z=(x,t)\in Q(r)} |x||v(z)|,$$

$$G_2(v;r) = \sup_{z=(x,t)\in Q(r)} \sqrt{-t}|v(z)|.$$

A proof of the corresponding statements has been presented in [Seregin and Zajaczkowski (2006)], see also [Takahashi (1990)], [Kim and Kozono (2004)], and [Chen and Price (2001)] for similar results.

6.2 Bounded Ancient Solutions

Definition 6.2. A bounded divergence free field $u \in L_\infty(Q_-;\mathbb{R}^n)$ is called a weak bounded ancient solution (or simply bounded ancient solution) to the Navier-Stokes equations if

$$\int\limits_{Q_-} (u\cdot\partial_t w + u\otimes u : \nabla w + u\cdot\Delta w)dz = 0$$

for any $w \in C^\infty_{0,0}(Q_-)$.

Without loss of generality, we may assume that $|u(z)| \leq 1$ a.e. in Q_-. If not, the function $u^\lambda(x,t) = \lambda u(\lambda x, \lambda^2 t)$ with $\lambda = 1/\|u\|_{\infty,Q_-}$ will be a bounded ancient solution satisfying the condition $|u^\lambda(z)| \leq 1$ a.e. in Q_-.

Our aim is to analyze differentiability properties of an arbitrary bounded ancient solution. Before stating and proving the main result, let us formulate several auxiliary lemmata.

Lemma 6.5. *For any $F = L_\infty(\mathbb{R}^n; \mathbb{M}^{n\times n})$, there exists a unique function $q_F \in BMO(\mathbb{R}^n)$ that $[q_F]_{B(1)} = 0$ and*

$$\Delta q_F = -\operatorname{div}\operatorname{div} F = -F_{ij,ij} \qquad in \quad \mathbb{R}^3$$

in the sense of distributions. Moreover, the following estimate is valid

$$\|q_F\|_{BMO(\mathbb{R}^n)} \leq c(n)\|F\|_{\infty,\mathbb{R}^n}.$$

Here, the space $BMO(\mathbb{R}^n)$ consists of all functions $f \in L_{1,\mathrm{loc}}(\mathbb{R}^n)$ with bounded mean oscillation, i.e.,

$$\sup\Big\{\frac{1}{|B(R)|}\int\limits_{B(x_0,R)} |f - [f]_{B(x_0,R)}|dx : \ \forall x_0 \in \mathbb{R}^n, \ \forall R > 0\Big\} < \infty.$$

$[f]_\Omega$ is the mean value of a function f over a spatial domain $\Omega \in \mathbb{R}^n$. The mean value of a function g over a space-time domain Q is denoted by $(g)_Q$.

Lemma 6.6. *Assume that functions $f \in L_m(B(2))$ and $q \in L_m(B(2))$ satisfy the equation*

$$\Delta q = -\operatorname{div} f \qquad in \quad B(2).$$

Then

$$\int\limits_{B(1)} |\nabla q|^m dx \leq c(m,n)\Big(\int\limits_{B(2)} |f|^m dx + \int\limits_{B(2)} |q - [q]_{B(2)}|^m dx\Big).$$

Lemma 6.7. *Assume that functions $f \in L_m(Q(2))$ and $u \in W^{1,0}_m(Q(2))$ satisfy the equation*

$$\partial_t u - \Delta u = f \qquad in \quad Q(2).$$

Then $u \in W^{2,1}_m(Q(1))$ and the following estimate is valid:

$$\|\partial_t u\|_{m,Q(1)} + \|\nabla^2 u\|_{m,Q(1)} \leq c(m,n)\Big[\|f\|_{m,Q(2)} + \|u\|_{W^{1,0}_m(Q(2))}\Big].$$

Lemma 6.5 is proved with the help of the singular integral theory, see [Stein (1970)]. Proof of Lemmata 6.6 and 6.7 can be found, for example, in [Ladyzhenskaya and Uraltseva (1973)] and [Ladyzhenskaya *et al.* (1967)].

If we let

$$F(\cdot,t) = u(\cdot,t)\otimes u(\cdot,t),$$

then, by Lemma 6.5, there exists a unique function

$$p_{u\otimes u} \in L_\infty(-\infty,0;BMO(\mathbb{R}^n))$$

which satisfies the condition $[p_{u\otimes u}]_{B(1)}(t)=0$ and the equation

$$\Delta p_{u\otimes u}(\cdot,t) = -\mathrm{div}\,\mathrm{div} F(\cdot,t) \qquad \text{in} \quad \mathbb{R}^n$$

for all $t\leq 0$.

To state the main result of this section, we introduce the space

$$\mathcal{L}_m(Q_-) := \{\sup_{z_0\in Q_-} \|f\|_{m,Q(z_0,1)} < \infty\}.$$

Theorem 2.6. *Let u be an arbitrary bounded ancient solution. For any number $m>1$,*

$$|\nabla u| + |\nabla^2 u| + |\nabla p_{u\otimes u}| \in \mathcal{L}_m(Q_-).$$

Moreover, for each $t_0\leq 0$, there exists a function $b_{t_0}\in L_\infty(t_0-1,t_0)$ with the following property

$$\sup_{t_0\leq 0} \|b_{t_0}\|_{L_\infty(t_0-1,t_0)} \leq c(n) < +\infty.$$

If we let $u^{t_0}(x,t) = u(x,t)+b_{t_0}(t)$ in $Q^{t_0} = \mathbb{R}^n\times]t_0-1,t_0[$, then, for any number $m>1$ and for any point $x_0\in\mathbb{R}^n$, the uniform estimate

$$\|u^{t_0}\|_{W^{2,1}_m(Q(z_0,1))} \leq c(m,n) < +\infty, \qquad z_0=(x_0,t_0),$$

is valid and, for a.a. $z=(z,t)\in Q^{t_0}$, functions u and u^{t_0} obey the system of equations

$$\partial_t u^{t_0} + \mathrm{div} u\otimes u - \Delta u = -\nabla p_{u\otimes u}, \qquad \mathrm{div} u = 0.$$

Remark 6.2. The first equation of the above system can be rewritten in the following way

$$\partial_t u + \mathrm{div} u\otimes u - \Delta u = -\nabla p_{u\otimes u} - b'_{t_0}, \qquad b'_{t_0}(t) = db_{t_0}(t)/dt,$$

in Q^{t_0} in the sense of distributions. So, the real pressure field in Q^{t_0} is the following distribution $p_{u\otimes u} + b'_{t_0}\cdot x$.

Remark 6.3. We can find a measurable vector-valued function b defined on $]-\infty,0[$ and having the following property. For any $t_0 \leq 0$, there exists a constant vector c_{t_0} such that

$$\sup_{t_0\leq 0} \|b - c_{t_0}\|_{L_\infty(t_0-1,t_0)} < +\infty.$$

Moreover, the Navier-Stokes system takes the form

$$\partial_t u + \mathrm{div} u \otimes u - \Delta u = -\nabla(p_{u\otimes u} + b' \cdot x), \qquad \mathrm{div} u = 0$$

in Q_- in the sense of distributions.

Remark 6.4. In most of our applications, we shall have some additional global information about the pressure field, which will make it possible to conclude that $b' = 0$. For example, it is true if the pressure field belongs to $L_\infty(-\infty, 0; BMO(\mathbb{R}^n))$, i.e., u is a mild bounded ancient solution, see the next section for details and definitions.

We can exclude the pressure field completely by considering the equation for vorticity $\omega = \nabla \wedge u$. Differentiability properties of ω are described by the following theorem.

Theorem 2.7. *Let u be an arbitrary bounded ancient solution. For any $m > 1$, we have the following statements. If $n = 2$, then*

$$\omega = \nabla^\perp u = u_{2,1} - u_{1,2} \in \mathcal{W}^{2,1}_m(Q_-) := \{\omega, \nabla\omega, \nabla^2\omega, \partial_t\omega \in \mathcal{L}_m(Q_-)\}$$

and

$$\partial_t\omega + u \cdot \nabla\omega - \Delta\omega = 0 \qquad a.e. \ in \quad Q_-.$$

If $n = 3$, then

$$\omega = \nabla \wedge u \in \mathcal{W}^{2,1}_m(Q_-; \mathbb{R}^3)$$

and

$$\partial_t\omega + u \cdot \nabla\omega - \Delta\omega = \omega \cdot \nabla u \qquad a.e. \ in \quad Q_-.$$

Remark 6.5. We could analyze smoothness of solutions to the vorticity equations further and it would be a good exercise. However, regularity results stated in Theorem 2.7 are sufficient for our purposes.

Remark 6.6. By the embedding theorems, see [Ladyzhenskaya *et al.* (1967)], functions ω and $\nabla\omega$ are Hölder continuous in Q_- and uniformly bounded there.

Proof of Theorem 2.6: Step 1. Energy estimate. Fix an arbitrary number $t_0 < 0$. Let $k_\varepsilon(z)$ be a standard smoothing kernel (mollifier). We use the following notation for mollified functions:

$$F^\varepsilon(z) = \int\limits_{Q_-} k_\varepsilon(z - z')F(z')dz', \qquad F = u \otimes u,$$

$$u^\varepsilon(z) = \int\limits_{Q_-} k_\varepsilon(z - z')u(z')dz'.$$

Assume that $w \in \overset{\circ}{C}{}_0^\infty(Q_-^{t_0})$, where $Q_-^{t_0} = \mathbb{R}^n \times]-\infty, t_0[$. For sufficiently small ε $(0 < \varepsilon < \varepsilon(t_0))$, w^ε belongs to $\overset{\circ}{C}{}_0^\infty(Q_-)$ as well. Then using known properties of smoothing kernel and Definition 6.2, we find

$$\int\limits_{Q_-} w \cdot (\partial_t u^\varepsilon + \mathrm{div} F^\varepsilon - \Delta u^\varepsilon)dz = 0, \qquad \forall w \in \overset{\circ}{C}{}_0^\infty(Q_-^{t_0}).$$

It is easy to see that in our case there exists a smooth function p_ε with the following property

$$\partial_t u^\varepsilon + \mathrm{div} F^\varepsilon - \Delta u^\varepsilon = -\nabla p_\varepsilon, \qquad \mathrm{div} u^\varepsilon = 0 \tag{6.2.1}$$

in $Q_-^{t_0}$. Let us decompose p_ε so that

$$p_\varepsilon = p_{F^\varepsilon} + \widetilde{p}_\varepsilon. \tag{6.2.2}$$

It is not difficult to show that the function ∇p_{F^ε} is bounded in $Q_-^{t_0}$ (exercise). So, it follows from (6.2.1) and (6.2.2) that

$$\Delta \widetilde{p}_\varepsilon = 0 \quad \text{in} \quad Q_-^{t_0}, \qquad \nabla \widetilde{p}_\varepsilon \in L_\infty(Q_-^{t_0}; \mathbb{R}^n).$$

By the Liouville theorem for harmonic functions, there exists a function $a_\varepsilon : [-\infty, t_0[\to \mathbb{R}^n$ such that

$$\nabla \widetilde{p}_\varepsilon(x,t) = a_\varepsilon(t), \qquad x \in \mathbb{R}^n, \quad -\infty < t \le t_0.$$

So, we have

$$\partial_t u^\varepsilon + \mathrm{div} F^\varepsilon - \Delta u^\varepsilon = -\nabla p_{F^\varepsilon} - a_\varepsilon, \qquad \mathrm{div} u^\varepsilon = 0 \tag{6.2.3}$$

in $Q_-^{t_0}$.

Now, let us introduce new functions

$$b_{\varepsilon t_0}(t) = \int\limits_{t_0 - 1}^{t} a_\varepsilon(\tau)d\tau, \qquad t_0 - 1 \le t \le t_0,$$

$$v_\varepsilon(x,t) = u^\varepsilon(x,t) + b_{\varepsilon t_0}(t), \qquad z=(x,t)\in Q^{t_0}.$$

Using them, we may rewrite system (6.2.3) so that

$$\partial_t v_\varepsilon - \Delta v_\varepsilon = -\mathrm{div} F^\varepsilon - \nabla p_{F^\varepsilon}, \qquad \mathrm{div} v_\varepsilon = 0 \tag{6.2.4}$$

in $Q^{t_0}_-$.

Fix an arbitrary cut-off function φ so that

$$0\le \varphi\le 1, \qquad \varphi\equiv 1 \quad \text{in} \quad B(1), \qquad \mathrm{supp}\varphi\subset B(2).$$

And then let $\varphi_{x_0}(x) = \varphi(x-x_0)$.

Now, we can derive the energy identity from (6.2.4), multiplying the latter by $\varphi^2_{x_0}v_\varphi$ and integrating the product by parts. As a result, we have

$$I(t) = \int\limits_{\mathbb{R}^n} \varphi^2_{x_0}(x)|v_\varepsilon(x,t)|^2dx + 2\int\limits_{t_0-1}^{t}\int\limits_{\mathbb{R}^n} \varphi^2_{x_0}|\nabla v_\varepsilon|^2dxdt' =$$

$$= \int\limits_{\mathbb{R}^n} \varphi^2_{x_0}(x)|v_\varepsilon(x,t_0-1)|^2dx + \int\limits_{t_0-1}^{t}\int\limits_{\mathbb{R}^n} \Delta\varphi^2_{x_0}|v_\varepsilon|^2dxdt' +$$

$$+ \int\limits_{t_0-1}^{t}\int\limits_{\mathbb{R}^n}(p_{F^\varepsilon} - [p_{F^\varepsilon}]_{B(x_0,2)})v_\varepsilon\cdot\nabla\varphi^2_{x_0}dxdt' +$$

$$+ \int\limits_{t_0-1}^{t}\int\limits_{\mathbb{R}^n}(F^\varepsilon - [F^\varepsilon]_{B(x_0,2)}) : \cdot\nabla(\varphi^2_{x_0}v_\varepsilon)dxdt'.$$

Introducing the quantity

$$\alpha_\varepsilon(t) = \sup_{x_0\in\mathbb{R}^n}\int\limits_{B(x_0,1)} |v_\varepsilon(x,t)|^2dx$$

and taking into account that $v_\varepsilon(\cdot,t_0-1) = u^\varepsilon(\cdot,t_0-1)$ and $|u^\varepsilon(\cdot,t_0-1)|\le 1$, we can estimate the right-hand side of the energy identity in the following way

$$I(t)\le c(n) + c(n)\int\limits_{t_0-1}^{t}\alpha_\varepsilon(t')dt' +$$

$$+c(n)\Big(\int\limits_{t_0-1}^{t_0}\int\limits_{B(x_0,2)} |p_{F^\varepsilon} - [p_{F^\varepsilon}]_{B(x_0,2)}|^2dxdt\Big)^{\frac12}\Big(\int\limits_{t_0-1}^{t}\alpha_\varepsilon(t')dt'\Big)^{\frac12} +$$

$$+c(n)\Big(\int\limits_{t_0-1}^{t_0}\int\limits_{B(x_0,2)}|F^\varepsilon-[F^\varepsilon]_{B(x_0,2)}|^2dxdt\Big)^{\frac{1}{2}}\Big(\int\limits_{t_0-1}^{t}\int\limits_{\mathbb{R}^n}\varphi^2_{x_0}|\nabla v_\varepsilon|^2dxdt'+ \tag{6.2.5}$$

$$+\int\limits_{t_0-1}^{t}\alpha_\varepsilon(t')dt'\Big)^{\frac{1}{2}},\qquad t_0-1\leq t\leq t_0.$$

Next, since $|F^\varepsilon|\leq c(n)$, we find two estimates

$$\int\limits_{t_0-1}^{t_0}\int\limits_{B(x_0,2)}|F^\varepsilon-[F^\varepsilon]_{B(x_0,2)}|^2dxdt\leq c(n)$$

and

$$\int\limits_{t_0-1}^{t_0}\int\limits_{B(x_0,2)}|p_{F^\varepsilon}-[p_{F^\varepsilon}]_{B(x_0,2)}|^2dxdt\leq c(n)\|p_{F^\varepsilon}\|^2_{L_\infty(-\infty,t_0;BMO(\mathbb{R}^n))}$$

$$\leq c(n)\|F^\varepsilon\|^2_{L_\infty(Q^{t_0}_-)}\leq c(n).$$

The latter estimates, together with (6.2.5), implies two inequalities:

$$\alpha_\varepsilon(t)\leq c(n)\Big(1+\int\limits_{t_0-1}^{t}\alpha_\varepsilon(t')dt'\Big),\qquad t_0-1\leq t\leq t_0$$

and

$$\sup_{x_0\in\mathbb{R}^n}\int\limits_{t_0-1}^{t_0}\int\limits_{B(x_0,1)}|\nabla v_\varepsilon|^2dxdt\leq c(n)\Big(1+\int\limits_{t_0-1}^{t_0}\alpha_\varepsilon(t)dt\Big).$$

Usual arguments allows us to conclude that:

$$\sup_{t_0-1\leq t\leq t_0}\alpha_\varepsilon(t)+\sup_{x_0\in\mathbb{R}^n}\int\limits_{t_0-1}^{t_0}\int\limits_{B(x_0,1)}|\nabla u^\varepsilon|^2dxdt\leq c(n). \tag{6.2.6}$$

It should be emphasized that the right-hand size in (6.2.6) is independent of t_0. In particular, estimate (6.2.6) gives:

$$\sup_{t_0-1\leq t\leq t_0}b_{\varepsilon t_0}(t)\leq c(n).$$

Now, let us see what happens if $\varepsilon \to 0$. Selecting a subsequence if necessary and taking the limit as $\varepsilon \to 0$, we get the following facts:

$$b_{\varepsilon t_0} \overset{\star}{\rightharpoonup} b_{t_0} \qquad \text{in} \qquad L_\infty(t_0-1,t_0;\mathbb{R}^n);$$

the estimate

$$\|b_{t_0}\|_{L_\infty(t_0-1,t_0)} + \sup_{x_0\in\mathbb{R}^n} \int\limits_{t_0-1}^{t_0} \int\limits_{B(x_0,1)} |\nabla u|^2 dxdt \leq c(n) < +\infty \tag{6.2.7}$$

is valid for all $t_0 < 0$;
the system

$$\partial_t u^{t_0} + \mathrm{div} u\otimes u - \Delta u = -\nabla p - u\otimes u, \qquad \mathrm{div} u = 0$$

holds in Q^{t_0} in the sense of distributions.

The case $t_0 = 0$ can be treated by passing to the limit as $t_0 \to 0$.

STEP 2. BOOTSTRAP ARGUMENTS. By (6.2.7),

$$f = \mathrm{div} F = u\cdot\nabla u \in \mathcal{L}_2(Q_-;\mathbb{R}^n).$$

Then Lemma 6.6 in combination with shifts shows that

$$\nabla p_{u\otimes u} \in \mathcal{L}_2(Q_-;\mathbb{R}^n).$$

Next, obviously, the function u^{t_0} satisfies the system of equations

$$\partial_t u^{t_0} - \Delta u^{t_0} = -u\cdot\nabla u - \nabla p_{u\otimes u} \in \mathcal{L}_2(Q_-;\mathbb{R}^n).$$

Using the invariance with respect to shifts and Lemma 6.7, one can conclude that

$$u^{t_0} \in W^{2,1}_2(Q(z_0,\tau_2);\mathbb{R}^n), \qquad 1/2 < \tau_2 < \tau_1 = 1,$$

and, moreover, the estimate

$$\|u^{t_0}\|_{W^{2,1}_2(Q(z_0,\tau_2))} \leq c(n,\tau_2)$$

holds for any $z_0 = (x_0,t_0)$, where $x_0 \in \mathbb{R}^n$ and $t_0 \leq 0$. A parabolic embedding theorem, see [Ladyzhenskaya *et al.* (1967)], ensures that:

$$\nabla u^{t_0} = \nabla u \in W^{1,0}_{m_2}(Q(z_0,\tau_2);\mathbb{R}^n)$$

for

$$\frac{1}{m_2} = \frac{1}{m_1} - \frac{1}{n+2}, \qquad m_1 = 2.$$

By Lemma 6.6, by shifts, and by scaling, for $1/2 < \tau_3' < \tau_2$, we have the following estimate

$$\int\limits_{B(x_0,\tau_3')} |\nabla p_{u\otimes u}(\cdot,t)|^{m_2} dx \leq c(n,\tau_2,\tau_3')\Big[\int\limits_{B(x_0,\tau_3')} |\nabla u(\cdot,t)|^{m_2} dx + 1\Big].$$

In turn, Lemma 6.7 implies two statements:

$$u^{t_0} \in W^{2,1}_{m_2}(Q(z_0,\tau_3);\mathbb{R}^n), \qquad 1/2 < \tau_3 < \tau_3'$$

and

$$\|u^{t_0}\|_{W^{2,1}_{m_2}(Q(z_0,\tau_3))} \le c(n,\tau_3,\tau_3').$$

Then, again, by the embedding theorem, we find

$$\nabla u^{t_0} = \nabla u \in W^{1,0}_{m_3}(Q(z_0,\tau_3);\mathbb{R}^n)$$

provided

$$\frac{1}{m_3} = \frac{1}{m_2} - \frac{1}{n+2}.$$

Now, let us take an arbitrary large number $m > 2$ and fix it. Find α as an unique solution to the equation

$$\frac{1}{m} = \frac{1}{2} - \frac{\alpha}{n+2}.$$

Next, we let $k_0 = [\alpha]+1$, where $[\alpha]$ is the entire part of the number α. And then we determine the number m_{k_0+1} satisfying the identity

$$\frac{1}{m_{k_0+1}} = \frac{1}{2} - \frac{k_0}{n+2}.$$

Obviously, $m_{k_0+1} > m$. Setting

$$\tau_{k+1} = \tau_k - \frac{1}{4}\frac{1}{2^k}, \qquad \tau_1 = 1, \qquad k = 1,2,,,$$

and repeating our previous arguments k_0 times, we conclude that:

$$u^{t_0} \in W^{2,1}_{m_{k_0+1}}(Q(z_0,\tau_{k_0+1});\mathbb{R}^n)$$

and

$$\|u^{t_0}\|_{W^{2,1}_{m_{k_0+1}}(Q(z_0,\tau_{k_0+1}))} \le c(n,m).$$

Thanks to the inequality $\tau_k > 1/2$ for any natural numbers k, we complete the proof of Theorem 2.6. □

PROOF OF THEOREM 2.7 Let us consider the case $n = 3$. The case $n = 2$ is in fact easier. So, we have

$$\partial_t \omega - \Delta\omega = \omega\cdot\nabla u - u\cdot\nabla\omega \equiv f.$$

Take an arbitrary number $m > 2$ and fix it. By Theorem 2.6, the right-hand side has the following property

$$|f| \le c(n)(|\nabla^2 u| + |\nabla u|^2) \in L_m(Q(z_0,2))$$

and the norm of f in $L_m(Q(z_0,2))$ is dominated by a constant depending only on m and being independent of z_0. It remains to apply Lemma 6.7 and complete the proof of Theorem 2.7. □

6.3 Mild Bounded Ancient Solutions

In this section, we assume that $z = 0$ is a singular point. Making use of the space-time shift and the Navier-Stokes scaling, we can reduce the general problem of local regularity to a particular one that in a sense mimics the first time singularity.

Proposition 3.8. *Let v and q be a suitable weak solution to the Navier-Stokes equations in Q and $z = 0$ be a singular point of v. There exist two functions $\widetilde{v}$ and $\widetilde{q}$ having the following properties:*

(i) $\widetilde{v} \in L_3(Q)$ and $\widetilde{q} \in L_{\frac{3}{2}}(Q)$ obey the Navier-Stokes equations in Q in the sense of distributions;

(ii) $\widetilde{v} \in L_\infty(B\times]-1,-a^2[)$ for all $a \in]0,1[$;

(iii) there exists a number $0 < r_1 < 1$ such that $\widetilde{v} \in L_\infty(\{(x,t) : r_1 < |x| < 1, -1 < t < 0\})$.

Moreover, functions $\widetilde{v}$ and $\widetilde{q}$ are obtained from v and q with the help of the space-times shift and the Navier-Stokes scaling and the origin remains to be a singular point of $\widetilde{v}$.

We recall $z = 0$ is a regular point of v if there exists a positive number r such that v is Hölder continuous in the closure $Q(r)$. A point $z = 0$ is a singular point if it is not a regular one.

PROOF Consider now an arbitrary suitable weak solution v and q in Q. Let $S \subset B\times]-1,0]$ be a set of singular points of v. It is closed in Q. As it was shown in [Caffarelli *et al.* (1982)], $\mathcal{P}^1(S) = 0$, where $\mathcal{P}^1$ is the one-dimensional parabolic Hausdorff measure. By assumptions, $S \neq \emptyset$. We can choose number R_1 and R_2 satisfying $0 < R_2 < R_1 < 1$ such that $S \cap \overline{Q(R_1) \setminus Q(R_2)} = \emptyset$ and $S \cap B(R_2)\times]-R_2^2, 0] \neq \emptyset$. We put

$$t_0 = \inf\{t \ : \ (x,t) \in S \cap B(R_2)\times]-R_2^2, 0]\}.$$

Clearly, $(x_0, t_0) \in S$ for some $x_0 \in B(R_2)$. In a sense, t_0 is the instant of time when singularity of our suitable weak solution v and q appears in $Q(R_1)$. Next, the one-dimensional Hausdorff measure of the set

$$S_{t_0} = \{x_* \in B(R_2) \ : \ (x_*, t_0) \text{ is a singular point}\}$$

is zero as well. Therefore, given $x_0 \in S_{t_0}$, we can find sufficiently small $0 < r < \sqrt{R_2^2 + t_0}$ such that $B(x_0, r) \Subset B(R_2)$ and $\partial B(x_0, r) \cap S_{t_0} = \emptyset$. Since the velocity field v is Hölder continuous at regular points, we can ensure that all statements of Proposition 3.8 hold in the parabolic ball $Q(z_0, r)$ with $z_0 = (x_0, t_0)$. We may shift and re-scale our solution if $z_0 \neq 0$ and $r \neq 1$. $\square$

In what follows, it is always deemed that such a replacement of v and q with $\widetilde{v}$ and $\widetilde{q}$ has been already made. Coming back to the original notation, we assume that functions v and q satisfy all the properties listed in Proposition 3.8 and $z = 0$ is a singular point of v.

One of the most powerful methods to study possible singularities is a blowup technique based on the Navier-Stokes scaling

$$u^{(k)}(y,s) = \lambda_k v(x,t), \qquad p^{(k)}(y,s) = \lambda_k^2 q(x,t)$$

with

$$x = x^{(k)} + \lambda_k y, \qquad x = t_k + \lambda_k^2 s,$$

where $x^{(k)} \in \mathbb{R}^3$, $-1 < t_k \leq 0$, and $\lambda_k > 0$ are parameters of the scaling and $\lambda_k \to 0$ as $k \to +\infty$. It is supposed that functions v and q are extended by zero to the whole $\mathbb{R}^3 \times \mathbb{R}$. A particular selection of scaling parameters $x^{(k)}$, t_k, and λ_k depends upon a problem under consideration.

Now, our goal is to describe a universal method that makes it possible to reformulate the local regularity problem as a classical Liouville type problem for the Navier-Stokes equations. To see how things work, let us introduce the function

$$M(t) = \sup_{-1<\tau\leq t} \|v(\cdot,\tau)\|_{\infty,\overline{B}(r)}$$

for some $r \in]r_1, 1[$. It tends to infinity as time t goes to zero from the left since the origin is a singular point of v. Thanks to the obvious properties of the function M, one can choose parameters of the scaling in a particular way letting $\lambda_k = 1/M_k$, where a sequence M_k is defined as

$$M_k = \|v(,t_k)\|_{\infty,\overline{B}(r)} = |v(x^{(k)},t_k)|$$

with $x^{(k)} \in \overline{B}(r_1)$ for sufficiently large k. Before discussing what happens if k tends to infinity, let us introduce a subclass of *bounded ancient (backward)* solutions playing an important role in the regularity theory of the Navier-Stokes equations.

Definition 6.3. A bounded vector field u, defined on $\mathbb{R}^3 \times]-\infty,0[$, is called a mild bounded ancient solution to the Navier-Stokes equation if there exists a function p in $L_\infty(-\infty,0;BMO(\mathbb{R}^3))$ such that u and p satisfy the Navier-Stokes system

$$\partial_t u + \operatorname{div} u \otimes u - \Delta u + \nabla p = 0,$$
$$\operatorname{div} u = 0$$

in $\mathbb{R}^3 \times]-\infty,0[$ in the sense of distributions.

The notion of mild bounded ancient solutions has been introduced in [Koch *et al.* (2009)]. It has been proved there that u has continuous derivatives of any order in both spatial and time variables. Actually, the definition accepted here is different but equivalent to the one given in [Koch *et al.* (2009)]. We follow [Seregin and Šverák (2009)].

Our first observation is that all mild bounded ancient solutions are very smooth.

Proposition 3.9. *Let u be an arbitrary mild bounded ancient solution. The u is of class C^∞ and moreover*

$$\sup_{(x,t)\in Q_-}\big(|\partial_t^k\nabla^l u(x,t)|+|\partial_t^k\nabla^{l+1}p(x,t)|\big)+$$

$$+\|\partial_t^k p\|_{L_\infty(BMO)}\leq C(k,l,\|p\|_{L_\infty(BMO)})<\infty$$

for any $k,l=0,1,\ldots$.

PROOF STEP 1 Let us show that

$$\nabla u\in\mathcal{L}_2(Q_-). \tag{6.3.1}$$

Here, $Q_-=\mathbb{R}^3\times]-\infty,0[$ and

$$\mathcal{L}_2(Q_-)=L_{2,\mathrm{unif}}(Q_-):=\{\|f\|_{L_{2,\mathrm{loc}}(Q_-)}:=\sup_{z_0\in Q_-}\|f\|_{L_2(Q(z_0,1))}<\infty\}.$$

Using a standard mollification kernel ω_ϱ, let us introduce

$$f_\varrho(z)=\int\limits_{\mathbb{R}^n}\omega_\varrho(z-z')f(z')dz',$$

where $z=(x,t)$. Then from Definition 6.3, it follows that

$$\partial_t u_\varrho+\mathrm{div}\,(u\otimes u)_\varrho-\Delta u_\varrho=-\nabla p_\varrho,\qquad \mathrm{div}\,u_\varrho=0. \tag{6.3.2}$$

Let us test the first equation in (6.3.2) with φu_ϱ. Then after integration by parts, we have

$$\int\limits_{\mathbb{R}^n}\varphi(x,t_0)|u_\varrho(x,t_0)|^2dx+2\int\limits_{t_1}^{t_0}\int\limits_{\mathbb{R}^n}\varphi|\nabla u_\varrho|^2dxdt=$$

$$=\int\limits_{\mathbb{R}^n}\varphi(x,t_1)|u_\varrho(x,t_1)|^2dx+\int\limits_{t_1}^{t_0}\int\limits_{\mathbb{R}^n}|u_\varrho|^2(\Delta\varphi+\partial_t\varphi)dxdt+$$

$$+2\int\limits_{t_1}^{t_0}\int\limits_{\mathbb{R}^n}(\varphi(u\otimes u)_\varrho:\nabla u_\varrho+(u\otimes u)_\varrho:u_\varrho\otimes\nabla\varphi+(p_\varrho-a(t))u_\varrho\cdot\nabla\varphi)dxdt$$

with an arbitrary function $a=a(t)$. Choosing an appropriate non-negative cut-off function φ, we can deduce from the above identity

$$\sup_{z_0\in Q_-}\|\nabla u_\varrho\|_{2,Q(z_0,1)}\leq c<\infty$$

with a constant c independent of ϱ. This certainly implies (6.3.1).

STEP 2 Let our non-negative function φ belong to the space $C_0^\infty(\mathbb{R}^n\times\mathbb{R})$. Then from the above identity, we can derive the local energy inequality by passing to the limit as $\varrho\to 0$:

$$\int\limits_{\mathbb{R}^n}\varphi(x,t_0)|u(x,t_0)|^2dx+2\int\limits_{-\infty}^{t_0}\int\limits_{\mathbb{R}^n}\varphi|\nabla u|^2dxdt\leq$$

$$\leq\int\limits_{-\infty}^{t_0}\int\limits_{\mathbb{R}^n}|u|^2(\Delta\varphi+\partial_t\varphi)dxdt+\int\limits_{-\infty}^{t_0}\int\limits_{\mathbb{R}^n}(|u|^2+2p)u\cdot\nabla\varphi dxdt.$$

This makes it possible to apply ε-regularity theory to the mild bounded ancient solution u. Indeed, restricting ourselves to the case $n=3$, we have

$$\frac{1}{R^2}\int\limits_{Q(z_0,R)}(|u|^3+|p-[p]_{B(x_0,R)}|^{\frac{3}{2}})dz\leq cR^3<\varepsilon,$$

where ε is an absolute constant and c depends on $\|p\|_{L_\infty(BMO)}$ only. So, a number R for which the above inequality is satisfied depends on the same norm only. Then, there are positive constants c_k with $k=1,2,...$ such that

$$|\nabla^k u(z_0)|\leq\frac{c_k}{R^{k+1}}$$

for any $z_0\in Q_-$.

Estimates for the pressure are coming from the pressure equation:

$$\Delta p=-u_{i,j}u_{j,i}.$$

Local regularity theory gives us:

$$\int\limits_{B(x_0,1)}|\nabla^k p|^2dx\leq c(k)\Big[\int\limits_{B(x_0,2)}|\nabla^k(u\otimes u)|^2dx+\int\limits_{B(x_0,2)}|p-[p]_{B(x_0,2)}|^2dx\Big]$$

for any $k=1,2,...$, and thus

$$|\nabla^k p(z_0)|\leq C(k,\|p\|_{L_\infty(BMO)})$$

for any $z_0 \in Q_-$ and for any $k = 1, 2, \ldots$.

STEP 3. Now, we wish to estimate derivative in time. Directly, from the equations and the above estimates, we deduce that

$$|\nabla^k \partial_t u(z_0)| \leq C(k, \|p\|_{L_\infty(BMO)})$$

for any $z_0 \in Q_-$ and for any $k = 0, 1, \ldots$. To get higher derivatives of p in t, we should estimate $\partial_t p$. To achieve this goal, let us use the pressure equations

$$\Delta \partial_t p = -\mathrm{div}\,\mathrm{div}\,(\partial_t u \otimes u + u \otimes \partial_t u).$$

This equation leads to the estimate

$$\|\partial_t p\|_{L_\infty(BMO)} \leq c(\|p\|_{L_\infty(BMO)}).$$

Repeating the same arguments as in Step 2, we establish

$$|\nabla^k \partial_t p(z_0)| \leq C(k, \|p\|_{L_\infty(BMO)})$$

for any $z_0 \in Q_-$ and for any $k = 1, 2, \ldots$. In turn, from the equation, we find that

$$|\nabla^k \partial_t^2 u(z_0)| \leq C(k, \|p\|_{L_\infty(BMO)})$$

for any $z_0 \in Q_-$ and for any $k = 0, 1, \ldots$. Then we again use the pressure equation to estimate first $L_\infty(BMO)$-norm of $\partial_t^2 p$ and afterwards L_∞-norm of $\nabla^k \partial_t^2 p$ with $k = 1, 2, \ldots$ And so on. $\Box$

The statement below proved in [Seregin and Šverák (2009)] shows how mild bounded ancient solutions occur in the regularity theory of the Navier-Stokes equations.

Proposition 3.10. *There exist a subsequence of $u^{(k)}$ (still denoted by $u^{(k)}$) and a mild bounded ancient solution u such that, for any $a > 0$, the sequence $u^{(k)}$ converges uniformly to u on the closure of the set $Q(a) = B(a) \times]-a^2, 0[$. The function u has the additional properties: $|u| \leq 1$ in $\mathbb{R}^3 \times]-\infty, 0[$ and $|u(0)| = 1$.*

PROOF OF PROPOSITION 3.10 Our solution v and q has good properties inside $Q_1 = B_1 \times]-1, 0[$ with $B_1 = \{r_1 < |x| < 1\}$. Let us list them. Let $Q_2 = B_2 \times]-\tau_2^2, 0[$, where $0 < \tau_2 < 1$, $B_2 = \{r_1 < r_2 < |x| < a_2 < 1\}$. Then, for any natural k,

$$z = (x, t) \mapsto \nabla^k v(z) \text{ is Hölder continuous in } \overline{Q}_2;$$

$$q \in L_{\frac{3}{2}}(-\tau_2^2, 0; C^k(\overline{B}_2)).$$

The corresponding norms are estimated by constants depending on $\|v\|_{3,Q}$, $\|q\|_{\frac{3}{2},Q}$, $\|v\|_{\infty,Q_1}$, and numbers k, r_1, r_2, a_2, τ_2. In particular, we have

$$\max_{x\in\overline{B}_2}\int\limits_{-\tau_2^2}^{0}|\nabla q(x,t)|^{\frac{3}{2}}dt \leq c_1 < \infty. \tag{6.3.3}$$

Proof of the first statement can be done by induction and found in [Escauriaza *et al.* (2003)], [Ladyzhenskaya and Seregin (1999)], and [Necas *et al.* (1996)]. The second statement follows directly from the first one and the pressure equation: $\Delta q = -v_{i,j}v_{j,i}$.

Now, let us decompose the pressure $q = q_1 + q_2$. For q_1, we have

$$\Delta q_1(x,t) = -\operatorname{div}\operatorname{div}\Big[\chi_B(x)v(x,t)\otimes v(x,t)\Big], \quad x\in\mathbb{R}^3, \quad -1<\tau<0,$$

where $\chi_B(x) = 1$ if $x \in B$ and $\chi_B(x) = 0$ if $x \notin B$. Obviously, the estimate

$$\int\limits_{-1}^{0}\int\limits_{\mathbb{R}^3}|q_1(x,t)|^{\frac{3}{2}}dxdt \leq c\int\limits_{Q}|v|^3dz$$

holds and it is a starting point for local regularity of q_1. Using differentiability properties of v, we can show

$$\max_{x\in\overline{B}_3}\int\limits_{-\tau_2^2}^{0}|\nabla q_1(x,t)|^{\frac{3}{2}}dt \leq c_2 < \infty, \tag{6.3.4}$$

where $B_3 = \{r_2 < r_3 < |x| < a_3 < a_2\}$. From (6.3.3) and (6.3.4), it follows that

$$\max_{x\in\overline{B}_3}\int\limits_{-\tau_2^2}^{0}|\nabla q_2(x,t)|^{\frac{3}{2}}dt \leq c_3 < \infty. \tag{6.3.5}$$

However, q_2 is a harmonic function in B, and thus, by the maximum principle, we have

$$\max_{x\in\overline{B}(r_4)}\int\limits_{-\tau_2^2}^{0}|\nabla q_2(x,t)|^{\frac{3}{2}}dt \leq c_3 < \infty, \tag{6.3.6}$$

where $r_4 = (r_3 + a_3)/2$.

Let us re-scale each part of the pressure separately, i.e.,

$$p_i^k(y,s) = \lambda_k^2 q_i(x,t), \quad i = 1,2,$$

so that $p^k = p_1^k + p_2^k$. As it follows from (6.3.6), for p_2^k, we have

$$\sup_{y\in B(-x^k/\lambda_k, r_4/\lambda_k)} \int\limits_{-(\tau_2^2-t_k)/\lambda_k^2}^{0} |\nabla_y p_2^k(y,s)|^{\frac{3}{2}} ds \leq c_3 \lambda_k^{\frac{5}{2}}. \qquad (6.3.7)$$

The first component of the pressure satisfies the equation

$$\Delta_y p_1^k(y,s) = -\mathrm{div}_y \mathrm{div}_y (\chi_{B(-x^k/\lambda_k, 1/\lambda_k)}(y) u^{(k)}(y,s) \otimes u^{(k)}(y,s)), \qquad y \in \mathbb{R}^3,$$

for all possible values of s. For such a function, we have the standard estimate

$$\|p_1^k(\cdot,s)\|_{BMO(\mathbb{R}^3)} \leq c \qquad (6.3.8)$$

for all $s \in]-(1-t_k)/\lambda_k^2, 0[$. It is valid since $|u^{(k)}| \leq 1$ in $B(-x^k/\lambda_k, 1/\lambda_k) \times]-(1-t_k)/\lambda_k^2, 0[$.

We slightly change p_1^k and p_2^k setting

$$\overline{p}_1^k(y,s) = p_1^k(y,s) - [p_1^k]_{B(1)}(s) \qquad \overline{p}_2^k(y,s) = p_2^k(y,s) - [p_2^k]_{B(1)}(s)$$

so that $[\overline{p}_1^k]_{B(1)}(s) = 0$ and $[\overline{p}_2^k]_{B(1)}(s) = 0$.

Now, we pick up an arbitrary positive number a and fix it. Then from (6.3.7) and (6.3.8) it follows that for sufficiently large k we have

$$\int\limits_{Q(a)} |\overline{p}_1^k|^{\frac{3}{2}} de + \int\limits_{Q(a)} |\overline{p}_2^k|^{\frac{3}{2}} de \leq c_4(c_2, c_3, a).$$

Using the same bootstrap arguments, we can show that the following estimate is valid:

$$\|u^{(k)}\|_{C^\alpha(\overline{Q}(a/2)} \leq c_5(c_2, c_3, c_4, a)$$

for some positive number $\alpha < 1/3$. Indeed, the norm $\|u^{(k)}\|_{C^\alpha(\overline{Q}(a/2))}$ is estimated with the help of norms $\|u^{(k)}\|_{L_\infty(Q(a)))}$ and $\|\overline{p}^k\|_{L_{\frac{3}{2}}(Q(a))}$, where $\overline{p}^k = \overline{p}_1^k + \overline{p}_2^k$. Hence, using the diagonal Cantor procedure, we can select subsequences such that for some positive α and for any positive a

$$u^{(k)} \to u \qquad \text{in } C^\alpha(Q(a)),$$

$$\overline{p}_1^k \rightharpoonup \overline{p}_1, \qquad \text{in } L_{\frac{3}{2}}(Q(a)), \qquad [\overline{p}_1]_{B(1)}(s) = 0,$$

$$\overline{p}_2^k \rightharpoonup \overline{p}_2 \qquad \text{in } L_{\frac{3}{2}}(Q(a)), \qquad [\overline{p}_2]_{B(1)}(s) = 0.$$

So, $|u| \leq 1$ in Q_- and u and $\overline{p} = \overline{p}_1 + \overline{p}_2$ satisfy the Navier-Stokes system in Q_- in the sense of distributions. Moreover, at it is follows from (6.3.8), $\overline{p}_1 \in L_\infty(-\infty, 0; BMO(\mathbb{R}^3))$.

Next, for sufficiently large k, we get from (6.3.7) that

$$\int\limits_{Q(a)} |\nabla p_2^k(y,s)|^{\frac{3}{2}} ds \leq c_3 \lambda_k^{\frac{5}{2}}.$$

Hence, $\nabla p_2 = 0$ in $Q(a)$ for any $a > 0$. So, $\overline{p}_2(y,s)$ is identically zero. This allows us to conclude that the pair u and $\overline{p}_1$ is a solution to the Navier-Stokes equations in the sense of distributions and thus u is a nontrivial mild bounded ancient solution satisfying the condition $|u(0,0)| = 1$ and the estimate $|u| \leq 1$ in Q_-. □

It is worthy to notice that the trivial bounded ancient solution of the form

$$u(x,t) = c(t), \qquad p(x,t) = -c'(t) \cdot x,$$

with arbitrary bounded function $c(t)$, is going to be a mild bounded ancient solution if and only if $c(t) \equiv constant$. This allows us to make the following plausible conjecture, see [Seregin and Šverák (2009)].

Conjecture *Any mild bounded ancient solution is a constant.*

To explain what consequences of the conjecture could be for regularity theory of the Navier-Stokes equations, let us formulate a question which can be raised in connection with the ε-regularity theory: *what happens if we drop the condition on smallness of scale-invariant quantities, assuming their uniform boundedness only, i.e,* $\sup_{0<r<1} F(v,r) < +\infty$. For Ladyzhenskaya-Prodi-Serrin type quantities with $s > 3$, the answer is still positive, i.e., $z = 0$ is a regular point. It follows from scale-invariance and the fact that the assumption $M_{s,l}(v;1) = \sup_{0<r<1} M_{s,l}(v;r) < +\infty$ implies $M_{s,l}(v;r) \to 0$ as $r \to 0$ if $s > 3$. Although in the marginal case $s = 3$ and $l = +\infty$, the answer remains positive, the known proof is more complicated and will be outlined later.

Let us recall certain definitions and make some general remarks about relationships between some scale-invariant quantities. Boundedness of

$$\sup_{0<r<1} G_2(v;r) = G_2(v,1) = G_{20} < +\infty$$

can be rewritten in the form

$$|v(z)| \leq \frac{G_{20}}{\sqrt{-t}}$$

for all $z = (x,t) \in Q$. If v satisfies the above inequality and $z = 0$ is still a singular point of v, we say that a *singularity of Type I* or *Type I blowup* takes place at $t = 0$. All other singularities are of Type II. The main feature of Type I singularities is that they have the same rate as potential self-similar solutions. The important properties connected with possible singularities of Type I have been proved in [Seregin (2007)], [Seregin and Zajaczkowski (2006)], and [Seregin and Šverák (2009)] and are as follows.

Proposition 3.11. *Let functions v and q be a suitable weak solution to the Navier-Stokes equations in Q.*

(i) If $\min\{G_1(v;1), G_2(v;1)\} < +\infty$, *then*

$$g = \sup_{0<r<1}\{A(v;r) + C(v;r) + D(q;r) + E(v;r)\} < +\infty.$$

(ii) If

$$g' = \min\{\sup_{0<r<1} A(v;r), \sup_{0<r<1} C(v;r), \sup_{0<r<1} E(v;r)\} < +\infty,$$

then $g < +\infty$.

This proposition admits many obvious generalizations.

If we assume that v possesses uniformly bounded energy scale-invariant quantities, then, by Proposition 3.10, the same type of quantities will be bounded for the ancient solution, which is not trivial if $z = 0$ is a singular point of v. However, by the conjecture, the above ancient solution must be zero. So, the origin $z = 0$ cannot be a singular point of v. This would be a positive answer to the question formulated above. In particular, according to Proposition 3.11, validity of the conjecture would rule out Type I blowups.

6.4 Liouville Type Theorems

6.4.1 *LPS Quantities*

Theorem 4.12. *Let u be a mild bounded ancient solution to the Navier-Stokes equations, i.e., $u \in L_\infty(Q_-)$ is divergence free and satisfies the identity*

$$\int\limits_{Q_-} (u\cdot\partial_t w + u\otimes u : \nabla w + u\cdot\Delta w)dz = 0 \qquad (6.4.9)$$

for any divergence free function w from $C_0^\infty(Q_-)$. Assume that

$$\sup_{0<r<\infty} M_{s,l}(u;r) = \int\limits_{-\infty}^{0}\Big(\int\limits_{\mathbb{R}^3}|u(x,t)|^s dx\Big)^{\frac{l}{s}} < \infty$$

with $3/s + 2/l = 1$ and $l < \infty$. Then $u \equiv 0$ in Q_-.

PROOF Let us consider the simplest case of the regular LPS quantity $M_{5,5}$. By the pressure equation, we may assume

$$\int\limits_{\infty}^{0}\int\limits_{\mathbb{R}^3}(|u|^5+|p|^{\frac{5}{2}})dxdt<+\infty.$$

Given $\varepsilon>0$, we can find $T<0$ such that

$$\int\limits_{\infty}^{T}\int\limits_{\mathbb{R}^3}(|u|^5+|p|^{\frac{5}{2}})dxdt<\varepsilon.$$

Then, by Hölder inequality, we have

$$\frac{1}{R^2}\int\limits_{t_0-R^2}^{t_0}\int\limits_{B(x_0,R)}(|u|^3+|p|^{\frac{3}{2}})dxdt<c\varepsilon^{\frac{3}{5}}$$

for any $x_0\in\mathbb{R}^3$, any $R>0$, and any $t_0\leq T$ with some universal constant c. In turn, the ε-regularity theory ensures the inequality

$$|u(x_0,t_0)|<\frac{c}{R}$$

with another universal constant c. Tending $R\to\infty$, we get $u(\cdot,t)=0$ as $t\leq T$. One can repeat more or less the same arguments in order to show that in fact u is identically zero on $\mathbb{R}^3\times]-\infty,0]$.

6.4.2 *2D case*

In two-dimensional case, we have the following Liouville type theorem.

Theorem 4.13. *Assume that $n=2$ and u is an arbitrary bounded ancient solution. Then $u(x,t)=b(t)$ for any $x\in\mathbb{R}^2$.*

To prove the above statement, we start with an auxiliary lemma.

Lemma 6.8. *Let functions*

$$\omega\in\mathcal{W}^{2,1}_m(Q_-)=\{u\in W^{2,1}_{m,\mathrm{loc}}(Q_-):\ \sup_{z_0\in Q_-}\|u\|_{W^{2,1}_m(Q(z_0,1))}<\infty\},$$

with $m>3$, and $u\in L_\infty(Q_-)$ satisfy the equation

$$\partial_t\omega+u\cdot\nabla\omega-\Delta\omega=0\qquad in\quad Q_-$$

and the inequality

$$|u|\leq 1\qquad in\quad Q_-.$$

Then, for any positive numbers ε *and* R, *there exists a point* $z_0 = (x_0, t_0)$, $x_0 \in \mathbb{R}^2$ *and* $t_0 \leq 0$, *such that*

$$\omega(z) \geq M - \varepsilon, \qquad z \in Q(z_0, R),$$

where $M = \sup\limits_{z \in Q_-} \omega(z)$.

Remark 6.7. By the embedding theorem, $M < +\infty$.

In order to prove Lemma 6.8, we need a strong maximum principle. Here, it is.

Theorem 4.14. STRONG MAXIMUM PRINCIPLE *Let functions* $w \in W^{2,1}_m(Q(z_0, R))$ *with* $m > n+1$ *and* $a \in L_\infty(Q(z_0, R); \mathbb{R}^n)$ *satisfy the equation*

$$\partial_t w + a \cdot \nabla w - \Delta w = 0 \qquad in \quad Q(z_0, R).$$

Let, in addition,

$$w(z_0) = \sup_{z \in Q(z_0, R)} w(z).$$

Then

$$w(z) = w(z_0) \qquad in \quad Q(z_0, R).$$

PROOF OF LEMMA 6.8 ([Koch *et al.* (2009)]) In fact, we shall prove even a stronger result. Let z_k be a sequence of points in Q_- such that

$$\omega(z_k) \to M.$$

We state that

$$\inf_{z \in Q(z_k, R)} \omega(z) \to M.$$

Indeed, assume that this statement is false. Then, we can find a number $\varepsilon > 0$ and a sequence of points $z_k' \in Q(z_k, R)$ such that

$$\omega(z_k') \leq M - \varepsilon.$$

Now, let us consider shifted functions

$$\omega^k(x, t) = \omega(x_k + x, t_k + t), \qquad u^k(x, t) = u(x_k + x, t_k + t),$$

for $z = (x, t) \in Q(R)$. By the definition of the space $\mathcal{W}^{2,1}_m(Q_-)$, these new functions are subject to the estimates

$$\|\omega^k\|_{W^{2,1}_m(Q(R))} \leq c,$$

$$|u^k| \leq 1 \qquad \text{in} \quad Q(R),$$

with a constant c that is independent of k. Moreover, we have

$$\partial_t \omega^k + u^k \cdot \nabla \omega^k - \Delta \omega^k = 0 \qquad \text{in} \quad Q(R),$$

$$\omega^k(z''_k) \leq M - \varepsilon, \qquad z''_k \in Q(R),$$

$$\omega^k(z) \leq M \qquad z \in Q(R).$$

Using standard compactness arguments, we show

$$\omega^k \rightharpoonup \overline{\omega} \qquad \text{in} \quad W^{2,1}_m(Q(R)),$$

$$u^k \overset{\star}{\rightharpoonup} \overline{u} \qquad \text{in} \quad L_\infty(Q(R);\mathbb{R}^2),$$

$$\omega^k \to \overline{\omega} \qquad \text{in} \quad C(\overline{Q}(R)),$$

$$\overline{\omega}(z) \leq \overline{\omega}(0) = M \qquad z \in Q(R), \tag{6.4.10}$$

$$\overline{\omega}(z_*) \leq M - \varepsilon, \tag{6.4.11}$$

where $z_* \in \overline{Q}(R)$. Clearly, $\overline{\omega} \in W^{2,1}_m(Q(R))$ and

$$\partial_t \overline{\omega} + \overline{u} \cdot \nabla \overline{\omega} - \Delta \overline{\omega} = 0 \qquad \text{in} \quad Q(R).$$

By (6.4.10) and by the above strong maximum principle,

$$\overline{\omega}(z) = M \qquad z \in \overline{Q}(R),$$

which is in a contradiction with (6.4.11). □

Proof of Theorem 4.13 We are going to apply Lemma 6.8 to the vorticity equation. Let us show first that

$$\sup_{z \in Q_-} \omega(z) = M \leq 0.$$

To this end, assume that the latter statement is wrong and in fact

$$M > 0.$$

Take a cut-off function $\varphi \in C^\infty_0(B(R))$ with the following properties:

$$0 \leq \varphi \leq 1, \qquad |\nabla \varphi| \leq c/R \qquad \text{in} \quad B(R),$$

$$\varphi \equiv 1 \qquad \text{in} \quad B(R/2).$$

By Lemma 6.8, for an arbitrary number $R > 0$, there exists a point $z_{0R} = (x_{0R}, t_{0R})$ with $t_{0R} \leq 0$ such that

$$|\omega(z)| \geq M - M/2 = M/2 > 0, \qquad z \in Q(z_{0R}, R).$$

If we let $\varphi_{x_{0R}}(x) = \varphi(x - x_{0R})$, then

$$A(R) = \int\limits_{Q(z_{0R},R)} \varphi_{x_{0R}}(x)\omega(z)dz \geq \frac{M}{2}R^2|B(x_{0R},R)| = \frac{M}{2}\pi R^4. \quad (6.4.12)$$

On the other hand, since $\omega = u_{2,1} - u_{1,2}$, we have after integration by parts

$$A(R) = \int\limits_{Q(z_{0R},R)} (\varphi_{x_{0R},2}u_1 - \varphi_{x_{0R},1}u_2)dz \leq$$

$$\leq cR^3,$$

where c is a universal constant. The latter inequality contradicts (6.4.12) for sufficiently large R. So, $M \leq 0$. In the same way, one can show that $m \geq 0$, where

$$-\infty < m = \inf_{z \in Q_-} \omega(z).$$

So, $\omega \equiv 0$ in Q_-. Since $u(\cdot, t)$ is a divergence free function in $\mathbb{R}^2$, we can state that $u(\cdot, t)$ is a bounded harmonic function in $\mathbb{R}^2$. Therefore, $u(x,t) = b(t)$, $x \in \mathbb{R}^2$. Theorem 4.13 is proved.

6.4.3 *Axially Symmetric Case with No Swirl*

In the case of axial symmetry, it is convenient to introduce the cylindrical coordinates ϱ, φ, x_3 so that $x_1 = \varrho\cos\varphi$, $x_2 = \varrho\sin\varphi$, $x_3 = x_3$. The velocity components are going to be u_ϱ, u_φ, u_3. By the definition of axial symmetry,

$$u_{\varrho,\varphi} = \partial u_\varrho/\partial\varphi = 0, \quad u_{\varphi,\varphi} = 0, \quad u_{3,\varphi} = 0, \quad p_{,\varphi} = 0.$$

For the vorticity components, we have simple formulae

$$\omega_\varrho = u_{\varphi,3}, \quad \omega_\varphi = u_{\varrho,3} - u_{3,\varrho}, \quad \omega_3 = u_{\varphi,\varrho} + u_\varphi/\varrho.$$

Now, assume that vector field u is an arbitrary axially symmetric bounded ancient solution with zero swirl, i.e., $u_\varphi = 0$. This, in particular, leads to the representation

$$\nabla\omega = -\frac{1}{\varrho}\omega_\varphi e_\varrho \otimes e_\varphi + \omega_{\varphi,\varrho}e_\varphi \otimes e_\varrho + \omega_{\varphi,3}e_\varphi \otimes e_3.$$

We know that $|\nabla\omega|$ is a bounded function, which implies boundedness of functions $\omega_{\varphi,\varrho}$, $\omega_{\varphi,3}$, and $\frac{1}{\varrho}\omega_\varphi$. Regarding $\nabla^2\omega$, we can state

$$\int\limits_{-T}^{0}\int\limits_{\mathcal{C}(a)} \Big[|\omega_{\phi,\varrho\varrho}|^2 + 2|\omega_{\varphi,\varrho\varphi}|^2 + |\omega_{\varphi,33}|^2 + 2|(\omega_\varphi/\varrho)_{,\varrho}|^2 + 2|(\omega_\varphi/\varrho)_{,3}|^2\Big]^{\frac{m}{2}} \varrho d\varrho d\varphi \leq$$

$$\leq c(a,T,p) < +\infty,$$

for any $a > 0$, for any $T > 0$, and for any $m > 1$. Here,

$$\mathcal{C}(a) = \{x = (x', x_3) \in \mathbb{R}^3 : \ |x'| < a, \quad |x_3| < a\}$$

and $x' = (x_1, x_2)$ so that $|x'| = \varrho$.

Theorem 4.15. *Let u be an arbitrary axially symmetric bounded ancient solution with zero swirl. Then $u(x,t) = b(t)$ for any $x \in \mathbb{R}^3$ and for any $t \leq 0$. Moreover, $u_1(x,t) = 0$ and $u_2(x,t) = 0$ for the same x and t or, equivalently, $u_\varrho(\varrho, x_3, t) = 0$ for any $\varrho > 0$, for any $x_3 \in \mathbb{R}$, and for any $t \leq 0$.*

PROOF We let $\eta = \omega_\varphi/\varrho$. It is not difficult to verify that η satisfies the equation

$$\partial_t \eta + u_\varrho \eta_{,\varrho} + u_3 \eta_{,3} - (\Delta \eta + \frac{2}{\varrho}\eta_{,\varrho}) = 0, \quad \varrho > 0, -\infty < x_3 < +\infty, t < 0,$$

where

$$\Delta \eta = \frac{1}{\varrho}(\varrho \eta_{,\varrho})_{,\varrho} + \eta_{,33} = \eta_{,\varrho\varrho} + \eta_{,33} + \frac{1}{\varrho}\eta_{,\varrho}.$$

Let us make the change of variables

$$y = (y', y_5) \in \mathbb{R}^5, \qquad y' = (y_1, y_2, y_3, y_4),$$

$$\varrho = |y'| = \sqrt{y_1^2 + y_2^2 + y_3^2 + y_4^2}, \qquad y_5 = x_3.$$

Then after simple calculations, we see that a new function

$$f(y,t) = f(y_1, y_2, y_3, y_4, y_5, t) = \eta(\varrho, \varphi, t)$$

obeys the equation

$$\partial_t f + U \cdot \nabla_5 f - \Delta_5 f = 0 \tag{6.4.13}$$

in $Q^5_- = \mathbb{R}^5 \times]-\infty, 0[$. Here, ∇_5 and Δ_5 are usual nabla and Laplacian operators with respect to the Cartesian coordinates in $\mathbb{R}^5$ and

$$U(y,t) = (U_1(y,t), U_2(y,t), U_3(y,t), U_4(y,t), U_5(y,t))$$

with

$$U_i(y,t) = \frac{u_\varrho(\varrho, x_3, t)}{\varrho} y_i, \quad i = 1,2,3,4, \qquad U_5(y,t) = u_3(\varrho, x_3,).$$

Obviously, the function U is bounded in Q^5_-. However, previous arguments show that $\nabla_5 U$ is a bounded function as well. Indeed,we have

$$|\nabla_5 U(y,t)| \leq c(|\nabla u(x,t)| + |u_\varrho(\varrho, x_3, t)|/\varrho) \leq c|\nabla u(x,t)| \leq c < +\infty$$

for any $y \in \mathbb{R}^5$ and any $t \leq 0$. So,

$$|U|,\, |\nabla_5 U| \in L_\infty(Q^5_-).$$

For bounded f, weak solution to (6.4.13) can be defined as follows

$$\int\limits_{Q^5_-} \Big[f\partial_t g + fU\cdot\nabla_5 g + fg\,\mathrm{div}_5 U + f\Delta_5 g \Big]\,dy\,dt = 0$$

for any $g \in C_0^\infty(Q^5_-)$. In the way, explained in the previous section, one can show that, for any $m > 1$,

$$f \in W^{2,1}_m(Q^5_-)$$

and the norm can be dominated by a positive constant, depending on m, $\sup_{Q^5_-}|f|$, and $\sup_{Q^5_-}(|U| + |\nabla_5 U|)$ only.

We let

$$M = \sup_{y\in\mathbb{R}^5}\sup_{t\leq 0} f(y,t) = \sup_{x\in\mathbb{R}^3}\sup_{t\leq 0}\eta(|x'|, x_3, t).$$

Our goal is to show that $M \leq 0$. Assume it is not so, i.e., $M > 0$. Now, let us apply Lemma 6.8 in our five-dimensional setting. Then, for any $R > 0$, there exists a point y_R in $\mathbb{R}^5$ and a moment of time $t_R \leq 0$ such that

$$f(y,t) \geq M/2, \qquad (y,t) \in Q((y_R, t_R), R) = B(y_R, R)\times]t_R - R^2, t_R[,$$

where $B(y_R, R) = \{|y - y_R| < R\}$.

By our assumptions,

$$0 < M_0 = \sup_{x\in\mathbb{R}^3,\, t\leq 0} \omega_\varphi(|x'|, x_3, t) < +\infty.$$

We may choose a number R so big as

$$R > 100\frac{2M_0}{M}$$

and then let

$$y_* = (y'_*, y_{5R}), \qquad y'_* = 50\frac{2M_0}{M}l + y'_R,$$

where

$$l \in \mathbb{R}^4, \qquad |l| = 1, \qquad (l, y'_R) = l_1 y_{1R} + l_2 y_{2R} + l_3 y_{3R} + l_4 y_{4R} = 0.$$

It is not difficult to check that $y_* \in B(y_R, R)$ and, moreover,

$$|y'_*| \geq 50\frac{2M_0}{M}.$$

Then we find

$$\frac{M}{2} \leq f(y_*, t_R - R^2/2) \leq \frac{M_0}{|y'_*|} < \frac{M_0}{50\frac{2M_0}{M}} = \frac{M}{100}.$$

This means that in fact $M \leq 0$. In the same way, one can show that $m \geq 0$ and then conclude that $f \equiv 0$ in Q^5_-, which in turn implies

$$\omega_\varphi(|x'|, x_3, t) = 0, \qquad \forall (x,t) \in Q_-,$$

and therefore

$$\omega \equiv 0 \quad \text{in} \quad Q_-.$$

The rest of the proof is the same as in Theorem 4.13.

6.4.4 *Axially Symmetric Case*

We are going to prove the following statement.

Theorem 4.16. *Let u be an arbitrary axially symmetric bounded ancient solution satisfying assumption*

$$|u(x,t)| \leq \frac{A}{|x'|}, \qquad x = (x', x_3) \in \mathbb{R}^3,\ -\infty < t \leq 0, \tag{6.4.14}$$

where A is a positive constant independent of x and t. Then $u \equiv 0$ in Q_-.

Proof Let us explain our strategy. First, we are going to show that, under condition (6.4.14), the swirl is zero, i.e., $u_\varphi = 0$. Then we apply Theorem 4.15 and state that $u(x,t) = b(t)$. But condition (6.4.14) says $b(t) = 0$ for all $t \leq 0$. So, our aim now is to show that $u_\varphi \equiv 0$ in Q_-.

Let us introduce the additional notation:

$$\widetilde{\mathbb{R}} = \mathbb{R}_+ \times \mathbb{R}, \qquad \mathbb{R}_+ = \{\varrho \in \mathbb{R}, \quad \varrho > 0\}, \qquad \widetilde{Q}_- = \widetilde{\mathbb{R}} \times]-\infty, 0[,$$

$$\Pi(\varrho_1, \varrho_2; h_1, h_2) = \{\varrho_1 < \varrho < \varrho_2, \quad h_1 < x_3 < h_2\},$$

$$\widetilde{Q}(\varrho_1, \varrho_2; h_1, h_2; t_1, t_2) = \Pi(\varrho_1, \varrho_2; h_1, h_2) \times]t_1, t_2[.$$

Now, our aim is to show that

$$M = \sup_{\widetilde{Q}_-} \varrho u_\varphi \leq 0.$$

Assume that it is false, i.e., $M > 0$, and let

$$g = \varrho u_\varphi / M.$$

The new scaled function g satisfies the equation

$$\partial_t g + u_\varrho g_{,\varrho} + u_3 g_{,3} - (\Delta g - 2 g_{,\varrho}/\varrho) = 0 \qquad \text{in} \quad \widetilde{Q}_-.$$

By the assumptions,

$$\sup_{\widetilde{Q}_-} g = 1, \qquad \sqrt{u_\varrho^2 + u_3^2} \leq A/\varrho, \quad |g| \leq A/M \quad \text{in} \quad \widetilde{Q}_- \tag{6.4.15}$$

and

$$(\varrho u_\varrho)_{,\varrho} + (\varrho u_3)_{,3} = 0 \qquad \text{in} \quad \widetilde{Q}_-.$$

To formulate the lemma below, we abbreviate

$$\Pi = \Pi(\varrho_1, \varrho_2; h_1, h_2), \qquad \widetilde{Q} = \Pi \times]\bar{t} - t_1, \bar{t}[.$$

Lemma 6.9. *For any $\varepsilon > 0$, there exists a positive number*

$$\delta = \delta(\Pi, \bar{t}, t_1, A, M, \varepsilon) \leq \varepsilon$$

such that if

$$\sup_{x \in \Pi} g(x, \bar{t}) > 1 - \delta,$$

then

$$\inf_{z \in \widetilde{Q}} g(z) > 1 - \varepsilon.$$

PROOF If we assume that the statement of the lemma is false, then there must exist a number $\varepsilon_0 > 0$ such that, for any natural k, one can find sequences with the following properties:

$$\delta_k > \delta_{k+1}, \quad \delta_k \to 0, \quad \sup_{x \in \Pi} g^k(x, \bar{t}) > 1 - \delta_k, \quad \inf_{z \in \widetilde{Q}} g^k(z) \leq 1 - \varepsilon_0, \quad (6.4.16)$$

functions u^k and g^k satisfy the equations

$$(\varrho u^k_\varrho)_{,\varrho} + (\varrho u^k_3)_{,3} = 0, \qquad \partial_t g^k + u^k_\varrho g^k_{,\varrho} + u^k_3 g^k_{,3} - (\Delta g^k - 2g^k_{,\varrho}/\varrho) = 0$$

in $\widetilde{Q}_-$ and the relations

$$\sup_{\widetilde{Q}_-} g^k = 1, \qquad \sqrt{|u^k_\varrho|^2 + |u^k_3|^2} \leq A/\varrho, \quad |g^k| \leq A/M \qquad \text{in} \quad \widetilde{Q}_-.$$

By (6.4.16), there are points $(\varrho_k, x_{k3}, \bar{t})$, with $(\varrho_k, x_{k3}) \in \Pi$, and $(\varrho'_k, x'_{k3}, t'_k) \in \widetilde{Q}$ such that

$$g^k(\varrho_k, x_{k3}, \bar{t}) > 1 - 2\delta_k, \qquad g^k(\varrho'_k, x'_{k3}, t'_k) \leq 1 - \varepsilon_0/2. \qquad (6.4.17)$$

Weak form of the equations for u^k and g^k is as follows:

$$\int\limits_{\widetilde{Q}_-} \Big[g^k \partial_t f + g^k (u^k_\varrho f_{,\varrho} + u^k_3 f_{,3}) + g^k (\Delta f + 2 f_{,\varrho}/\varrho) \Big] \varrho d\varrho dx_3 dt = 0$$

for any $f \in C^\infty_0(\widetilde{Q}_-)$. Routine arguments show

$$u^k \stackrel{\star}{\rightharpoonup} \bar{u} \qquad \text{in} \quad L_\infty(\widetilde{Q}_-; \mathbb{R}^2),$$

$$g^k \rightharpoonup \bar{g} \qquad \text{in} \quad W^{2,1}_m(\widetilde{Q}_2),$$

where $\widetilde{Q}_2 = \Pi_2 \times]\bar{t} - t_2, \bar{t}[$, $\Pi_2 = \Pi(\varrho_1^2, \varrho_2^2; h_1^2, h_2^2) \Supset \Pi$, $t_2 > t_1$, and $m >> 1$. Then we have

$$g^k \to \bar{g} \qquad \text{in} \quad C(\widetilde{Q}_2) \qquad (6.4.18)$$

and

$$\sup_{\widetilde{Q}_2} \overline{g} \leq 1, \qquad \sqrt{|\overline{u}_\varrho|^2 + |\overline{u}_3|^2} \leq A/\varrho, \quad |\overline{g}| \leq A/M \qquad \text{in} \quad \widetilde{Q}_2, \quad (6.4.19)$$

and

$$\partial_t \overline{g} + (\overline{u}_\varrho + 1/\varrho)\overline{g}_{,\varrho} + \overline{u}_3 \overline{g}_{,3} - \overline{g}_{,\varrho\varrho} - \overline{g}_{,33} = 0 \qquad \text{in} \quad \widetilde{Q}_2.$$

According to (6.4.17) and (6.4.18),

$$\overline{g}(\varrho_0, x_{03}, \overline{t}) = 1, \qquad \overline{g}(\varrho_0', x_{03}', t_0') \leq 1 - \varepsilon_0/2, \quad t_0' \leq \overline{t}, \qquad (6.4.20)$$

where

$$(\varrho_k, x_{k3}, \overline{t}) \to (\varrho_0, x_{03}, \overline{t}), \qquad (\varrho_k', x_{k3}', t_k') \to (\varrho_0', x_{03}', t_0')$$

and points $(\varrho_0, x_{03}, \overline{t})$ and $(\varrho_0', x_{03}', t_0')$ belong to the closure of the set $\widetilde{Q}$. Clearly, by (6.4.19),

$$\overline{g}(\varrho_0, x_{03}, \overline{t}) = \sup_{z \in \widetilde{Q}_2} \overline{g}(z) = 1.$$

By the strong maximum principle, $\overline{g} \equiv 1$ in $\widetilde{Q}_2$. But this contradicts (6.4.20). □

Now, we proceed with the proof of Theorem 4.16[1]. Take arbitrary positive numbers R, L, T, and $0 < \varepsilon \leq 1/2$. We can always assume that

$$1 - \varepsilon \leq g \leq 1 \qquad \text{on} \quad \widetilde{Q}_0 = \Pi_0 \times]-T, 0[, \qquad (6.4.21)$$

where $\Pi_0 = \Pi(1, R; -L, L)$. To explain this, we let $\delta_* = \delta(\Pi_0, 0, -T, A, M, \varepsilon)$. Obviously, there exists a point $(\varrho_0, x_{03}, t_0) \in \widetilde{Q}_-$ such that

$$1 - g(\varrho_0, x_{03}, t_0) < \delta_* \leq \varepsilon \leq 1/2.$$

It is easy to see

$$1/2 \leq \varrho_0 u_\varphi(\varrho_0, x_{03}, t_0)/M = g(\varrho_0, x_{03}, t_0) \leq \varrho_0/M$$

and, therefore, $\varrho_0 > M/2 > 0$. Then one can scale our functions so that

$$g_\lambda(r, y_3, s) = g(\lambda r, x_{03} + \lambda y_3, t_0 + \lambda^2 s), \qquad \lambda = \varrho_0,$$

$$u_r^\lambda(r, y_3, s) = \lambda u_\varrho(\lambda r, x_{03} + \lambda y_3, t_0 + \lambda^2 s),$$

$$u_3^\lambda(r, y_3, s) = \lambda u_3(\lambda r, x_{03} + \lambda y_3, t_0 + \lambda^2 s).$$

[1]The idea of the proof belongs to V. Sverak

For scaled functions, we have

$$1 - g_\lambda(1,0,0) < \delta_*,$$

$$\partial_s g_\lambda + u_r^\lambda g_{\lambda,r} + u_3^\lambda g_{\lambda,3} - (\Delta g_\lambda - 2g_{\lambda,r}/r) = 0 \qquad \text{in} \quad \widetilde{Q}_-,$$

$$(ru_r^\lambda)_{,r} + (ru_3^\lambda)_{,3} = 0 \qquad \text{in} \quad \widetilde{Q}_-,$$

$$\sup_{\widetilde{Q}_-} g_\lambda = 1, \qquad \sqrt{|u_r^\lambda|^2 + |u_3^\lambda|^2} \le A/r, \quad |g_\lambda| \le A/M \qquad \text{in} \quad \widetilde{Q}_-.$$

By Lemma 6.9,

$$1 - \varepsilon \le g_\lambda \le 1 \qquad \text{on} \quad \widetilde{Q}_0.$$

It is always deemed that this operation has been already made and script λ is dropped. It is important to notice two things. Numbers R, L, T, and ε are in our hands and we cannot use the fact $|u| \le 1$ any more since after scaling $|u| \le \varrho_0(R, T, L, A, M, \varepsilon)$.

We choose a cut-off function

$$\Phi(\varrho, x_3, t) = \psi(\varrho)\eta(x_3)\chi(t),$$

where functions ψ, η, and χ have the following properties:

$$\psi(\varrho) = 1 \quad 0 \le \varrho \le R-1, \qquad \psi(\varrho) = 0 \quad \varrho \ge R,$$

$$|\psi'(\varrho)| + |\psi''(\varrho)| \le c \quad 0 \le \varrho < +\infty;$$

$$\eta(x_3) = 1 \quad |x_3| \le L-1, \qquad \eta(x_3) = 0 \quad |x_3| \ge L,$$

$$|\eta'(x_3)| + |\eta''(x_3)| \le c \quad |x_3| < +\infty;$$

$$\chi(t) = 1 \quad -T+1 < t \le -1, \qquad \chi(t) = 0 \quad t < -T,$$

$$\chi(t) = t + T \quad -T \le t \le -T+1,$$

$$\chi(t) = -t \qquad -1 < t \le 0.$$

So, we have

$$I_0 = \int\limits_{\widetilde{Q}_-} \Big(\partial_t g + u_\varrho g_{,\varrho} + u_3 g_{,3} - \Delta g\Big)\Phi \varrho d\varrho dx_3 dt = I_0' =$$

$$= -2 \int\limits_{\widetilde{Q}_-} \frac{g_{,\varrho}}{\varrho} \Phi \varrho d\varrho dx_3 dt. \tag{6.4.22}$$

We replace g with $g-1$ in the left-hand side of (6.4.22) and, after integration by parts, have

$$I_0 = \frac{1}{4\pi} \int\limits_{Q_-} \Big(\partial_t \Phi + u_\varrho \Phi_{,\varrho} + u_3 \Phi_{,3} + \Delta \Phi \Big)(1-g) dx dt.$$

We know that $1 - g \leq \varepsilon$ in $\widetilde{Q}_0$. Then, by 6.4.15,

$$I_0 \geq \int\limits_{-T}^{0} \int\limits_{-L}^{L} \int\limits_{0}^{1} (1-g) \Big(\partial_t \Phi + u_3 \Phi_{,3} + \Phi_{,33} \Big) \varrho d\varrho dx_3 dt + \varepsilon C_0(R,T,L,A,M)$$

$$\geq -(L+T) C_1(A,M) + \varepsilon C_0(R,T,L,A,M). \tag{6.4.23}$$

Next, let us start with evaluation of the right-hand side in (6.4.22). Integration by parts gives:

$$I_0' = -2 \int\limits_{-T}^{0} \int\limits_{-L}^{L} g(0,x_3,t) \Phi(0,x_3,t) dx_3 dt + 2 \int\limits_{-T}^{0} \int\limits_{-L}^{L} \int\limits_{0}^{R} \Phi_{,\varrho} g d\varrho dx_3 dt.$$

The first term on the right-hand side of the above identity is equal to zero. An upper bound of the second one is derived as follows:

$$I_0' = 2 \int\limits_{-T}^{0} \int\limits_{-L}^{L} \int\limits_{0}^{R} \Phi_{,\varrho} d\varrho dx_3 dt + 2 \int\limits_{-T}^{0} \int\limits_{-L}^{L} \int\limits_{0}^{R} \Phi_{,\varrho} (g-1) d\varrho dx_3 dt \leq$$

$$\leq 2 \int\limits_{-T}^{0} \int\limits_{-L}^{L} \int\limits_{0}^{R} \Phi_{,\varrho} d\varrho dx_3 dt + \varepsilon C_0'(R,T,L,A,M) =$$

$$= -2 \int\limits_{-T}^{0} \int\limits_{-L}^{L} \Phi(0,x_3,t) dx_3 dt + \varepsilon C_0'(R,T,L,A,M) <$$

$$< -2(L-1)(T-2) + \varepsilon C_0'(R,T,L,A,M).$$

The latter, together with identity (6.4.22) and (6.4.23), implies the following inequality

$$2(L-1)(T-2) \leq (L+T) C_1(A,M) + \varepsilon C_0''(R,T,L,A,M).$$

This leads to contradiction for large L and T and sufficiently small ε.

So, the assumption $M > 0$ is wrong. In the same way, one shows that

$$\inf_{\widetilde{Q}_-} \varrho u_\varphi = m \geq 0.$$

This means that the swirl is zero. □

6.5 Axially Symmetric Suitable Weak Solutions

In this section, just for convenience, we replace balls $B(r)$ with cylinders $\mathcal{C}(r) = \{x = (x', x_3),\, x' = (x_1, x_2),\, |x'| < r,\, |x_3| < r\}$, $\mathcal{C} = \mathcal{C}(1)$, and then $Q(r) = \mathcal{C}(r)\times]-r^2, 0[$. As usual, let us set

$$\overline{v} = v_\varrho e_\varrho + v_3 e_3 \qquad \widehat{v} = v_\varphi e_\varphi$$

for $v = v_\varrho e_\varrho + v_\varphi e_\varphi + v_3 e_3$.

Here, we follow paper [Seregin and Šverák (2009)], where results are stated for the canonical domain $Q = Q(1)$. The general case can be done by re-scaling.

Theorem 5.17. *Assume that functions $v \in L_3(Q)$ and $q \in L_{\frac{3}{2}}(Q)$ are an axially symmetric weak solution to the Navier-Stokes equations in Q. Let, in addition, a positive constant C exists such that*

$$|\overline{v}(x,t)| \leq \frac{C}{\sqrt{-t}} \tag{6.5.1}$$

for almost all points $z = (x,t) \in Q$. Then $z = 0$ is a regular point of v.

Theorem 5.18. *Assume that functions $v \in L_3(Q)$ and $q \in L_{\frac{3}{2}}(Q)$ are an axially symmetric weak solution to the Navier-Stokes equations in Q. Let, in addition,*

$$v \in L_\infty(\mathcal{C}\times]-1, -a^2[) \tag{6.5.2}$$

for each $0 < a < 1$ and

$$|\overline{v}(x,t)| \leq \frac{C}{|x'|} \tag{6.5.3}$$

for almost all points $z = (x,t) \in Q$ with some positive constant C. Then $z = 0$ is a regular point of v.

According to the Caffarelli-Kohn-Nirenberg theorem if v and q are an axially symmetric suitable weak solution and $z = (x,t)$ is a singular (i.e., not regular) point of v, then there must be $x' = 0$. In other words, all singular points must seat on the axis of symmetry, which in our case is the axis x_3.

The following estimate is obtained with help of Mozer's iterations. A proof is not complicated, see, for example, [Seregin and Šverák (2009)].

Lemma 6.10. *Assume that functions $v \in L_3(Q)$ and $q \in L_{\frac{3}{2}}(Q)$ are an axially symmetric weak solution to the Navier-Stokes equations in Q. Let, in addition, condition (6.5.2) hold. Then following estimate is valid:*

$$\text{ess}\sup_{z\in Q(1/2)} |\varrho v_\varphi(z)| \leq C(M)\Big(\int\limits_{Q(3/4)} |\varrho v_\varphi|^{\frac{10}{3}}dz\Big)^{\frac{3}{10}}, \tag{6.5.4}$$

where

$$M = \Big(\int\limits_{Q(3/4)} |\overline{v}|^{\frac{10}{3}}dz\Big)^{\frac{3}{10}} + 1.$$

Remark 6.8. Under the assumptions of Lemma 6.10, the pair v and q is a suitable weak solution to the Navier-Stokes equations in Q. Hence, the right-hand side of (6.5.4) is bounded from above.

With some additional notation

$$\mathcal{C}(x_0,R) = \{x \in \mathbb{R}^3 : x = (x',x_3),\ x' = (x_1,x_2),\ |x'-x'_0| < R,\ |x_3-x_{03}| < R\},$$

$$Q(z_0,R) = \mathcal{C}(x_0,R)\times]t_0-R^2,t_0[,$$

we recall the definition of certain scaled energy quantities:

$$A(z_0,r;v) = \text{ess}\sup_{t_0-r^2<t<t_0}\frac{1}{r}\int\limits_{\mathcal{C}(x_0,r)} |v(x,t)|^2dx,$$

$$E(z_0,r;v) = \frac{1}{r}\int\limits_{Q(z_0,r)} |\nabla v|^2dz, \qquad D(z_0,r;q) = \frac{1}{r^2}\int\limits_{Q(z_0,r)} |q|^{\frac{3}{2}}dz,$$

$$C(z_0,r;v) = \frac{1}{r^2}\int\limits_{Q(z_0,r)} |v|^3dz, \qquad H(z_0,r;v) = \frac{1}{r^3}\int\limits_{Q(z_0,r)} |v|^2dz,$$

$$M_{s,l}(z_0,r;v) = \frac{1}{r^\kappa}\int\limits_{t_0-r^2}^{t_0}\Big(\int\limits_{\mathcal{C}(x_0,r)} |v|^s dx\Big)^{\frac{l}{s}}dt,$$

where $\kappa = l(\frac{3}{s}+\frac{2}{l}-1)$ and $s \geq 1$, $l \geq 1$.

The following statement is proven in a similar way as Proposition 3.11, see details in [Seregin and Šverák (2009)].

Lemma 6.11. *Under assumptions of Theorem 5.17, we have the estimate*

$$A(z_b,r;v) + E(z_b,r;v) + C(z_b,r;v) + D(z_b,r;q) \leq C_1 < +\infty \tag{6.5.5}$$

for all z_b and for all r satisfying conditions

$$z_b = (be_3, 0), \quad b \in \mathbb{R}, \quad |b| \leq \frac{1}{4}, \qquad 0 < r < \frac{1}{4}. \tag{6.5.6}$$

A constant C_1 depends only on the constant C in (6.5.1), $\|v\|_{L_3(Q)}$, and $\|q\|_{L_{\frac{3}{2}}(Q)}$.

To prove Theorem 5.18, we need an analogue of Lemma 6.11. Here, it is.

Lemma 6.12. *Under assumptions of Theorem 5.18, estimate (6.5.5) is valid as well with constant C_1 depending only on the constant C in (6.5.3), $\|v\|_{L_3(Q)}$, and $\|q\|_{L_{\frac{3}{2}}(Q)}$.*

Lemma 6.12 is proved along the same lines as Lemma 6.11.

As it follows from conditions of Theorem 5.18 and the statement of Lemma 6.10, the module of the velocity field grows not faster than $C/|x'|$ as $|x'| \to 0$. Moreover, the corresponding estimate is uniform in time. However, it turns out that the same is true under conditions of Theorem 5.17. More precisely, we have the following.

Proposition 5.19. *Assume that all conditions of Theorem 5.17 hold. Then*

$$|v(x,t)| \leq \frac{C_1}{|x'|} \tag{6.5.7}$$

for all $z = (x,t) \in Q(1/8)$. A constant C_1 depends only on the constant C in (6.5.1), $\|v\|_{L_3(Q)}$, and $\|q\|_{L_{\frac{3}{2}}(Q)}$.

PROOF In view of (6.11), we can argue essentially as in [Seregin and Zajaczkowski (2007)].

Let us fix a point $x_0 \in \mathcal{C}(1/8)$ and put $r_0 = |x_0'|$, $b_0 = x_{03}$. So, we have $r_0 < \frac{1}{8}$ and $|b_0| < \frac{1}{8}$. Further, we introduce the following cylinders:

$$\mathcal{P}^1_{r_0} = \{r_0 < |x'| < 2r_0, \, |x_3| < r_0\}, \quad \mathcal{P}^2_{r_0} = \{r_0/4 < |x'| < 3r_0, \, |x_3| < 2r_0\}.$$

$$\mathcal{P}^1_{r_0}(b_0) = \mathcal{P}^1_{r_0} + b_0 e_3, \quad \mathcal{P}^2_{r_0}(b_0) = \mathcal{P}^2_{r_0} + b_0 e_3,$$

$$Q^1_{r_0}(b_0) = \mathcal{P}^1_{r_0}(b_0) \times]-r_0^2, 0[, \quad Q^2_{r_0}(b_0) = \mathcal{P}^2_{r_0}(b_0) \times]-(2r_0)^2, 0[.$$

Now, let us scale our functions so that

$$x = r_0 y + b_0 e_3, \quad t = r_0^2 s, \quad u(y,s) = r_0 v(x,t), \quad p(y,s) = r_0^2 q(x,t).$$

As it was shown in [Seregin and Zajaczkowski (2007)], there exists a continuous nondecreasing function $\Phi : \mathbb{R}_+ \to \mathbb{R}_+$, $\mathbb{R}_+ = \{s > 0\}$, such that

$$\sup_{(y,s)\in Q^1_1(0)} |u(u,s)| + |\nabla u(y,s)| \leq \Phi\Big(\sup_{-2^2<s<0} \int\limits_{\mathcal{P}^2_1(0)} |u(y,s)|^2 dy$$

$$+ \int\limits_{Q^2_1(0)} |\nabla u|^2 dy\, ds + \int\limits_{Q^2_1(0)} |u|^3 dy\, ds + \int\limits_{Q^2_1(0)} |p|^{\frac{3}{2}} dy\, ds\Big). \tag{6.5.8}$$

After making inverse scaling in (6.5.8), we find

$$\sup_{z\in Q^1_{r_0}(b_0)} r_0|u(x,t)| + r_0^2|\nabla u(x,t)| \leq \Phi\Big(cA(z_{b_0}, 3r_0; v) + cE(z_{b_0}, 3r_0; v)+$$

$$+cC(z_{b_0}, 3r_0; v) + cD(z_{b_0}, 3r_0; q)\Big) \leq \Phi\Big(4cC_1\Big).$$

It remains to apply Lemma 6.11 and complete the proof of the proposition. □

Now, we proceed with proof of Theorems 5.17 and 5.18. Using Lemmata 6.10, 6.11, 6.12, Remark 6.8, Proposition 5.19 and scaling arguments, we may assume (without loss of generality) that our solution v and q have the following properties:

$$\sup_{0<r\leq 1} \Big(A(0,r;v) + E(0,r;v) + C(0,r;v) + D(0,r;q)\Big) = A_1 < +\infty, \tag{6.5.9}$$

$$\text{ess} \sup_{z=(x,t)\in Q} |x'||v(x,t)| = A_2 < +\infty. \tag{6.5.10}$$

We may also assume that the function v is Hölder continuous in the closure of the set $\mathcal{C}\times]-1,-a^2[$ for any $0 < a < 1$.

Introducing functions

$$H(t) = \sup_{x\in \mathcal{C}} |v(x,t)|, \quad h(t) = \sup_{-1<\tau\leq t} H(\tau),$$

let us suppose that our statement is wrong, i.e., $z = 0$ is a singular point. Then there are sequences $x_k \in \overline{\mathcal{C}}$ and $-1 < t_k < 0$, having the following properties:

$$h(t_k) = H(t_k) = M_k = |v(x_k,t_k)| \to +\infty \qquad \text{as} \quad k \to +\infty.$$

We scale our functions v and q so that scaled functions keep axial symmetry:

$$u^k(y,s) = \lambda_k v(\lambda_k y', x_{3k} + \lambda_k y_3, t_k + \lambda_k^2 s), \qquad \lambda_k = \frac{1}{M_k},$$

$$p^k(y,s) = \lambda_k^2 q(\lambda_k y', x_{3k} + \lambda_k y_3, t_k + \lambda_k^2 s).$$

These functions satisfy the Navier-Stokes equations in $Q(M_k)$. Moreover,

$$|u^k(y_k', 0, 0)| = 1, \qquad y_k' = M_k x_k'. \tag{6.5.11}$$

According to (6.5.10),

$$|y_k'| \leq A_2$$

for all $k \in \mathbb{N}$. Thus, without loss of generality, we may assume that

$$y_k' \to y_*' \qquad \text{as} \quad k \to +\infty. \tag{6.5.12}$$

Now, let us see what happens as $k \to +\infty$. By the identity

$$\sup_{e=(y,s)\in C(M_k)} |u^k(e)| = 1 \tag{6.5.13}$$

and by (6.5.9), we can select subsequences (still denote as the entire sequence) such that

$$u^k \overset{\star}{\rightharpoonup} u \qquad \text{in} \qquad L_\infty(Q(a)), \tag{6.5.14}$$

and

$$p^k \rightharpoonup p \qquad \text{in} \qquad L_{\frac{3}{2}}(Q(a))$$

for any $a > 0$. Functions u and p are defined on $Q_- = \mathbb{R}^3 \times]-\infty, 0[$. Obviously, they possess the following properties:

$$\operatorname*{ess\,sup}_{e\in Q_-} |u(e)| \leq 1,$$

$$\sup_{0<r<+\infty} \Big(A(0,r;u) + E(0,r;u) + C(0,r;u) + D(0,r;p) \Big) \leq A_1,$$

$$\operatorname*{ess\,sup}_{e=(y,s)\in Q_-} |y'||u(y,s)| \leq A_2. \tag{6.5.15}$$

Now, our aim is to show that u and p satisfy the Navier-Stokes equations Q_- and u is smooth enough to obey the identity

$$|u(y_*', 0, 0)| = 1. \tag{6.5.16}$$

To this end, we fix an arbitrary positive number $a > 0$ and consider numbers k so big that $a < M_k/4$. We know that u^k satisfies the non-homogeneous heat equation of the form

$$\partial_t u^k - \Delta u^k = -\operatorname{div} F^k \qquad \text{in} \quad Q(4a),$$

where $F^k = u^k \otimes u^k + p^k \mathbb{I}$ and

$$\|F^k\|_{\frac{3}{2},Q(4a)} \leq c_1(a) < \infty.$$

This implies the following fact, see [Ladyzhenskaya *et al.* (1967)],

$$\|\nabla u^k\|_{\frac{3}{2},Q(3a)} \leq c_2(a) < \infty.$$

Now, we can interpret the pair u^k and p^k as a solution to the non-homogeneous Stokes system

$$\partial_t u^k - \Delta u^k + \nabla p^k = f^k, \quad \operatorname{div} u^k = 0 \qquad \text{in} \quad Q(3a), \tag{6.5.17}$$

where $f^k = -u^k \cdot \nabla u^k$ is the right-hand side having the property

$$\|f^k\|_{\frac{3}{2},Q(3a)} \leq c_2(a).$$

Then, according to the local regularity theory for the Stokes system, see Chapter 4, we can state that

$$\|\partial_t u^k\|_{\frac{3}{2},Q(2a)} + \|\nabla^2 u^k\|_{\frac{3}{2},Q(2a)} + \|\nabla p^k\|_{\frac{3}{2},Q(2a)} \leq c_3(a).$$

The latter, together with the embedding theorem, implies

$$\|\nabla u^k\|_{3,\frac{3}{2},(Q(2a))} + \|p^k\|_{3,\frac{3}{2},Q(2a)} \leq c_4(a).$$

In turn, this improves integrability of the right-hand side in (6.5.17)

$$\|f^k\|_{3,\frac{3}{2},Q(2a)} \leq c_4(a).$$

Therefore, by the local regularity theory,

$$\|\partial_t u^k\|_{3,\frac{3}{2},Q(2a)} + \|\nabla^2 u^k\|_{3,\frac{3}{2},Q(2a)} + \|\nabla p^k\|_{3,\frac{3}{2},Q(2)} \leq c_5(a).$$

Applying the imbedding theorem once more, we find

$$\|\nabla u^k\|_{6,\frac{3}{2},Q(2a)} + \|p^k\|_{6,\frac{3}{2},Q(2a)} \leq c_6(a).$$

The local regularity theory leads then to the estimate

$$\|\partial_t u^k\|_{6,\frac{3}{2},Q(a)} + \|\nabla^2 u^k\|_{6,\frac{3}{2},Q(a)} + \|\nabla p^k\|_{6,\frac{3}{2},Q(a)} \leq c_7(a).$$

By the embedding theorem, sequence u^k is uniformly bounded in a Hölder space, for example, in $C^{\frac{1}{2}}(\overline{Q}(a/2))$. Hence, without loss of generality, one may assume that

$$u^k \to u \qquad \text{in} \quad C^{\frac{1}{4}}(\overline{Q}(a/2)).$$

This means that the pair u and p obeys the Navier-Stokes system and (6.5.16) holds. So, the function u is the so-called bounded ancient solution to the Navier-Stokes system which is, in addition, axially symmetric and satisfies the decay estimate (6.5.15). As it has been shown in Section 5 of this chapter, such a solution must be identically zero. But this contradicts (6.1.43). Theorems (5.17) and (5.18) are proved.

6.6 Backward Uniqueness for Navier-Stokes Equations

In this section, we deal with another subclass of ancient solutions u possessing the following property: *there exists a function p defined on $\mathbb{R}^3\times]-\infty,0[$ such that functions u and p are a suitable weak solution to the Navier-Stokes equations in $\mathbb{R}^3\times]-\infty,0[$, i.e., they are a suitable weak solution on each parabolic ball of the form $Q(a)=B(a)\times]-a^2,0[$ with $< a < +\infty$.* We call u a *local energy ancient* solution. Certainly, mild bounded ancient solutions belong to this subclass.

Local energy ancient solutions can be obtained from a given suitable weak solution v and q defined in Q with the help of the scaling mentioned in the previous section provided boundedness of g' takes place, see the definition of g' in Proposition 3.11.

Proposition 6.20. *Let v and q be a suitable weak solution to the Navier-Stokes equations in Q with $g' < +\infty$ and let $u^{(k)}(y,s)=\lambda_k v(\lambda_k y,\lambda_k^2 s)$ and $p^{(k)}(y,s)=\lambda_k^2 q(\lambda_k y,\lambda_k^2 s)$ with $\lambda_k\to 0$ as $k\to+\infty$. Then there exist subsequences of $u^{(k)}$ and $p^{(k)}$ still denoted by $u^{(k)}$ and $p^{(k)}$ such that, for each $a>0$,*

$$u^{(k)}\to u$$

in $L_3(Q(a))\cap C([-a^2,0];L_{\frac{9}{8}}(B(a)))$ and

$$p^{(k)}\rightharpoonup p$$

in $L_{\frac{3}{2}}(Q(a))$, where u is a local energy ancient solution with the corresponding pressure p. For them, the energy scale-invariant quantities are uniformly bounded, i.e.,

$$\sup_{0<a<+\infty}\{A(u;a)+C(u;a)+D(p;a)+E(u;a)\}<+\infty.$$

Moreover, if $z=0$ is a singular point of the velocity field v, then

$$\int\limits_{Q(3/4)}|u|^3dz>c \tag{6.6.1}$$

with a positive universal constant c, i.e., u is not identically equal to zero.

A proof of this proposition and similar facts can be found in [Escauriaza *et al.* (2003)], [Seregin (2007)], [Seregin and Šverák (2009)], and [Seregin

(2011)]. Let us comment the last statement of Proposition 6.20. Indeed, if $z=0$ is a singular point of v, the ε-regularity theory gives us

$$\frac{1}{r^2}\int\limits_{Q(r)}(|v|^3+|q|^{\frac{3}{2}})dz>\varepsilon>0$$

for all $0<r<1$ and for some universal constant ε. Making the inverse change of variables, we find

$$\frac{1}{a^2}\int\limits_{Q(a)}(|u^{(k)}|^3+|p^{(k)}|^{\frac{3}{2}})dyds=$$

$$=\frac{1}{\lambda_k^2a^2}\int\limits_{Q(\lambda_k a)}(|v|^3+|q|^{\frac{3}{2}})dxds>\varepsilon>0$$

for each fixed radius $a>0$ and for sufficiently large natural number k. We cannot simply pass to the limit in the latter identity since it is not clear whether the pressure $p^{(k)}$ converges strongly. This is a typical issue for those who work with sequences of weak solutions to the Navier-Stokes equations. In order to treat this case, let us split the pressure $p^{(k)}$ into two parts. The first part is completely controlled by the velocity field $u^{(k)}$ while the second one is a harmonic function with respect to the spatial variables. This, together with a certain boundedness of the sequence $p^{(k)}$, implies (6.6.1). For more details, we recommend papers [Seregin (2007)] and [Seregin (2011)].

We do not know whether local energy ancient solutions with bounded scaled energy quantities are identically equal to zero. However, there are some interesting cases for which the answer is positive. Let us describe them.

Our additional standing assumption of this section can be interpreted as a restriction on the blowup profile of v and has the form

$$\frac{1}{r^{\frac{15}{8}}}\int\limits_{B(r)}|v(x,0)|^{\frac{9}{8}}dx\to 0 \tag{6.6.2}$$

as $r\to 0$. The most important consequence of (6.6.2) is that

$$u(\cdot,0)=0, \tag{6.6.3}$$

where u is a local energy ancient solution that is generated by the scaling described by Proposition 6.20. Indeed, for any $a>0$, we have

$$\frac{1}{a^{\frac{15}{8}}}\int\limits_{B(a)}|u(y,0)|^{\frac{9}{8}}dy\leq$$

$$\leq c\frac{1}{a^{\frac{15}{8}}}\int\limits_{B(a)}|u(y,0)-u^{(k)}(y,0)|^{\frac{9}{8}}dy+c\frac{1}{a^{\frac{15}{8}}}\int\limits_{B(a)}|u^{(k)}(y,0)|^{\frac{9}{8}}dy=$$

$$=\alpha_k(a)+c\frac{1}{(\lambda_k a)^{\frac{15}{8}}}\int\limits_{B(\lambda_k a)}|v(x,0)|^{\frac{9}{8}}dx.$$

Now, by Proposition 6.20 and by (6.6.2), the right-hand side of the latter inequality tends to zero and this completes the proof of (6.6.3).

In a view of (6.6.3), one could expect that our local energy ancient solution is identically equal to zero. We call this phenomenon a backward uniqueness for the Navier-Stokes equations. So, if the backward uniqueness takes place or at least our ancient solution is zero on the time interval $]-3/4,0[$, then (6.6.1) cannot be true and thus, by Proposition 6.20, the origin $z=0$ is not a singular point of the velocity field v.

The crucial point for understanding the backward uniqueness for the Navier-Stokes equations is a similar phenomenon for the heat operator with lower order terms. The corresponding statement for the partial differential inequality involving the backward heat operator with lower order terms has been proved in [Escauriaza *et al.* (2003)] and reads:

Theorem 6.21. *Assume that we are given a function ω defined on $\mathbb{R}^n_+\times]0,1[$, where $\mathbb{R}^n_+=\{x=(x_i)\in\mathbb{R}^n,\ x_n>0\}$. Suppose further that they have the properties:*

ω and the generalized derivatives $\nabla\omega$, $\partial_t\omega$, and $\nabla^2\omega$ are square integrable over any bounded subdomain of $\mathbb{R}^n_+\times]0,1[$;

$$|\partial_t\omega+\Delta\omega|\leq c(|\omega|+|\nabla\omega|) \tag{6.6.4}$$

on $\mathbb{R}^n_+\times]0,1[$ with a positive constant c;

$$|\omega(x,t)|\leq\exp\{M|x|^2\} \tag{6.6.5}$$

for all $x\in\mathbb{R}^n_+$, for all $0<t<1$, and for some $M>0$;

$$\omega(x,0)=0 \tag{6.6.6}$$

for all $x\in\mathbb{R}^n_+$.

Then ω is identically zero in $\mathbb{R}^n_+\times]0,1[$.

The interesting feature of Theorem 6.21 is that there has been made no assumption on ω on the boundary $x_n=0$. In order to prove the theorem, two Carleman's inequalities have been established, see details in [Escauriaza *et al.* (2003)] and [Escauriaza *et al.* (2003)] and Appendix A. For the further improvements of the above backward uniqueness result, we refer to the interesting paper [Escauriaza *et al.* (2006)].

Theorem 6.21 clearly indicates what should be added to (6.6.3) in order to get the backward uniqueness for ancient solutions to the Navier-Stokes equations. Obviously, we need more regularity for sufficiently large x and a decay at infinity. One can hope then to apply Theorem 6.21 to the vorticity equation

$$\partial_t \omega - \Delta \omega = \omega \cdot \nabla u - u \cdot \nabla \omega, \qquad \omega = \nabla \wedge u,$$

which could be interpreted as a perturbation of the heat equation by lower order terms. To make it possible, it is sufficient to show boundedness of u and ∇u outside of the Cartesian product of some spatial ball and some time interval. The rest of the section will be devoted to a certain situation, for which it is really true.

Let us assume that

$$|u(x,t)| + |\nabla u(x,t)| \leq c < +\infty \tag{6.6.7}$$

for all $|x| > R$, for all $-1 < t < 0$, and for some constant c and try to figure out what follows from (6.6.7). It is not difficult to see that (6.6.3) and (6.6.7) implies (6.6.6) and (6.6.4), (6.6.5), respectively. At last, the linear theory ensures the validity of first condition in Theorem 6.21, see details in [Seregin (2007)]. So, Theorem 6.21 is applicable and by it, $\omega(x,t) = 0$ for all $|x| > R$ and for $-1 < t < 0$. Using unique continuation across spatial boundaries, see, for instance, [Escauriaza *et al.* (2003)] or Appendix A, we deduce $\omega(x,t) = \nabla \wedge u(x,t) = 0$ for all $x \in \mathbb{R}^3$ and, say, for $-5/6 < t < 0$. Since u is divergence free, it is a harmonic function in $\mathbb{R}^3$ depending on $t \in]-5/6, 0[$ as a parameter. Therefore, for any $a > \sqrt{5/6}$ and for any $x_0 \in \mathbb{R}^3$, by the mean value theorem for harmonic functions, we have

$$\sup_{-5/6<t<0} |u(x_0,t)|^2 \leq c \sup_{-5/6<t<0} \frac{1}{a^3} \int\limits_{B(x_0,a)} |u(x,t)|^2 dx$$

$$\leq c \sup_{-5/6<t<0} \frac{1}{a^3} \int\limits_{B(|x_0|+a)} |u(x,t)|^2 dx \leq c \frac{a+|x_0|}{a^3} A(u, a+|x_0|).$$

Thanks to boundedness of scaled energy quantities stated in Proposition 6.20, the right-hand side of the above inequality tends to zero as a goes to infinity. By arbitrariness of x_0, we conclude that $u(x,t) = 0$ for all $x \in \mathbb{R}^3$ and for $-5/6 < t < 0$, which contradicts (6.6.1). Hence, the origin $z = 0$ cannot be a singular point of v.

Coming back to a marginal case of Ladyzhenskaya-Prodi-Serrin condition, which is called $L_{3,\infty}$-case, and show that it can be completely embedded into the above scheme. So, we assume that functions v and q are a suitable weak solution to the Navier-Stokes equations in Q and satisfy the additional condition

$$\|v\|_{3,\infty,Q} < +\infty. \tag{6.6.8}$$

With the help of Proposition 3.11, it is not so difficult to show that $g' < +\infty$. So, for v, all the assumptions of Proposition 6.20 hold and thus our blowup procedure produces a local energy ancient solution u with the properties listed in that proposition. Exploited the ε-regularity theory once more, we can show further that $v(\cdot,0) \in L_3(B(2/3))$, which in turn implies (6.6.2). Now, in order to prove regularity of the velocity v at the point $z=0$, it is sufficient to verify the validity of (6.6.7). Indeed, by scale-invariance,

$$\|u\|_{3,\infty,\mathbb{R}^3\times]-\infty,0[} < +\infty.$$

Applying Proposition 6.20 again and taking into account properties of harmonic functions, one can conclude that

$$\|p\|_{\frac{3}{2},\infty,\mathbb{R}^3\times]-\infty,0[} < +\infty.$$

Combining the latter estimates, we show that for any $T>0$

$$\int\limits_{-T}^{0}\int\limits_{\mathbb{R}^3}(|u|^3+|p|^{\frac{3}{2}})dxdt < +\infty. \tag{6.6.9}$$

Our further arguments rely upon the ε-regularity theory. Indeed, letting, say, $T=4$, one can find $R>4$ so that

$$\int\limits_{-4}^{0}\int\limits_{\mathbb{R}^3\setminus B(R/2)}(|u|^3+|p|^{\frac{3}{2}})dxdt < \varepsilon.$$

The rest of the proof of (6.6.7) is easy.

6.7 Comments

The section is essentially the context of my lectures on the local regularity theory given in Summer School, Cetraro, Italy, 2010, see [Seregin (2013)]. It contains an introduction to the so-called ε-regularity theory in the spirit of the paper [Escauriaza *et al.* (2003)], see also [Seregin (2007)] for some generalizations. A big part of this section is an alternative approach to derivation of mild bounded ancient solutions and Liouville type theorems for them presented in [Koch *et al.* (2009)]. Here, we follow the paper [Seregin and Šverák (2009)] although proofs of Liouville type theorems is essentially the same as in [Koch *et al.* (2009)].

Chapter 7

Behavior of L_3-Norm

7.1 Main Result

Let us consider the Cauchy problem for the classical Navier-Stokes system

$$\partial_t v + v \cdot \nabla v - \Delta v = -\nabla q, \qquad \mathrm{div} v = 0 \tag{7.1.1}$$

with the initial condition

$$v|_{t=0} = v_0 \tag{7.1.2}$$

in $\mathbb{R}^3$. For simplicity, assume

$$v_0 \in C^\infty_{0,0}(\mathbb{R}^3) \equiv \{v \in C^\infty_0(\mathbb{R}^3) : \ \mathrm{div} v = 0\}. \tag{7.1.3}$$

In 1934, J. Leray proved certain necessary conditions for T to be a blow up time. They can be stated as follows. Suppose that T is a blow up time, then, for any $3 < m \leq \infty$, there exists a constant c_m, depending on m only, such that

$$\|v(\cdot,t)\|_m \equiv \|v(\cdot,t)\|_{m,\mathbb{R}^3} \equiv \Big(\int\limits_{\mathbb{R}^3} |v(x,t)|^m dx\Big)^{\frac{1}{m}} \geq \frac{c_m}{(T-t)^{\frac{m-3}{2m}}} \tag{7.1.4}$$

for all $0 < t < T$.

However, for the scale-invariant L_3-norm, a weaker statement

$$\limsup_{t \to T-0} \|v(\cdot,t)\|_3 = \infty \tag{7.1.5}$$

has been proven in the previous chapter. The aim of this chapter is to improve (7.1.5). At the moment, the best improvement of (7.1.5) is given by the following theorem.

Theorem 1.1. *Let v be an energy solution to the Cauchy problem (7.1.1) and (7.1.2) with the initial data satisfying (7.1.3). Suppose that $T > 0$ is a finite blow up time. Then*

$$\lim_{t \to T-0} \|v(\cdot,t)\|_3 = \infty \tag{7.1.6}$$

holds true.

Let us briefly outline our proof of Theorem 1.1 that relays upon ideas developed in [Seregin (2007)]-[Seregin (2010)]. In particular, in [Seregin (2007)], a certain type of scaling has been invented, which, after passing to the limit, gives a special non-trivial solution to the Navier-Stokes equations provided there is a finite time blow up. In [Seregin (2011)] and [Seregin (2010)], it has been shown that the same type of scaling and blowing-up can produce the so-called Lemarie-Rieusset local energy solutions, introduced and carefully studied in the monograph [Lemarie-Riesset (2002)], see Appendix B for details. It turns out to be that the backward uniqueness technique is still applicable to those solutions. Although the theory of backward uniqueness itself is relatively well understood, its realization is not an easy task and based on delicate regularity results for the Navier-Stokes equations. Actually, there are two main points to verify: solutions, produced by scaling and blowing-up, vanish at the last moment of time and have a certain spatial decay. The first property is easy for solutions with bounded L_3-norm while the second one is harder. However, under certain restrictions, the required decay is a consequence of the Lemarie-Rieusset theory. So, the main technical part of the whole procedure is to show that scaling and blowing-up lead to local energy solutions. On that way, a lack of compactness of initial data of scaled solutions in $L_{2,\rm loc}$ is the main obstruction. This is why the same theorem for a stronger scale-invariant norm of the space $H^{\frac{1}{2}}$ is easier. The reason for that is a compactness of the corresponding embedding, see [Rusin and Sverak (2011)] and [Seregin (2011)].

In this chapter, we are going to show that, despite of a lack of compactness in L_3-case, the limit of the sequence of scaled solutions is still a local energy solution, for which a spatial decay takes place. Technically, this can be done by splitting each scaled solution into two parts. The first one is a solution to a non-linear problem but with zero initial data while the second one is a solution of a linear problem with weakly converging non-homogeneous initial data.

We also prove (7.1.4) as a by-product of the proof of Theorem 1.1, see Section 4.

7.2 Estimates of Scaled Solutions

Assume that our statement is false and there exists an increasing sequence t_k converging to T as $k \to \infty$ such that

$$\sup_{k\in\mathbb{N}} \|v(\cdot, t_k)\|_3 = M < \infty. \tag{7.2.1}$$

By the definition of a blow up time for energy solutions, there exists at least one singular point at time T. Without loss of generality, we may assume that it is $(0, T)$. Moreover, the blow-up profile has the finite L_3-norm, i.e.,

$$\|v(\cdot, T)\|_3 < \infty. \tag{7.2.2}$$

Indeed, by the ε-regularity theory, one-dimensional Hausdorff's measure of singular points at the blow up time T is equal to zero. Therefore, $v(x, t) \to v(x, T)$ as $t \to T - 0$ for a.a. x and thus (7.2.2) follows from Fatou's lemma.

Let us scale v and q so that

$$u^{(k)}(y, s) = \lambda_k v(x, t), \quad p^{(k)}(y, s) = \lambda_k^2 q(x, t), \tag{7.2.3}$$

for $(y, s) \in \mathbb{R}^3\times] - \lambda_k^{-2} T, 0[$, where

$$x = \lambda_k y, \qquad t = T + \lambda_k^2 s,$$

$$\lambda_k = \sqrt{\frac{T - t_k}{S}}$$

and a positive parameter $S < 10$ will be defined later.

By the scale invariance of L_3-norm, $u^{(k)}(\cdot, -S)$ is uniformly bounded in $L_3(\mathbb{R}^3)$, i.e.,

$$\sup_{k\in\mathbb{N}} \|u^{(k)}(\cdot, -S)\|_3 = M < \infty. \tag{7.2.4}$$

Let us decompose our scaled solution $u^{(k)}$ into two parts: $u^{(k)} = v^{(k)} + w^{(k)}$. Here, $w^{(k)}$ is a solution to the Cauchy problem for the Stokes system:

$$\partial_t w^{(k)} - \Delta w^{(k)} = -\nabla r^{(k)}, \quad \operatorname{div} w^{(k)} = 0 \quad \text{in} \quad \mathbb{R}^3\times] - S, 0[,$$
$$w^{(k)}(\cdot, -S) = u^{(k)}(\cdot, -S). \tag{7.2.5}$$

Obviously, (7.2.5) can be reduced to the Cauchy problem for the heat equation so that the pressure $r^{(k)} = 0$ and $w^{(k)}$ can be worked out with the help of the heat potential. The estimate below is well-known, see, for example [Kato (1984)],

$$\sup_k \{\|w^{(k)}\|_{L_5(\mathbb{R}^3\times]-S,0[} + \|w^{(k)}\|_{L_{3,\infty}(\mathbb{R}^3\times]-S,0[}\} \leq c(M) < \infty. \tag{7.2.6}$$

It is worthy to note that, by the scale invariance, $c(M)$ in (7.2.6) is independent of S.

As to $v^{(k)}$, it is a solution to the Cauchy problem for the following perturbed Navier-Stokes system

$$\partial_t v^{(k)} + \operatorname{div}(v^{(k)} + w^{(k)}) \otimes (v^{(k)} + w^{(k)}) - \Delta v^{(k)} = -\nabla p^{(k)},$$
$$\operatorname{div} v^{(k)} = 0 \quad \text{in} \quad \mathbb{R}^3 \times]-S, 0[, \tag{7.2.7}$$
$$v^{(k)}(\cdot, -S) = 0.$$

Now, our aim is to show that, for a suitable choice of $-S$, we can prove uniform estimates of $v^{(k)}$ and $p^{(k)}$ in certain spaces, pass to the limit as $k \to \infty$, and conclude that the limit functions u and p are a local energy solution to the Cauchy problem for the Navier-Stokes system in $\mathbb{R}^3 \times]-S, 0[$ associated with the initial data, generated by the weak L_3-limit of the sequence $u^{(k)}(\cdot, -S)$.

Let us start with estimates of solution to (7.2.7). First of all, we know the formula for the pressure:

$$p^{(k)}(x,t) = -\frac{1}{3}|u^{(k)}(x,t)|^2 + \frac{1}{4\pi}\int\limits_{\mathbb{R}^3} K(x-y) : u^{(k)}(y,t) \otimes u^{(k)}(y,t)dy, \tag{7.2.8}$$

where $K(x) = \nabla^2(1/|x|)$.

Next, we may decompose the pressure in the same way as it has been done in [Kikuchi and Seregin (2007)], see Appendix B. For $x_0 \in \mathbb{R}^3$ and for $x \in B(x_0, 3/2)$, we let

$$p^{(k)}_{x_0}(x,t) \equiv p^{(k)}(x,t) - c^{(k)}_{x_0}(t) = p^{1(k)}_{x_0}(x,t) + p^{2(k)}_{x_0}(x,t), \tag{7.2.9}$$

where

$$p^{1(k)}_{x_0}(x,t) = -\frac{1}{3}|u^{(k)}(x,t)|^2 + \frac{1}{4\pi}\int\limits_{B(x_0,2)} K(x-y) : u^{(k)}(y,t) \otimes u^{(k)}(y,t)dy,$$

$$p^{2(k)}_{x_0}(x,t) = \frac{1}{4\pi}\int\limits_{\mathbb{R}^3 \setminus B(x_0,2)} (K(x-y) - K(x_0-y)) : u^{(k)}(y,t) \otimes u^{(k)}(y,t)dy,$$

$$c^{(k)}_{x_0}(t) = \frac{1}{4\pi}\int\limits_{\mathbb{R}^3 \setminus B(x_0,2)} K(x_0-y) : u^{(k)}(y,t) \otimes u^{(k)}(y,t)dy.$$

Using the similar arguments as in [Lemarie-Riesset (2002)], see also Appendix B, one can derive estimates of $p^{1(k)}_{x_0}$ and $p^{2(k)}_{x_0}$. Here, they are:

$$\|p^{1(k)}_{x_0}(\cdot,t)\|_{L_{\frac{3}{2}}(B(x_0,3/2))} \leq c(M)(\|v^{(k)}(\cdot,t)\|^2_{L_3(B(x_0,2))} + 1), \tag{7.2.10}$$

$$\sup_{B(x_0,3/2)} |p^{2(k)}_{x_0}(x,t)| \leq c(M)(\|v^{(k)}(\cdot,t)\|^2_{L_{2,\text{unif}}} + 1), \tag{7.2.11}$$

where

$$\|g\|_{L_{2,\text{unif}}} = \sup_{x_0 \in \mathbb{R}^3} \|g\|_{L_2(B(x_0,1))}.$$

We further let

$$\alpha(s) = \alpha(s; k, S) = \|v^{(k)}(\cdot, s)\|^2_{2,\text{unif}},$$

$$\beta(s) = \beta(s; k, S) = \sup_{x \in R^3} \int_{-S}^{s} \int_{B(x,1)} |\nabla v^{(k)}|^2 dy d\tau.$$

From (7.2.10), (7.2.11), we find the estimate of the scaled pressure

$$\delta(0) \leq c(M)\Big[\gamma(0) + \int_{-S}^{0} (1 + \alpha^{\frac{3}{2}}(s))ds\Big], \tag{7.2.12}$$

with some positive constant $c(M)$ independent of k and S. Here, γ and δ are defined as

$$\gamma(s) = \gamma(s; k, S) = \sup_{x \in R^3} \int_{-S}^{s} \int_{B(x,1)} |v^{(k)}(y,\tau)|^3 dy d\tau$$

and

$$\delta(s) = \delta(s; k, S) = \sup_{x \in R^3} \int_{-S}^{s} \int_{B(x,3/2)} |p^{(k)}(y,\tau) - c^{(k)}_x(\tau)|^{\frac{3}{2}} dy\, d\tau,$$

respectively. It is known that an upper bound for γ can be given by the known multiplicative inequality

$$\gamma(s) \leq c\Big(\int_{-S}^{s} \alpha^3(\tau)d\tau\Big)^{\frac{1}{4}} \Big(\beta(s) + \int_{-S}^{s} \alpha(\tau)d\tau\Big)^{\frac{3}{4}}. \tag{7.2.13}$$

Fix $x_0 \in \mathbb{R}^3$ and a smooth non-negative function φ such that

$$\varphi = 1 \quad \text{in} \quad B(1), \qquad \text{spt} \subset B(3/2)$$

and let $\varphi_{x_0}(x) = \varphi(x - x_0)$.

Since the function $v^{(k)}$ is smooth on $[-S,0[$, we may write down the following energy identity

$$\int\limits_{\mathbb{R}^3}\varphi^2_{x_0}(x)|v^{(k)}(x,s)|^2dx+2\int\limits_{-S}^{s}\int\limits_{\mathbb{R}^3}\varphi^2_{x_0}|\nabla v^{(k)}|^2dxd\tau=$$

$$=\int\limits_{-S}^{s}\int\limits_{\mathbb{R}^3}\Big[|v^{(k)}|^2\Delta\varphi^2_{x_0}+v^{(k)}\cdot\nabla\varphi^2_{x_0}(|v^{(k)}|^2+2p^{(k)}_{x_0})\Big]dxd\tau+$$

$$+\int\limits_{-S}^{s}\int\limits_{\mathbb{R}^3}\Big[w^{(k)}\cdot\nabla\varphi^2_{x_0}|v^{(k)}|^2+2\varphi^2_{x_0}w^{(k)}\otimes(w^{(k)}+v^{(k)}):\nabla v^{(k)}+$$

$$+2w^{(k)}\cdot v^{(k)}(w^{(k)}+v^{(k)})\cdot\nabla\varphi^2_{x_0}\Big]dxd\tau=I_1+I_2.$$

The first term I_1 is estimated with the help of the Hölder inequality, multiplicative inequality (7.2.13), and bounds (7.2.10), (7.2.11). So, we find

$$I_1\leq c(M)\Big[\int\limits_{-S}^{s}(1+\alpha(\tau)+\alpha^{\frac32}(\tau))d\tau+$$

$$+\Big(\int\limits_{-S}^{s}\alpha^3(\tau)d\tau\Big)^{\frac14}\Big(\beta(s)+\int\limits_{-S}^{s}\alpha(\tau)d\tau\Big)^{\frac34}\Big].$$

Now, let us evaluate the second term

$$I_2\leq c\int\limits_{-S}^{s}\|v^{(k)}(\cdot,\tau)\|^2_{L_3(B(x_0,3/2))}\|w^{(k)}(\cdot,\tau)\|_{L_3(B(x_0,3/2))}d\tau+$$

$$+c\int\limits_{-S}^{s}\Big(\int\limits_{B(x_0,3/2)}|w^{(k)}|^5dx\Big)^{\frac15}\Big(\int\limits_{B(x_0,3/2)}|v^{(k)}|^{\frac54}|\nabla v^{(k)}|^{\frac54}dx\Big)^{\frac45}d\tau+$$

$$+c\beta^{\frac12}(s)\Big(\int\limits_{-S}^{s}\int\limits_{B(x_0,3/2)}|w^{(k)}|^4dxd\tau\Big)^{\frac12}d\tau+$$

$$+c\int\limits_{-S}^{s}\|v^{(k)}(\cdot,\tau)\|_{L_3(B(x_0,3/2))}\|w^{(k)}(\cdot,\tau)\|^2_{L_3(B(x_0,3/2))}d\tau.$$

Taking into account (7.2.6) and applying Hölder inequality several times, we find

$$I_2 \leq c(M)\gamma^{\frac{2}{3}}(s)(s+S)^{\frac{1}{3}}+$$

$$+c\int\limits_{-S}^{s}\Big(\int\limits_{B(x_0,3/2)}|w^{(k)}|^5dx\Big)^{\frac{1}{5}}\Big(\int\limits_{B(x_0,3/2)}|\nabla v^{(k)}|^2dx\Big)^{\frac{1}{2}}\times$$

$$\times\Big(\int\limits_{B(x_0,3/2)}|v^{(k)}|^{\frac{10}{3}}dx\Big)^{\frac{3}{10}}d\tau + c(M)\beta^{\frac{1}{2}}(s)(s+S)^{\frac{1}{10}}+$$

$$+c(M)\gamma^{\frac{1}{3}}(s)(s+S)^{\frac{2}{3}}.$$

It remains to use another known multiplicative inequality

$$\Big(\int\limits_{B(x_0,3/2)}|v^{(k)}(x,s)|^{\frac{10}{3}}dx\Big)^{\frac{3}{10}} \leq c\Big(\int\limits_{B(x_0,3/2)}|v^{(k)}(x,s)|^2dx\Big)^{\frac{1}{5}}\times$$

$$\times\Big(\int\limits_{B(x_0,3/2)}(|\nabla v^{(k)}(x,s)|^2+|v^{(k)}(x,s)|^2dx\Big)^{\frac{3}{10}}$$

and to conclude that

$$I_2 \leq c(M)\gamma^{\frac{2}{3}}(s)(s+S)^{\frac{1}{3}} + c(M)\beta^{\frac{1}{2}}(s)(s+S)^{\frac{1}{10}} + c(M)\gamma^{\frac{1}{3}}(s)(s+S)^{\frac{2}{3}}+$$

$$+c\Big(\beta(s)+\int\limits_{-S}^{s}\alpha(\tau)d\tau\Big)^{\frac{4}{5}}\times\Big(\int\limits_{-S}^{s}\alpha(\tau)\|w^{(k)}(\cdot,\tau)\|^5_{L_{5,\mathrm{unif}}}d\tau\Big)^{\frac{1}{5}}.$$

Finally, we find

$$\alpha(s)+\beta(s) \leq c(M)\Big[(s+S)^{\frac{1}{5}}+$$

$$+\int\limits_{-S}^{s}\Big(\alpha(\tau)(1+\|w^{(k)}(\cdot,\tau)\|^5_{L_{5,\mathrm{unif}}})+\alpha^3(\tau)\Big)d\tau\Big], \tag{7.2.14}$$

which is valid for any $s \in [-S,0[$ and for some positive constant $c(M)$ independent of k, s, and S.

It is not so difficult to show that there is a positive constant $S(M)$ such that

$$\alpha(s) \leq \frac{1}{10} \tag{7.2.15}$$

for any $s \in]-S(M), 0[$. In turn, the latter will also imply that

$$\alpha(s) \leq c(M)(s+S)^{\frac{1}{5}} \tag{7.2.16}$$

for any $s \in]-S(M), 0[$.

To see how this can be worked out, let us assume

$$\alpha(s) \leq 1 \tag{7.2.17}$$

for $-S \leq s < s_0 \leq 0$. Then (7.2.14) yields

$$\alpha(s) \leq c(M)((s+S)^{\frac{1}{5}} + y(s)) \tag{7.2.18}$$

for the same s. Here,

$$y(s) = \int\limits_{-S}^{s} \alpha(\tau)(2+g(\tau))d\tau, \qquad g(s) = \|w^{(k)}(\cdot,s)\|^2_{L_5(\mathbb{R}^3)}.$$

The function $y(s)$ obeys the differential inequality

$$y'(s) \leq c(M)(2+g(s))((s+S)^{\frac{1}{5}} + y(s)) \tag{7.2.19}$$

for $-S \leq s < s_0 \leq 0$. After integrating (7.2.19), we find

$$y(s) \leq c(M) \int\limits_{-S}^{s} \Big((\tau+S)^{\frac{1}{5}}(2+g(\tau)) \exp\Big\{c(M) \int\limits_{\tau}^{s} (2+g(\vartheta))\Big\}d\vartheta\Big)d\tau \tag{7.2.20}$$

for $-S \leq s < s_0 \leq 0$. Taking into account estimate (7.2.6), we derive from (7.2.20) the following bound

$$y(s) \leq c_1(M)(s+S)^{\frac{1}{5}} \tag{7.2.21}$$

for $-S \leq s < s_0 \leq 0$ and thus

$$y(s) \leq c_1(M)S^{\frac{1}{5}} \tag{7.2.22}$$

for the same s.

Now, let us pick up $S(M) > 0$ so small that

$$c(M)(1+c_1(M))S^{\frac{1}{5}}(M) = \frac{1}{20}. \tag{7.2.23}$$

We claim that, for such a choice of S(M), statement (7.2.15) holds true. Indeed, assume that it is false. Then since $\alpha(s)$ is a continuous function on $[-S, 0[$ and $\alpha(0) = 0$, there exists $s_0 \in]-S, 0[$ such that $0 \leq \alpha(s) < \frac{1}{10}$ for all $s \in]-S, s_0[$ and $\alpha(s_0) = \frac{1}{10}$. In this case, we may use first (7.2.22) and then (7.2.18), (7.2.23) to get

$$\alpha(s) \leq c(M)(1+c_1(M))S^{\frac{1}{5}}(M) = \frac{1}{20}$$

for $s \in]-S, s_0[$. This leads to a contradiction and, hence, (7.2.15) has been proven. It remains to use (7.2.18) and (7.2.21) with $s_0 = 0$ in order to establish (7.2.16).

7.3 Limiting Procedure

As to $w^{(k)}$, it is defined by the solution formula

$$w^{(k)}(x,t) = \frac{1}{(4\pi(s+S))^{\frac{3}{2}}} \int\limits_{\mathbb{R}^3} \exp\Big(-\frac{|x-y|^2}{4(s+S)}\Big) u^{(k)}(y,-S)dy.$$

Moreover, by standard localization arguments, the following estimate can be derived:

$$\sup_{-S<s<0}\sup_{x_0\in\mathbb{R}^3} \|w^{(k)}(\cdot,s)\|^2_{L_2(B(x_0,1))}+$$

$$+\sup_{x_0\in\mathbb{R}^3}\int\limits_{-S}^{0}\int\limits_{B(x_0,1)} |\nabla w^{(k)}(y,s)|^2 dyds \leq c(M)<\infty.$$

Obviously, $w^{(k)}$ and all its derivatives converge to w and to its corresponding derivatives uniformly in sets of the form $\overline{B}(R)\times[\delta,0]$ for any $R>0$ and for any $\delta\in]-S,0[$. The limit function satisfies the same representation formula

$$w(x,t) = \frac{1}{(4\pi(s+S))^{\frac{3}{2}}} \int\limits_{\mathbb{R}^3} \exp\Big(-\frac{|x-y|^2}{4(s+S)}\Big) w_0(y)dy,$$

in which w_0 is the weak $L_3(\mathbb{R}^3)$-limit of the sequence $u^{(k)}(\cdot,-S)$. The function w satisfies the uniform local energy estimate

$$\sup_{-S<s<0}\sup_{x_0\in\mathbb{R}^3} \|w(\cdot,s)\|^2_{L_2(B(x_0,1))}+$$

$$+\sup_{x_0\in\mathbb{R}^3}\int\limits_{-S}^{0}\int\limits_{B(x_0,1)} |\nabla w(y,s)|^2 dyds \leq c(M)<\infty.$$

The important fact, coming from the solution formula, is as follows:

$$w\in C([-S,0];L_3(\mathbb{R}^3))\cap L_5(\mathbb{R}^3\times]-S,0[). \tag{7.3.1}$$

Next, the uniform local energy estimate for the sequence $u^{(k)}$ (with respect to k) can be deduced from the estimates above. This allows us to exploit the limiting procedure explained in [Kikuchi and Seregin (2007)], see Appendix B, in details. As a result, one can select a subsequence, still denoted by $u^{(k)}$, with the following properties:

for any $a>0$,

$$u^{(k)}\to u \tag{7.3.2}$$

weakly-star in $L_\infty(-S,0;L_2(B(a)))$ and strongly in $L_3(B(a)\times]-S,0[)$ and in $C([\tau,0];L_{\frac{9}{8}}(B(a)))$ for any $-S<\tau<0$;

$$\nabla u^{(k)}\to\nabla u \tag{7.3.3}$$

weakly in $L_2(B(a)\times]-S,0[)$;

$$t\mapsto\int\limits_{B(a)}u^{(k)}(x,t)\cdot w(x)dx\to t\mapsto\int\limits_{B(a)}u(x,t)\cdot w(x)dx \tag{7.3.4}$$

strongly in $C([-S,0])$ for any $w\in L_2(B(a))$. The corresponding sequences $v^{(k)}$ and $w^{(k)}$ converge to their limits v and w in the same sense and of course $u=v+w$. For the pressure p, we have the following convergence: for any $n\in\mathbb{N}$, there exists a sequences $c_n^{(k)}\in L_{\frac{3}{2}}(-S,0)$ such that

$$\widetilde{p}_n^{(k)}\equiv p^{(k)}-c_n^{(k)}\rightharpoonup p \tag{7.3.5}$$

in $L_{\frac{3}{2}}(-S,0;L_{\frac{3}{2}}(B(n)))$.

So, arguing in the same way as in [Kikuchi and Seregin (2007)], see Appendix B, one can show that u and p satisfy the following conditions:

$$\sup_{-S<s<0}\sup_{x_0\in\mathbb{R}^3}\|u(\cdot,s)\|^2_{L_2(B(x_0,1))}+\sup_{x_0\in\mathbb{R}^3}\int\limits_{-S}^{0}\int\limits_{B(x_0,1)}|\nabla u(y,s)|^2dyds<\infty; \tag{7.3.6}$$

$$p\in L_{\frac{3}{2}}(-S,0;L_{\frac{3}{2},\mathrm{loc}}(\mathbb{R}^3); \tag{7.3.7}$$

the function

$$s\mapsto\int\limits_{\mathbb{R}^3}u(y,s)\cdot w(y)dy \tag{7.3.8}$$

is continuous on $[-S,0]$ for any compactly supported $w\in L_2(\mathbb{R}^3)$;

$$\partial_t u+u\cdot\nabla u-\Delta u=-\nabla p,\quad \operatorname{div}u=0 \tag{7.3.9}$$

in $\mathbb{R}^3\times]-S,0[$ in the sense of distributions;

for any $x_0\in\mathbb{R}^3$, there exists a function $c_{x_0}\in L_{\frac{3}{2}}(-S,0)$ such that

$$p(x,t)-c_{x_0}(t)=p^1_{x_0}(x,t)+p^2_{x_0}(x,t) \tag{7.3.10}$$

for all $x\in B(x_0,3/2)$, where

$$p^1_{x_0}(x,t)=-\frac{1}{3}|u(x,t)|^2+\frac{1}{4\pi}\int\limits_{B(x_0,2)}K(x-y):u(y,t)\otimes u(y,t)dy,$$

$$p^2_{x_0}(x,t) = \frac{1}{4\pi}\int\limits_{\mathbb{R}^3\setminus B(x_0,2)} (K(x-y) - K(x_0-y)) : u(y,t)\otimes u(y,t)dy;$$

for any $s \in]-S, 0[$ and for $\varphi \in C^\infty_0(\mathbb{R}^3\times]-S,S[)$,

$$\int\limits_{\mathbb{R}^3} \varphi^2(y,s)|u(y,s)|^2dy + 2\int\limits_{-S}^{s}\int\limits_{\mathbb{R}^3} \varphi^2|\nabla u|^2dyd\tau \leq$$

$$\leq \int\limits_{-S}^{s}\int\limits_{\mathbb{R}^3}\Big(|u|^2(\Delta\varphi^2 + \partial\varphi^2) + u\cdot\nabla\varphi^2(|u|^2+2p)\Big)dyd\tau. \qquad (7.3.11)$$

Passing to the limit in (7.2.16), we find

$$\sup_{x_0\in\mathbb{R}^3} \|v(\cdot,s)\|^2_{L_2(B(x_0,1))} \leq c(M)(s+S)^{\frac{1}{5}}$$

for all $s \in [-S,0]$. And thus

$$v(\cdot,s) \to 0 \qquad \text{in} \quad L_{2,\mathrm{loc}}(\mathbb{R}^3)$$

as $s \downarrow -S$. Then, taking into account (7.3.1), we can conclude that

$$u(\cdot,s) \to w_0 \qquad \text{in} \quad L_{2,\mathrm{loc}}(\mathbb{R}^3). \qquad (7.3.12)$$

as $s \downarrow -S$.

By definition accepted in [Kikuchi and Seregin (2007)], see Apendix B, the pair u and p, satisfying (7.3.6)–(7.3.12), is a local energy solution to the Cauchy problem for the Navier-Stokes equations in $\mathbb{R}^3\times]-S,0[$ associated with the initial velocity w_0.

Now, our aim is to show that u is not identically zero. Using the inverse scaling, we observe that the following identity takes place:

$$\frac{1}{a^2}\int\limits_{Q(a)} (|u^{(k)}|^3 + |\widetilde{p}^{(k)}|^{\frac{3}{2}})dy\,ds = \frac{1}{(a\lambda_k)^2}\int\limits_{Q(z_T,a\lambda_k)} (|v|^3 + |q-b^{(k)}|^{\frac{3}{2}})dx\,dt$$

for all $0 < a < a_* = \inf\{1, \sqrt{S/10}, \sqrt{T/10}\}$ and for all $\lambda_k \leq 1$. Here, $z_T = (0,T)$, $\widetilde{p}^{(k)} \equiv \widetilde{p}^{(k)}_2$, and $b^{(k)}(t) = \lambda_k^{-2}c_2^{(k)}(s)$. Since the pair v and $q-b^{(k)}$ is a suitable weak solution to the Navier-Stokes equations in $Q(z_T,\lambda_k a_*)$, we find

$$\frac{1}{a^2}\int\limits_{Q(a)} (|u^{(k)}|^3 + |\widetilde{p}^{(k)}|^{\frac{3}{2}})dy\,ds > \varepsilon \qquad (7.3.13)$$

for all $0 < a < a_*$ with a positive universal constant ε.

Now, by (7.3.2) and (7.3.5),

$$\frac{1}{a^2}\int\limits_{Q(a)}|u^{(k)}|^3dy\,ds\to\frac{1}{a^2}\int\limits_{Q(a)}|u|^3dy\,ds \tag{7.3.14}$$

for all $0<a<a_*$ and

$$\sup_{k\in\mathbb{N}}\frac{1}{a_*^2}\int\limits_{Q(a_*)}(|u^{(k)}|^3+|\widetilde{p}^{(k)}|^{\frac{3}{2}})dy\,ds=M_1<\infty. \tag{7.3.15}$$

To treat the pressure $\widetilde{p}^{(k)}$, we do the usual decomposition of it into two parts, see similar arguments in [Seregin (2011)]. The first one is completely controlled by the pressure while the second one is a harmonic function in $B(a_*)$ for all admissible t. In other words, we have

$$\widetilde{p}^{(k)}=p_1^{(k)}+p_2^{(k)}$$

where $p_1^{(k)}$ obeys the estimate

$$\|p_1^{(k)}(\cdot,s)\|_{\frac{3}{2},B(a_*)}\leq c\|u^{(k)}(\cdot,s)\|^2_{3,B(a_*)}. \tag{7.3.16}$$

For the harmonic counterpart of the pressure, we have

$$\sup_{y\in B(a_*/2)}|p_2^{(k)}(y,s)|^{\frac{3}{2}}\leq c(a_*)\int\limits_{B(a_*)}|p_2^{(k)}(y,s)|^{\frac{3}{2}}dy$$

$$\leq c(a_*)\int\limits_{B(a_*)}(|\widetilde{p}^{(k)}(y,s)|^{\frac{3}{2}}+|u^{(k)}(y,s)|^3)dy \tag{7.3.17}$$

for all $-a_*^2<s<0$.

For any $0<a<a_*/2$,

$$\varepsilon\leq\frac{1}{a^2}\int\limits_{Q(a)}(|\widetilde{p}^{(k)}|^{\frac{3}{2}}+|u^{(k)}|^3)dy\,ds\leq$$

$$\leq c\frac{1}{a^2}\int\limits_{Q(a)}(|p_1^{(k)}|^{\frac{3}{2}}+|p_2^{(k)}|^{\frac{3}{2}}+|u^{(k)}|^3)dy\,ds\leq$$

$$\leq c\frac{1}{a^2}\int\limits_{Q(a)}(|p_1^{(k)}|^{\frac{3}{2}}+|u^{(k)}|^3)dy\,ds+$$

$$+ca^3\frac{1}{a^2}\int\limits_{-a^2}^{0}\sup_{y\in B(a_*/2)}|p_2^{(k)}(y,s)|^{\frac{3}{2}}ds.$$

From (7.3.15)–(7.3.17), it follows that

$$\varepsilon \leq c\frac{1}{a^2}\int\limits_{Q(a_*)}|u^{(k)}|^3 dy\,ds + ca\int\limits_{-a^2}^{0} ds\int\limits_{B(a_*)}(|\widetilde{p}^{(k)}(y,s)|^{\frac{3}{2}} + |u^{(k)}(y,s)|^3)dy \leq$$

$$\leq c\frac{1}{a^2}\int\limits_{Q(a_*)}|u^{(k)}|^3 dy\,ds + ca\int\limits_{Q(a_*)}(|\widetilde{p}^{(k)}|^{\frac{3}{2}} + |u^{(k)}|^3)dy\,ds \leq$$

$$\leq c\frac{1}{a^2}\int\limits_{Q(a_*)}|u^{(k)}|^3 dy\,ds + cM_1 a a_*^2$$

for all $0 < a < a_*/2$. After passing to the limit and picking up sufficiently small a, we find

$$0 < c\varepsilon a^2 \leq \int\limits_{Q(a_*)}|u|^3 dy\,ds \tag{7.3.18}$$

for some positive $0 < a < a_*/2$. So, the limit function u is non-trivial.

PROOF THEOREM 1.1 The limit function $w_0 \in L_3$ and, hence,

$$\|w_0\|_{2,B(x_0,1)} \to 0$$

as $|x_0| \to \infty$. The latter, together with Theorem 1.6 from Appendix B, and ε-regularity theory for the Navier-Stokes equations, gives a required decay at infinity. To be more precise, there are positive numbers R, $T \in]a_*, S[$, and c_k with $k = 0, 1, ...$ such that

$$|\nabla^k u(x,t)| \leq c_k \tag{7.3.19}$$

for any $x \in \mathbb{R}^3 \setminus B(R/2)$ and for any $t \in]-T, 0[$.

The second thing to be noticed is that the following important property holds true:

$$u(\cdot, 0) = 0. \tag{7.3.20}$$

This follows from (7.2.2) and (7.3.2), see the last statement in (7.3.2). Since vorticity $\omega = \nabla \wedge u$ vanishes at $t = 0$ as well, we can apply the backward uniqueness result from Appendix A to the vorticity equation and conclude that $\omega = 0$ in $(\mathbb{R}^3 \setminus B(R/2)) \times]-T, 0[$. Now, our aim is to show that in fact $\omega = 0$ in $\mathbb{R}^3 \times]-T, 0[$. If so, $u(\cdot, t)$ is going to be a bounded harmonic function with the additional property $\|u(\cdot,t)\|_{L_2(B(x_0,1))} \to 0$ as $|x_0| \to \infty$ and thus we may conclude that $u = 0$ in $(\mathbb{R}^3 \setminus B(R/2)) \times]-T, 0[$. The latter contradicts (7.3.18) and, hence, z_T is not a singular point.

The idea of the proof of the above claim is more or less the same as in paper [Escauriaza *et al.* (2003)]. However, in the present case, we have less regularity and no global finite norm for the pressure. The way out is to use decomposition (7.3.10) in order to get better estimates for the pressure, say, in the domain $(\mathbb{R}^3 \setminus B(R))\times]-T,0[$. Indeed, using estimates of type (7.2.10) and (7.2.11) for the parts of the pressure $p^1_{x_0}$ and $p^2_{x_0}$ in (7.3.10), we show

$$\|p(\cdot,t) - c_{x_0}(t)\|_{L_{3/2}(B(x_0,3/2))} < c$$

provided $B(x_0,2) \in \mathbb{R}^3 \setminus B(R/2)$. Here, a constant c is independent of x_0 and t. Then, local regularity theory, applied to the pressure equation $\Delta p = -\mathrm{div}\,\mathrm{div}\, u\otimes u$, together with estimate (7.3.19), implies

$$|\nabla^k p(x,t)| < c^1_k$$

for any $x \in \mathbb{R}^3 \setminus B(R)$, any $t \in]0,T[$, and any $k = 1,2,\ldots$. If we replace the pressure p with $p - [p]_{B(4R)\setminus B(R)}$, then from Poincare's inequality, from previous estimates, and from the equation $\partial_t u + u\cdot\nabla u - \Delta u = -\nabla p$, it follows that

$$|\nabla^k u(x,t)| + |\nabla^k p(x,t)| + |\nabla^k \partial_t u(x,t)| < c^2_k \tag{7.3.21}$$

for all $x \in B(4R)\setminus B(R)$, for all $t \in]-T,0[$, and for all $k = 0,1,\ldots$.

Next, we pick up a smooth cut-off function φ such that $\varphi = 0$ out of $B(3R)$ and $\varphi = 1$ in $B(2R)$ and introduce auxiliary functions $\widetilde{w}$ and $\widetilde{r}$ obeying the equations

$$\Delta\widetilde{w} = \nabla\widetilde{r}, \qquad \mathrm{div}\widetilde{w} = u\cdot\nabla\varphi$$

in $B(4R)$ and the additional conditions

$$\widetilde{w}|_{\partial B(4R)} = 0, \qquad \int\limits_{B(4R)} \widetilde{r}dx = 0.$$

In a view of (7.3.19), the regularity theory for the stationary Stokes system gives the estimates

$$|\nabla^k \widetilde{w}(x,t)| + |\nabla^k \widetilde{r}(x,t)| + |\nabla^k \partial_t \widetilde{w}(x,t)| < c^2_k \tag{7.3.22}$$

being valid for all $x \in B(4R)$, for all $t \in]-T,0[$, and for all $k = 0,1,\ldots$. Letting $U = w - \widetilde{w}$ and $P = r - \widetilde{r}$, where $w = \varphi u$ and $r = \varphi p$, we find

$$\partial_t U + \mathrm{div}(U\otimes U) - \Delta U + \nabla P = F = -\mathrm{div}(U\otimes\widetilde{w} + \widetilde{w}\otimes U) + G,$$

$$\mathrm{div}\, U = 0$$

in $Q_* = B(4R)\times]-T,0[$,

$$U|_{\partial B(4R)\times]-T,0[} = 0.$$

Here, $G = -\mathrm{div}(\widetilde{w}\otimes\widetilde{w}) + g - \partial_t\widetilde{w}$ and

$$g = (\varphi^2-\varphi)\mathrm{div}(u\otimes u) + uu\cdot\nabla\varphi^2 + p\nabla\varphi - 2\nabla\varphi\cdot\nabla u - u\Delta\varphi.$$

Since u and p are a local energy solution, it follows from its definition that there exists a set $\Sigma\subset]-T,0[$ of full measure, i.e., $|\Sigma| = T$, such that U is a weak Leray-Hopf solution to initial boundary problem for the above system in $B(4R)\times]t_0,0[$ for each $t_0\in\Sigma$. The rest of the proof is based upon estimates (7.3.21) and (7.3.22) and unique continuation across spatial boundaries for parabolic differential inequalities and goes along lines of arguments in the last section of Chapter 6. Theorem 1.1 is proved.

Let us outline the proof of (7.1.4), which is much easier than the proof of Theorem 1.1. Indeed, arguing as in the main case, we find a sequence $t_k\to T-0$ such that

$$\lim_{k\to\infty}\|v(\cdot,t_k)\|_m(T-t_k)^{\frac{m-3}{2m}} = 0.$$

The scaling implies $\|u^{(k)}(\cdot,-S)\|_m\to 0$ and thus

$$\|u^{(k)}(\cdot,-S)\|_{2,\mathrm{unif}}\to 0. \tag{7.3.23}$$

For solutions $u^{(k)}$, we may use local energy estimates proved in Appendix B. In particular, they give the estimate

$$\|u^{(k)}(\cdot,t)\|^2_{2,\mathrm{unif}} \leq 2c\|u^{(k)}(\cdot,-S)\|^2_{2,\mathrm{unif}}$$

for any $t\in]-S,0[$. And S should be chosen independently of k so that

$$0 < S < \frac{\ln 2}{c(1+(2c\|u^{(k)}(\cdot,-S)\|_{2,\mathrm{unif}})^2)}$$

for all $k\in\mathbb{N}$. It is possible because of (7.3.23).

So, we can claim that

$$\sup_{-S<t<0}\|u^{(k)}(\cdot,t)\|^2_{2,\mathrm{unif}} + \sup_{x_0\in\mathbb{R}^3}\int\limits_{-S}^{0}\int\limits_{B(x_0,1)}|\nabla u^{(k)}|^2dxdt\to 0$$

as $k\to\infty$. This means that the limit solution must be identically zero. However, using the same arguments as in the previous section, we can show that the limit solution is not a trivial one provided that the original solution blows up at time T.

7.4 Comments

This section is essentially based on my paper [Seregin (2012)], which in turn summarizes all previous attempts made in [Seregin (2007)]-[Seregin (2010)] to solve the problem about behavior of L_3-norm of the velocity field as time approaches possible blow up time.

Appendix A

Backward Uniqueness and Unique Continuation

A.1 Carleman-Type Inequalities

We start with the first Carleman type inequality which has been already used in [Escauriaza *et al.* (2002)] and [Escauriaza *et al.* (2003)] (see also [Escauriaza (2000)], [Escauriaza and Vega (2001)], and [Tataru (2000)]).

Proposition 1.2. *For any function $u \in C_0^\infty(\mathbb{R}^n \times]0,2[;\mathbb{R}^m)$ and for any positive number a, the inequality*

$$\int\limits_{\mathbb{R}^n\times]0,2[} h^{-2a}(t)e^{-\frac{|x|^2}{4t}}\left(\tfrac{a}{t}|u|^2+|\nabla u|^2\right)dxdt$$
$$\leq c_0 \int\limits_{\mathbb{R}^n\times]0,2[} h^{-2a}(t)e^{-\frac{|x|^2}{4t}}|\partial_t u+\Delta u|^2\,dxdt \tag{A.1.1}$$

is valid with an absolute positive constant c_0 and a function $h(t) = te^{\frac{1-t}{3}}$.

Proof of Proposition 1.2 Our approach is based on the L_2-theory of Carleman inequalities developed essentially in [Hörmander (1963)], see also [Tataru (2000)].

Let u be an arbitrary function from $C_0^\infty(\mathbb{R}^n\times]0,2[;\mathbb{R}^m)$. We let $\phi(x,t) = -\frac{|x|^2}{8t} - (a+1)\ln h(t)$ and $v = e^\phi u$. Then, we have
$Lv := e^\phi(\partial_t u+\Delta u) = \partial_t v - \operatorname{div}(v\otimes\nabla\phi) - \nabla v\nabla\phi + \Delta v + (|\nabla\phi|^2-\partial_t\phi)v.$
The main step in the above approach is the decomposition of operator tL into symmetric and skew symmetric parts, i.e.,

$$tL = S + A, \tag{A.1.2}$$

where

$$Sv := t(\Delta v + (|\nabla\phi|^2-\partial_t\phi)v) - \frac{1}{2}v \tag{A.1.3}$$

and

$$Av := \frac{1}{2}(\partial_t(tv) + t\partial_t v) - t(\mathrm{div}(v \otimes \nabla\phi) + \nabla v \nabla\phi). \tag{A.1.4}$$

Obviously,

$$\begin{aligned} \int t^2 e^{2\phi}|\partial_t u + \Delta u|^2\, dxdt &= \int t^2 |Lv|^2\, dxdt \\ &= \int |Sv|^2\, dxdt + \int |Av|^2\, dxdt + \int [S, A]v \cdot v\, dxdt, \end{aligned} \tag{A.1.5}$$

where $[S, A] = SA - AS$ is the commutator of S and A. Simple calculations show that

$$\begin{aligned} I &:= \int [S, A]v \cdot v\, dxdt = \\ &= 4\int t^2 \Big[\phi_{,ij} v_{,i} \cdot v_{,j} + \phi_{,ij}\phi_{,i}\phi_{,j}|v|^2\Big]\, dxdt \\ &+ \int t^2|v|^2(\partial_t^2\phi - 2\partial_t|\nabla\phi|^2 - \Delta^2\phi)\, dxdt \\ &+ \int t|\nabla v|^2\, dxdt - \int t|v|^2(|\nabla\phi|^2 - \partial_t\phi)\, dxdt. \end{aligned} \tag{A.1.6}$$

Given choice of function ϕ, we have

$$I = (a+1)\int t^2\Big[-\Big(\frac{h'(t)}{h(t)}\Big)' - \frac{h'(t))}{th(t)}\Big]|v|^2\, dxdt = \frac{a+1}{3}\int t|v|^2\, dxdt. \tag{A.1.7}$$

By the simple identity

$$|\nabla v|^2 = \frac{1}{2}(\partial_t + \Delta)|v|^2 - v \cdot (\partial_t v + \Delta v), \tag{A.1.8}$$

we find

$$\begin{aligned} \int t^2|\nabla v|^2\, dxdt &= -\int t|v|^2\, dxdt - \int t^2 v \cdot Lv\, dxdt \\ &+ \int t^2|v|^2(|\nabla\phi|^2 - \partial_t\phi)\, dxdt. \end{aligned} \tag{A.1.9}$$

In our case,

$$|\nabla\phi|^2 - \partial_t\phi = -|\nabla\phi|^2 + (a+1)\frac{h'(t)}{h(t)}.$$

The latter relation (together with (A.1.7)) implies the bound

$$\begin{aligned} &\int t^2(|\nabla v|^2 + |v|^2|\nabla\phi|^2)\, dxdt \\ &\le 3I - \int t^2 v \cdot Lv\, dxdt \le b_1 \int t^2|Lv|^2\, dxdt \end{aligned} \tag{A.1.10}$$

with an absolute positive constant b_1. Since

$$e^{\phi}|\nabla u| \le |\nabla v| + |v||\nabla\phi|, \tag{A.1.11}$$

it follows from (A.1.5)–(A.1.10) that

$$\int h^{-2a}(t)(th^{-1}(t))^2\Big((a+1)\frac{|u|^2}{t}+|\nabla u|^2\Big)e^{-\frac{|x|^2}{4t}}\,dxdt$$
$$\leq b_2\int h^{-2a}(t)(th^{-1}(t))^2|\partial_t u+\Delta u|^2 e^{-\frac{|x|^2}{4t}}\,dxdt.$$

Here, b_2 is an absolute positive constant. Inequality (A.1.1) is proved.

The second Carleman-type inequality is, in a sense, an anisotropic one.

Proposition 1.3. *Let*

$$\phi=\phi^{(1)}+\phi^{(2)},$$

where $\phi^{(1)}(x,t)=-\frac{|x'|^2}{8t}$ *and* $\phi^{(2)}(x,t)=a(1-t)\frac{x_n^{2\alpha}}{t^\alpha}$, $x'=(x_1,x_2,...,x_{n-1})$ *so that* $x=(x',x_n)$, *and* $e_n=(0,0,...,0,1)$. *Then, for any function* $u\in C_0^\infty((\mathbb{R}^n_+ + e_n)\times]0,1[;\mathbb{R}^m)$ *and for any number* $a>a_0(\alpha)$, *the following inequality is valid:*

$$\int\limits_{(\mathbb{R}^n_+ + e_n)\times]0,1[} t^2e^{2\phi(x,t)}\Big(a\frac{|u|^2}{t^2}+\frac{|\nabla u|^2}{t}\Big)\,dxdt$$
$$\leq c_\star\int\limits_{(\mathbb{R}^n_+ + e_n)\times]0,1[} t^2e^{2\phi(x,t)}|\partial_t u+\Delta u|^2\,dxdt. \tag{A.1.12}$$

Here, $c_\star=c_\star(\alpha)$ *is a positive constant and* $\alpha\in]1/2,1[$ *is fixed.*

Proof Let $u\in C_0^\infty(Q^1_+;\mathbb{R}^m)$, where $Q^1_+=(\mathbb{R}^n_+ + e_n)\times]0,1[$. We are going to use formulae (A.1.2)–(A.1.6) for new functions u, v, and ϕ. All integrals in those formulae are taken now over Q^1_+.

First, we observe that

$$\nabla\phi=\nabla\phi^{(1)}+\nabla\phi^{(2)}$$
$$\nabla\phi^{(1)}(x,t)=-\frac{x'}{4t},\qquad \nabla\phi^{(2)}(x,t)=2\alpha a\frac{1-t}{t^\alpha}x_n^{2\alpha-1}e_n. \tag{A.1.13}$$

Therefore,

$$\nabla\phi^{(1)}\cdot\nabla\phi^{(2)}=0,\qquad |\nabla\phi|^2=|\nabla\phi^{(1)}|^2+|\nabla\phi^{(2)}|^2. \tag{A.1.14}$$

Moreover,

$$\nabla^2\phi=\nabla^2\phi^{(1)}+\nabla^2\phi^{(2)},$$

$$\phi^{(1)}_{,ij}=\begin{cases}-\frac{\delta_{ij}}{4t} & \text{if}\quad 1\leq i,j\leq n-1\\ 0 & \text{if}\quad i=n\quad\text{or}\quad j=n\end{cases}, \tag{A.1.15}$$

$$\phi^{(2)}_{,ij}=\begin{cases}0 & \text{if}\quad i\neq n\quad\text{or}\quad j\neq n\\ 2\alpha(2\alpha-1)a\frac{1-t}{t^\alpha}x_n^{2\alpha-2} & \text{if}\quad i=n\quad\text{and}\quad j=n\end{cases}.$$

In particular, (A.1.15) implies

$$\phi_{,ij}\phi_{,i}\phi_{,j} = \phi^{(1)}_{,ij}\phi^{(1)}_{,i}\phi^{(1)}_{,j} + \phi^{(2)}_{,ij}\phi^{(2)}_{,i}\phi^{(2)}_{,j}. \tag{A.1.16}$$

Using (A.1.14)–(A.1.16), we present integral I in (A.1.6) in the following way:

$$I = I_1 + I_2 + \int t|\nabla v|^2\, dxdt, \tag{A.1.17}$$

where

$$I_s = 4\int t^2\Big[\phi^{(s)}_{,ij} v_{,i}\cdot v_{,j} + \phi^{(s)}_{,ij}\phi^{(s)}_{,i}\phi^{(s)}_{,j}|v|^2\Big]\, dxdt$$

$$+\int t^2|v|^2\Big(\partial^2_t\phi^{(s)} - 2\partial_t|\nabla\phi^{(s)}|^2 - \Delta^2\phi^{(s)}$$

$$-\tfrac{1}{t}|\nabla\phi^{(s)}|^2 + \tfrac{1}{t}\partial_t\phi^{(s)}\Big)\, dxdt, \qquad s = 1, 2.$$

Direct calculations give us

$$I_1 = -\int t(|\nabla v|^2 - |v_{,n}|^2)\, dxdt$$

and, therefore,

$$I = \int t|v_{,n}|^2\, dxdt + I_2. \tag{A.1.18}$$

Now, our aim is to estimate I_2 from below. Since $\alpha\in]1/2,1[$, we can drop the first integral in the expression for I_2. As a result, we have

$$I_2 \geq \int t^2|v|^2(A_1 + A_2 + A_3)\, dxdt, \tag{A.1.19}$$

where

$$A_1 = -\partial_t|\nabla\phi^{(2)}|^2,$$

$$A_2 = A_1 - \Delta^2\phi^{(2)} - \frac{1}{t}|\nabla\phi^{(2)}|^2,$$

$$A_3 = \partial^2_t\phi^{(2)} + \frac{1}{t}\partial_t\phi^{(2)}.$$

For A_2, we find

$$A_2 \geq \frac{1-t}{t^\alpha}x_n^{2\alpha-4}a(2\alpha-1)\Big[\frac{4\alpha^2 a x_n^{2\alpha+2}}{t^{\alpha+1}} - 2\alpha(2\alpha-2)(2\alpha-3)\Big].$$

Since $x_n \geq 1$ and $0 < t < 1$, we see that $A_2 \geq 0$ for all $a \geq 2$. Hence, it follows from (A.1.18) and (A.1.19) that

$$I \geq \int t^2|v|^2(A_1 + A_3)\,dxdt. \tag{A.1.20}$$

It is not difficult to check the following inequality

$$A_3 \geq a(2\alpha-1)\frac{x_n^{2\alpha}}{t^{\alpha+2}}. \tag{A.1.21}$$

On the other hand,

$$-\partial_t|\nabla\phi^{(2)}|^2 - \frac{1}{t}|\nabla\phi^{(2)}|^2 \geq (2\alpha-1)\frac{1-t}{t^{2\alpha+1}}4\alpha^2a^2x_n^{2(2\alpha-1)} \geq 0$$

and thus

$$A_1 \geq \frac{1}{t}|\nabla\phi^{(2)}|^2. \tag{A.1.22}$$

Combining (A.1.20)–(A.1.22), we deduce from (A.1.5) the estimate

$$\begin{aligned}\int t^2|Lv|^2\,dxdt &\geq I\\ &\geq a(2\alpha-1)\int \tfrac{x_n^{2\alpha}}{t^\alpha}|v|^2\,dxdt + \int t|v|^2|\nabla\phi^{(2)}|^2\,dxdt \qquad \text{(A.1.23)}\\ &\geq a(2\alpha-1)\int|v|^2\,dxdt + \int t|v|^2|\nabla\phi^{(2)}|^2\,dxdt.\end{aligned}$$

Using (A.1.8), we can find the following analog of (A.1.9)

$$\begin{aligned}\int t|\nabla v|^2\,dxdt &= -\tfrac{1}{2}\int|v|^2\,dxdt - \int tv\cdot Lv\,dxdt \qquad \text{(A.1.24)}\\ &\quad + \int t|v|^2(|\nabla\phi|^2 - \partial_t\phi)\,dxdt.\end{aligned}$$

Due to special structure of ϕ, we have

$$\begin{aligned}|\nabla\phi|^2 - \partial_t\phi &= |\nabla\phi^{(1)}|^2 - \partial_t\phi^{(1)} + |\nabla\phi^{(2)}|^2 - \partial_t\phi^{(2)}\\ &= -|\nabla\phi^{(1)}|^2 + |\nabla\phi^{(2)}|^2 - \partial_t\phi^{(2)}\end{aligned}$$

and, therefore, (A.1.24) can be reduced to the form

$$\begin{aligned}&\int\Big(t|\nabla v|^2 + t|v|^2(|\nabla\phi^{(1)}|^2 + |\nabla\phi^{(2)}|^2)\Big)\,dxdt\\ &= \int t\Big(|\nabla v|^2 + |v|^2|\nabla\phi|^2\Big)\,dxdt = -\tfrac{1}{2}\int|v|^2\,dxdt \qquad \text{(A.1.25)}\\ &\quad - \int tv\cdot Lv\,dxdt + 2\int t|v|^2|\nabla\phi^{(2)}|^2\,dxdt - \int t|v|^2\partial_t\phi^{(2)}\,dxdt.\end{aligned}$$

But

$$-t\partial_t\phi^{(2)} \leq a\frac{x_n^{2\alpha}}{t^\alpha}$$

and, by (A.1.11) and (A.1.25),

$$\begin{aligned}\tfrac{1}{2}\int te^{2\phi}|\nabla u|^2 &\leq -\int v\cdot(tLv)\,dxdt \qquad \text{(A.1.26)}\\ &\quad + 2\int t|v|^2|\nabla\phi^{(2)}|^2\,dxdt + a\int\tfrac{x_n^{2\alpha}}{t^\alpha}|v|^2\,dxdt.\end{aligned}$$

The classical Cauchy-Schwarz inequality, (A.1.23), and (A.1.26) yield required inequality (A.1.12). □

A.2 Unique Continuation Across Spatial Boundaries

We will work with the backward heat operator $\partial_t + \Delta$ rather than the more usual heat operator $\partial_t - \Delta$ since this will save us writing some minus signs in many formulae. In the space-time cylinder $Q(R,T) \equiv B(R) \times]0,T[\subset \mathbb{R}^n \times \mathbb{R}^1$, we consider a vector-valued function $u = (u_i) = (u_1, u_2, ..., u_m)$, satisfying three conditions:

$$u \in W^{2,1}_2(Q(R,T); \mathbb{R}^m); \tag{A.2.1}$$

$$|\partial_t u + \Delta u| \leq c_1(|u| + |\nabla u|) \qquad \text{a.e. in } Q(R,T) \tag{A.2.2}$$

for some positive constant c_1;

$$|u(x,t)| \leq C_k(|x| + \sqrt{t})^k \tag{A.2.3}$$

for all $k = 0, 1, ...,$ for all $(x,t) \in Q(R,T)$, and for some positive constants C_k. Here,

$$W^{2,1}_2(Q(R,T); \mathbb{R}^m) \equiv \{|u| + |\nabla u| + |\nabla^2 u| + |\partial_t u| \in L_2(Q(R,T))\}.$$

Condition (A.2.3) means that the origin is zero of infinite order for the function u.

Theorem 2.4. *Assume that a function u obeys conditions (A.2.1)–(A.2.3). Then, $u(x,0) = 0$ for all $x \in B(R)$.*

Without loss of generality, we may assume that $T \leq 1$. Theorem 2.4 is an easy consequence of the following lemma.

Lemma A.1. *Suppose that all conditions of Theorem 2.4 hold. Then, there exist a constant $\gamma = \gamma(c_1) \in]0, 3/16[$ and absolute constants β_1 and β_2 such that*

$$|u(x,t)| \leq c_2(c_1) A_0(R,T) e^{-\frac{|x|^2}{4t}} \tag{A.2.4}$$

for all $(x,t) \in Q(R,T)$ satisfying the following restrictions:

$$0 < t \leq \gamma T, \qquad |x| \leq \beta_1 R, \qquad \beta_2 t \leq |x|^2.$$

Here,

$$A_0 \equiv \max_{(x,t) \in Q(\frac{3}{4}R, \frac{3}{4}T)} |u(x,t)| + \sqrt{T}|\nabla u(x,t)|.$$

Remark A.1. According to the statement of Lemma A.1, $u(x,0) = 0$ if $|x| \leq \beta_1 R$.

Remark A.2. From the regularity theory for parabolic equations (see [Ladyzhenskaya *et al.* (1967)]), it follows that

$$A_0 \le c_3(c_1, R, T)\Big(\int\limits_{Q(R,T)} |u|^2\, dz\Big)^{\frac{1}{2}}.$$

PROOF OF LEMMA A.1 We let $\lambda = \sqrt{2t}$ and $\varrho = 2|x|/\lambda$. Suppose that $t \le \gamma T \le \gamma$, $|x| \le \frac{3}{8}R$, and $8t \le |x|^2$. Then, as it is easy to verify, we have $\varrho \ge 4$ and

$$\lambda y \in B(3R/4) \quad \text{if } y \in B(\varrho); \qquad \lambda^2 s \in]0, 3/4[\quad \text{if } s \in]0, 2[$$

under the condition $0 < \gamma \le 3/16$. Thus the function $v(y, s) = u(\lambda y, \lambda^2 s)$ is well defined on $Q(\varrho, 2) = B(\varrho)\times]0, 2[$. This function satisfies the conditions:

$$|\partial_s v + \Delta v| \le c_1 \lambda(|v| + |\nabla v|) \tag{A.2.5}$$

in $Q(\varrho, 2)$;

$$|v(y, s)| \le C'_k(|y| + \sqrt{s})^k \tag{A.2.6}$$

for all $k = 0, 1, \ldots$ and for all $(y, s) \in Q(\varrho, 2)$. Here, $C'_k = C_k \lambda^k$.

Given $\varepsilon > 0$, we introduce two smooth cut-off functions such that:

$$0 \le \varphi(y, s) = \begin{cases} 1, & (y, s) \in Q(\varrho - 1, 3/2) \\ 0, & (y, s) \notin B(\varrho)\times]-2, 2[\end{cases} \le 1,$$

$$0 \le \varphi_\varepsilon(s) = \begin{cases} 1, & s \in]2\varepsilon, 2[\\ 0, & s \in]0, \varepsilon[\end{cases} \le 1.$$

We let $w = \varphi v$ and $w_\varepsilon = \varphi_\varepsilon w$. Obviously, (A.2.5) implies the following inequality:

$$\begin{aligned} &|\partial_s w_\varepsilon + \Delta w_\varepsilon| \le c_1 \lambda(|w_\varepsilon| + |\nabla w_\varepsilon|) \\ &+ c_4(|\nabla \varphi||\nabla v| + |\nabla \varphi||v| + |\Delta \varphi||v| + |\partial_s \varphi||v|) + c_4|\varphi'_\varepsilon||v|. \end{aligned} \tag{A.2.7}$$

The crucial point is the application of the following Carleman-type inequality, see Proposition 1.2, to the function w_ε

$$\begin{aligned} &\int\limits_{Q(\varrho,2)} h^{-2a}(s) e^{-\frac{|y|^2}{4s}} (|\nabla w_\varepsilon| + |w_\varepsilon|)^2\, dyds \\ &\le c_5 \int\limits_{Q(\varrho,2)} h^{-2a}(s) e^{-\frac{|y|^2}{4s}} |\partial_s w_\varepsilon + \Delta w_\varepsilon|^2\, dyds. \end{aligned} \tag{A.2.8}$$

Here, c_5 is an absolute positive constant, a is an arbitrary positive number, and $h(t) = te^{\frac{1-t}{3}}$. We let

$$A = \max_{(y,s)\in Q(\varrho,2)\setminus\overline{Q}(\varrho-1,\frac{3}{2})} |v(y,s)| + |\nabla v(y,s)|$$

and choose γ sufficiently small in order to provide the condition

$$10c_5c_1^2\lambda^2 \leq 20c_5c_1^2\gamma < \frac{1}{2}. \tag{A.2.9}$$

Condition (A.2.9) makes it possible to hide the strongest term in the right-hand side of (A.2.8) into the left-hand side of (A.2.8). So, we derive from (A.2.7)–(A.2.9) the following relation

$$\begin{aligned}
&\int\limits_{Q(\varrho,2)} h^{-2a}(s)e^{-\frac{|y|^2}{4s}}(|\nabla w_\varepsilon| + |w_\varepsilon|)^2\, dyds \\
&\leq c_6A^2 \int\limits_{Q(\varrho,2)} h^{-2a}(s)e^{-\frac{|y|^2}{4s}}\chi(y,s)\, dyds \\
&+c_6\frac{1}{\varepsilon^2} \int\limits_{Q(\varrho,2\varepsilon)} h^{-2a}(s)e^{-\frac{|y|^2}{4s}}|v|^2\, dyds.
\end{aligned} \tag{A.2.10}$$

Here, χ is the characteristic function of the set $Q(\varrho,2)\setminus\overline{Q}(\varrho-1,3/2)$. We fix a and take into account (A.2.6). As a result of the passage to the limit as $\varepsilon\to 0$, we find from (A.2.10)

$$\begin{aligned}
D \equiv &\int\limits_{Q(\varrho-1,3/2)} h^{-2a}(s)e^{-\frac{|y|^2}{4s}}(|\nabla v| + |v|)^2\, dyds \\
&\leq c_6A^2 \int\limits_{Q(\varrho,2)} h^{-2a}(s)e^{-\frac{|y|^2}{4s}}\chi(y,s)\, dyds \\
&\leq c_6'A^2\Big(h^{-2a}(3/2) + \rho^{n-1}\int\limits_0^2 h^{-2a}(s)e^{-\frac{(\varrho-1)^2}{4s}}\, ds\Big).
\end{aligned} \tag{A.2.11}$$

Since $\varrho\geq 4$, it follows from (A.2.11) that:

$$D \leq c_7A^2\Big(h^{-2a}(3/2) + \rho^{n-1}\int\limits_0^2 h^{-2a}(s)e^{-\frac{\varrho^2}{8s}}\, ds\Big). \tag{A.2.12}$$

In (A.2.12), the constant c_7 depends on n and c_1 only.

Given positive number β, we can take a number a in the following way

$$a = \frac{\beta\varrho^2}{2\ln h(3/2)}. \tag{A.2.13}$$

This is legal, since $h(3/2) > 1$. Hence, by (A.2.13), inequality (A.2.12) can be reduced to the form

$$D \leq c_7 A^2 e^{-\beta\rho^2}\Big(1+\rho^{n-1}e^{-\beta\varrho^2}\int\limits_0^2 h^{-2a}(s)e^{2\beta\varrho^2-\frac{\varrho^2}{8s}}\,ds\Big).$$

We fix $\beta \in]0,1/64[$, say, $\beta = 1/100$. Then, the latter relation implies the estimate

$$D \leq c_7'(c_1,n)A^2 e^{-\beta\varrho^2}\Big(1+\int\limits_0^2 h^{-2a}(s)e^{-\frac{\varrho^2}{16s}}\,ds\Big). \tag{A.2.14}$$

It is easy to check that $\beta < \frac{\ln(3/2)}{12}$ and therefore $g'(s) \geq 0$ if $s \in]0,2[$, where $g(s) = h^{-2a}(s)e^{-\frac{\varrho^2}{16s}}$ and a and ϱ satisfy condition (A.2.13). So, we have

$$D \leq c_8(c_1,n)A^2 e^{-\beta\varrho^2}, \tag{A.2.15}$$

where β is an absolute positive constant.

By the choice of ϱ and λ, we have $B(\mu\frac{x}{\lambda},1) \subset B(\varrho-1)$ for any $\mu \in]0,1]$. Then, setting $\widetilde{Q} = B(\mu\frac{x}{\lambda},1)\times]1/2,1[$, we find

$$D \geq \int\limits_{\widetilde{Q}} e^{-\frac{|y|^2}{2}}|v|^2\,dyds. \tag{A.2.16}$$

Observing that $|y|^2 \leq 2\mu^2\frac{|x|^2}{\lambda^2}+2$ if $y \in B(\mu\frac{x}{\lambda},1)$ and letting $\mu = \sqrt{2\beta}$, we derive from (A.2.15) and (A.2.16) the following bound

$$\int\limits_{\widetilde{Q}} |v|^2\,dyds \leq c_8' A^2 e^{(-2\beta+\frac{\mu^2}{2})\frac{|x|^2}{t}} = c_8' A^2 e^{-\beta\frac{|x|^2}{t}}. \tag{A.2.17}$$

On the other hand, the regularity theory for linear parabolic equations give us:

$$|v(\mu x/\lambda,1/2)|^2 \leq c_9(c_1,n)\int\limits_{\widetilde{Q}} |v|^2\,dyds. \tag{A.2.18}$$

Combining (A.2.17) and (A.2.18), we show

$$|u(\sqrt{2\beta}x,t)|^2 = |u(\mu x,t)|^2 = |v(\mu x/\lambda,1/2)|^2 \leq c_9' A^2 e^{-\beta\frac{|x|^2}{t}}.$$

Changing variables $\widetilde{x} = \sqrt{2\beta}x$, we have

$$|u(\widetilde{x},t)| \leq \sqrt{c_9'}Ae^{-\frac{|\widetilde{x}|^2}{4t}}$$

for $|\widetilde{x}| \leq \beta_1 R$ and $|\widetilde{x}|^2 \geq \beta_2 t$ with $\beta_1 = 3/8\sqrt{2\beta}$ and $\beta_2 = 16\beta$. It remains to note that $\lambda \leq \sqrt{2T}$ and

$$A \leq \max_{(x,t)\in Q(\frac{3}{4}R,\frac{3}{4}T)} |u(x,t)| + \lambda|\nabla u(x,t)|.$$

Lemma A.1 is proved.

A.3 Backward Uniqueness for Heat Operator in Half Space

In this section, we deal with a backward uniqueness problem for the heat operator. Our approach is due to [Escauriaza *et al.* (2003)], see also [Escauriaza *et al.* (2002)].

Let $\mathbb{R}^n_+ = \{x = (x_i) \in \mathbb{R}^n : \ x_n > 0\}$ and $Q_+ = \mathbb{R}^n_+ \times]0, 1[$. We consider a vector-valued function $u : Q_+ \to \mathbb{R}^m$, which is "sufficiently regular" and satisfies

$$|\partial_t u + \Delta u| \leq c_1(|\nabla u| + |u|) \quad \text{in} \quad Q_+ \tag{A.3.1}$$

for some $c_1 > 0$ and

$$u(\cdot, 0) = 0 \quad \text{in} \quad \mathbb{R}^n_+. \tag{A.3.2}$$

Do (A.3.1) and (A.3.2) imply $u \equiv 0$ in Q_+? We prove that the answer is positive if we impose natural restrictions on the growth of the function u at infinity. For example, we can assume

$$|u(x,t)| \leq e^{M|x|^2} \tag{A.3.3}$$

for all $(x, t) \in Q_+$ and for some $M > 0$. Natural regularity assumptions, under which (A.3.1)–(A.3.3) may be considered are, for example, as follows:

$$\left.\begin{array}{c} u \text{ and weak derivatives } \partial_t u,\ \nabla u, \text{ and } \nabla^2 u \text{ are square} \\ \text{integrable over bounded subdomains of } Q_+. \end{array}\right\} \tag{A.3.4}$$

We can formulate the main result of this section.

Theorem 3.5. *Using the above notation introduced, assume that u satisfies conditions (A.3.1)–(A.3.4). Then $u \equiv 0$ in Q_+.*

We start with proofs of several lemmas. The first of them plays the crucial role in our approach. It enables us to apply powerful technique of Carleman's inequalities.

Lemma A.2. *Suppose that conditions (A.3.1), (A.3.2), and (A.3.4) hold. There exists an absolute positive constant $A_0 < 1/32$ with the following properties. If*

$$|u(x,t)| \leq e^{A|x|^2} \tag{A.3.5}$$

for all $(x,t) \in Q_+$ and for some $A \in [0, A_0]$, then there are constants $\beta(A) > 0$, $\gamma(c_1) \in]0, 1/12[$, and $c_2(c_1, A) > 0$ such that

$$|u(x,t)| \leq c_2 e^{4A|x'|^2} e^{-\beta \frac{x_n^2}{t}} \tag{A.3.6}$$

for all $(x,t) \in (\mathbb{R}^n_+ + 2e_n) \times]0, \gamma[$.

PROOF In what follows, we always assume that the function u is extended by zero to negative values of t.

According to the regularity theory of solutions to parabolic equations, see [Ladyzhenskaya *et al.* (1967)], we may assume

$$|u(x,t)| + |\nabla u(x,t)| \leq c_3 e^{2A|x|^2} \qquad \text{(A.3.7)}$$

for all $(x,t) \in (\mathbb{R}^n_+ + e_n)\times]0,1/2[$.

We fix $x_n > 2$ and $t \in]0,\gamma[$ and introduce the new function v by usual parabolic scaling

$$v(y,s) = u(x+\lambda y, \lambda^2 s - t/2).$$

The function v is well defined on the set $Q_\rho = B(\rho)\times]0,2[$, where $\rho = (x_n - 1)/\lambda$ and $\lambda = \sqrt{3t} \in]0,1/2[$. Then, relations (A.3.1), (A.3.2), and (A.3.7) take the form:

$$|\partial_s v + \Delta v| \leq c_1\lambda(|\nabla v| + |v|) \quad \text{a.e. in} \quad Q_\rho; \qquad \text{(A.3.8)}$$

$$|v(y,s)| + |\nabla v(y,s)| \leq c_3 e^{4A|x|^2} e^{4A\lambda^2|y|^2} \qquad \text{(A.3.9)}$$

for $(y,s) \in Q_\rho$;

$$v(y,s) = 0 \qquad \text{(A.3.10)}$$

for $y \in B(\rho)$ and for $s \in]0,1/6]$.

To apply inequality (A.1.1), we pick up two smooth cut-off functions:

$$\phi_\rho(y) = \begin{cases} 0 & |y| > \rho - 1/2 \\ 1 & |y| < \rho - 1 \end{cases},$$

$$\phi_t(s) = \begin{cases} 0 & 7/4 < s < 2 \\ 1 & 0 < s < 3/2 \end{cases}.$$

These functions take values in $[0,1]$. In addition, function ϕ_ρ obeys the inequalities: $|\nabla^k \phi_\rho| < C_k$, $k = 1,2$. We let $\eta(y,s) = \phi_\rho(y)\phi_t(s)$ and $w = \eta v$. It follows from (A.3.8) that

$$|\partial_s w + \Delta w| \leq c_1\lambda(|\nabla w| + |w|) + \chi c_4(|\nabla v| + |v|).$$

Here, c_4 is a positive constant depending on c_1 and C_k only, $\chi(y,s) = 1$ if $(y,s) \in \omega = \{\rho - 1 < |y| < \rho,\ 0 < s < 2\} \cup \{|y| \leq \rho - 1,\ 3/2 < s < 2\}$ and $\chi(y,s) = 0$ if $(y,s) \notin \omega$. Obviously, function w has the compact support in $\mathbb{R}^n\times]0,2[$ and we may use inequality (A.1.1), see Proposition 1.2. As a result, we have

$$I \equiv \int\limits_{Q_\rho} h^{-2a}(s) e^{-\frac{|y|^2}{4s}} (|w|^2 + |\nabla w|^2)\, dyds \leq c_0 10(c_1^2\lambda^2 I + c_6^2 I_1), \qquad \text{(A.3.11)}$$

where

$$I_1 = \int\limits_{Q_\rho} \chi(y,s)h^{-2a}(s)e^{-\frac{|y|^2}{4s}}(|v|^2 + |\nabla v|^2)\,dyds.$$

Choosing $\gamma = \gamma(c_1)$ sufficiently small, we may assume that the inequality $c_0 10 c_1^2 \lambda^2 \leq 1/2$ holds and then (A.3.11) implies

$$I \leq c_5(c_1) I_1.$$

On the other hand, if $A < 1/32$, then

$$8A\lambda^2 - \frac{1}{4s} < -\frac{1}{8s} \tag{A.3.12}$$

for $s \in]0,2]$. By (A.3.9) and (A.3.12), we have

$$\begin{aligned} I_1 &\leq c_3^2 e^{8A|x|^2} \int\limits_0^2 \int\limits_{B(\rho)} \chi(y,s)h^{-2a}(s)e^{-\frac{|y|^2}{8s}}\,dyds \\ &\leq c_6 e^{8A|x|^2}\Big[h^{-2a}(3/2) + \int\limits_0^2 h^{-2a}(s)e^{-\frac{(\rho-1)^2}{8s}}\,ds\Big]. \end{aligned} \tag{A.3.13}$$

Now, taking into account (A.3.13), we deduce the bound

$$\begin{aligned} D &\equiv \int\limits_{B(1)}\int\limits_{\frac{1}{2}}^{1} |w|^2\,dyds = \int\limits_{B(1)}\int\limits_{\frac{1}{2}}^{1} |v|^2\,dyds \\ &\leq c_7 \int\limits_{Q_\rho} h^{-2a}(s)e^{-\frac{|y|^2}{4s}}(|w|^2 + |\nabla w|^2)\,dyds \\ &\leq c_8(c_1)e^{8A|x|^2}\Big[h^{-2a}(3/2) + \int\limits_0^2 h^{-2a}(s)e^{-\frac{\rho^2}{32s}}\,ds\Big] \\ &= c_8 e^{8A|x|^2 - 2\beta\rho^2}\Big[h^{-2a}(3/2)e^{2\beta\rho^2} + \int\limits_0^2 h^{-2a}(s)e^{2\beta\rho^2 - \frac{\rho^2}{32s}}\,ds\Big]. \end{aligned}$$

We can take $\beta = 8A < 1/256$ and then choose

$$a = \beta\rho^2 / \ln h(3/2).$$

Since $\rho \geq x_n$, such a choice leads to the estimate

$$D \leq c_8 e^{8A|x'|^2} e^{-\beta\rho^2}\Big[1 + \int\limits_0^2 g(s)\,ds\Big],$$

where $g(s) = h^{-2a}(s)e^{-\frac{\rho^2}{64s}}$. It is easy to check that $g'(s) \geq 0$ for $s \in]0,2[$ if $\beta < \frac{1}{96}\ln h(3/2)$. So, we have

$$D \leq 2c_8 e^{8A|x'|^2} e^{-\beta\rho^2} \leq 2c_8 e^{8A|x'|^2} e^{-\frac{\beta x_n^2}{12t}}. \tag{A.3.14}$$

On the other hand, the regularity theory implies

$$|v(0,1/2)|^2 = |u(x,t)|^2 \leq c_8' D. \tag{A.3.15}$$

Combining (A.3.14) and (A.3.15), we complete the proof of the lemma. □

Next lemma is a consequence of Lemma A.2 and the second Carleman inequality (see (A.1.12)).

Lemma A.3. *Suppose that the function u obeys conditions (A.3.1), (A.3.2), (A.3.4), and (A.3.5). There exists a number $\gamma_1(c_1, c_\star) \in]0, \gamma/2]$ such that $u(x,t) = 0$ for all $x \in \mathbb{R}^n_+$ and for all $t \in]0, \gamma_1[$.*

PROOF As usual, by Lemma A.2 and by the regularity theory, we may assume

$$|u(x,t)| + |\nabla u(x,t)| \leq c_9(c_1, A) e^{8A|x'|^2} e^{-\beta\frac{x_n^2}{2t}} \tag{A.3.16}$$

for all $x \in \mathbb{R}^n_+ + 3e_n$ and for all $t \in]0, \gamma/2]$.

By scaling, we define function $v(y,s) = u(\lambda y, \lambda^2 s - \gamma_1)$ for $(y,s) \in Q_+$ with $\lambda = \sqrt{2\gamma_1}$. This function satisfies the relations:

$$|\partial_s v + \Delta v| \leq c_1\lambda(|\nabla v| + |v|) \quad \text{a.e. in} \quad Q_+; \tag{A.3.17}$$

$$v(y,s) = 0 \tag{A.3.18}$$

for all $y \in \mathbb{R}^n_+$ and for all $s \in]0, 1/2[$;

$$|\nabla v(y,s)| + |v(y,s)| \leq c_9 e^{8A\lambda^2|y'|^2} e^{-\frac{\beta\lambda^2 y_n^2}{2(\lambda^2 s - \gamma_1)}} \leq c_9 e^{8A\lambda^2|y'|^2} e^{-\beta\frac{y_n^2}{2s}} \tag{A.3.19}$$

for all $1/2 < s < 1$ and for all $y \in \mathbb{R}^n_+ + \frac{3}{\lambda}e_n$. Since $A < 1/32$ and $\lambda \leq \sqrt{\gamma} \leq 1/\sqrt{12}$, (A.3.19) can be reduced to the form

$$|\nabla v(y,s)| + |v(y,s)| \leq c_{11} e^{\frac{|y'|^2}{48}} e^{-\beta\frac{y_n^2}{2s}} \tag{A.3.20}$$

for the same y and s as in (A.3.19).

Let us fix two smooth cut-off functions:

$$\psi_1(y_n) = \begin{cases} 0 & y_n < \frac{3}{\lambda} + 1 \\ 1 & y_n > \frac{3}{\lambda} + \frac{3}{2} \end{cases},$$

and

$$\psi_2(r) = \begin{cases} 1 & r > -1/2 \\ 0 & r < -3/4 \end{cases}.$$

We set (see Proposition 1.3 for the definition of $\phi^{(1)}$ and $\phi^{(2)}$)

$$\phi_B(y_n, s) = \frac{1}{a}\phi^{(2)}(y_n, s) - B = (1-s)\frac{y_n^{2\alpha}}{s^\alpha} - B,$$

where $\alpha \in]1/2, 1[$ is fixed, $B = \frac{2}{a}\phi^{(2)}(\frac{3}{\lambda} + 2, 1/2)$, and

$$\eta(y_n, s) = \psi_1(y_n)\psi_2(\phi_B(y_n, s)/B), \qquad w(y, s) = \eta(y_n, s)v(y, s).$$

Although function w is not compactly supported in $Q^1_+ = (\mathbb{R}^3_+ + e_n)\times]0,1[$, but, by the statement of Lemma A.2 and by the special structure of the weight in (A.1.12), we can claim validity of (A.1.12) for w. As a result, we have

$$\int\limits_{Q^1_+} s^2 e^{2\phi^{(1)}} e^{2a\phi_B}(|w|^2 + |\nabla w|^2)\, dyds$$

$$\leq c_\star \int\limits_{Q^1_+} s^2 e^{2\phi^{(1)}} e^{2a\phi_B}|\partial_s w + \Delta w|^2\, dyds.$$

Arguing as in the proof of Lemma A.2, we can select $\gamma_1(c_1, c_\star)$ so small that

$$I \equiv \int\limits_{Q^1_+} s^2 e^{2a\phi_B}(|w|^2 + |\nabla w|^2)e^{-\frac{|y'|^2}{4s}}\, dyds$$

$$\leq c_{10}(c_1, c_\star) \int\limits_{(\mathbb{R}^n_+ + (\frac{3}{\lambda}+1)e_n)\times]1/2,1[} \chi(y_n, s)(sy_n)^2 e^{2a\phi_B}(|v|^2 + |\nabla v|^2)e^{-\frac{|y'|^2}{4s}}\, dyds,$$

where $\chi(y_n, s) = 1$ if $(y_n, s) \in \omega$, $\chi(y_n, s) = 0$ if $(y_n, s) \notin \omega$, and

$$\omega \equiv \{(y_n, s) : \; y_n > 1, \quad 1/2 < s < 1, \quad \phi_B(y_n, s) < -D/2\},$$

where $D = -2\phi_B(\frac{3}{\lambda} + \frac{3}{2}, \frac{1}{2}) > 0$. Now, we wish to estimate the right-hand side of the last inequality with the help of (A.3.20). We find

$$I \leq c_{11} e^{-Da} \int\limits_{\frac{3}{\lambda}+1}^{+\infty} \int\limits_{1/2}^{1} (y_n s)^2 e^{-\beta\frac{y_n^2}{s}}\, dy_n ds \int\limits_{\mathbb{R}^{n-1}} e^{(\frac{1}{24} - \frac{1}{4s})|y'|^2}\, dy'.$$

Passing to the limit as $a \to +\infty$, we see that $v(y,s) = 0$ if $1/2 \le s < 1$ and $\phi_B(y_n, s) > 0$. Using unique continuation across spatial boundaries, we show that $v(y,s) = 0$ if $y \in \mathbb{R}^n_+$ and $0 < s < 1$. □

Now, Theorem 3.5 follows from Lemmas A.2 and A.3 with the help of more or less standard arguments. We shall demonstrate them just for completeness.

Lemma A.4. *Suppose that the function u meets all conditions of Lemma A.3. Then $u \equiv 0$ in Q_+.*

PROOF By Lemma A.3, $u(x,t) = 0$ for $x \in \mathbb{R}^n_+$ and for $t \in]0, \gamma_1[$. By scaling, we introduce the function $u^{(1)}(y,s) = u(\sqrt{1-\gamma_1}y, (1-\gamma_1)s+\gamma_1)$. It is easy to check that function $u^{(1)}$ is well-defined in Q_+ and satisfies all conditions of Lemma A.3 with the same constants c_1 and A. Therefore, $u^{(1)}(y,s) = 0$ for $y_n > 0$ and for $0 < s < \gamma_1$. The latter means that $u(x,t) = 0$ for $x_n > 0$ and for $0 < t < \gamma_2 = \gamma_1 + (1-\gamma_1)\gamma_1$. Then, we introduce the function

$$u^{(2)}(y,s) = u(\sqrt{1-\gamma_2}y, (1-\gamma_2)s+\gamma_2), \qquad (y,s) \in Q_+,$$

and apply Lemma A.3. After k steps we shall see that $u(x,t) = 0$ for $x_n > 0$ and for $0 < t < \gamma_{k+1}$, where $\gamma_{k+1} = \gamma_k + (1-\gamma_k)\gamma_1 \to 1$. □

PROOF OF THEOREM 3.5 Assume that $A_0 < M$. Then $\lambda^2 \equiv \frac{A_0}{2M} < \frac{1}{2}$. Introducing function $v(y,s) = u(\lambda y, \lambda^2 s)$, $(y,s) \in Q_+$, we see that this function satisfies all conditions of Lemma A.4 with constants c_1 and $A = \frac{1}{2}A_0$. Therefore, $u(x,t) = 0$ for $x_n > 0$ and for $0 < t < \frac{A_0}{2M}$. Now, we repeat arguments of Lemma A.4, replacing γ_1 to $\frac{A_0}{2M}$ and A to M, and end up with the proof of the theorem. □

A.4 Comments

The whole chapter is essentially due to a part of the paper [Escauriaza *et al.* (2003)]

Appendix B

Lemarie-Riesset Local Energy Solutions

B.1 Introduction

In this chapter, we are going to construct solutions to the Cauchy problem for 3D Navier-Stokes equations with slow decaying initial data. For such initial data, we cannot expect the existence of energy solutions, i.e., weak Leray-Hopf solutions, but we can hope to construct solutions that satisfy the energy inequality at least locally. ε-regularity theory, including the Caffarelli-Kohn-Nirenberg theorem, would work for them as well. The right class of initial data is a certain subspace of the special Morrey space $L_{2,unif}$. This class contains slow decaying functions, for example, interesting homogeneous functions of order minus one.

The main difficulty is caused by the pressure field, which even does not appear in the definition of weak Leray-Hopf solutions. In the case of the Cauchy problem, one would hope to use a nice solution formula for the pressure that is well-defined for weak Leray-Hopf solutions, but it should be modified somehow in order to be useful for functions with very weak decay at the spatial infinity. The problem of the existence of weak solutions with the initial data from $L_{2,unif}$ has been essentially solved by P. G. Lemarie-Riesset [Lemarie-Riesset (2002)] and our aim is to give an interpretation of his interesting and important results.

Let us consider the classical Cauchy problem for the Navier-Stokes equations:

$$\begin{gathered}\partial_t v(x,t) + \operatorname{div} v(x,t)\otimes v(x,t) - \Delta\, v(x,t) = g(x,t) - \nabla\, p(x,t), \\ \operatorname{div} v(x,t) = 0\end{gathered} \tag{B.1.1}$$

for $(x,t)\in Q_T=\mathbb{R}^3\times]0,T[$ together with the initial condition

$$v(x,0)=a(x),\qquad x\in\mathbb{R}^3. \tag{B.1.2}$$

It is supposed that

$$a\in\overset{\circ}{E}_2,\qquad g\in\overset{\circ}{G}_2(0,T). \tag{B.1.3}$$

Here, spaces $\overset{\circ}{E}_m$ and $\overset{\circ}{G}_m(0,T)$ with $m\geq 1$ are defined as follows:

$$\overset{\circ}{E}_m=\{u\in E_m:\ \operatorname{div}u=0\quad\text{in}\quad\mathbb{R}^3\},$$

$$\overset{\circ}{G}_m(0,T)=\{u\in G_m(0,T):\ \operatorname{div}u=0\quad\text{in}\quad Q_T=\mathbb{R}^3\times]0,T[\},$$

$$E_m=\{u\in L_{m,unif}:\quad\int\limits_{B(x_0,1)}|u(x)|^m dx\to 0\quad\text{as}\quad|x_0|\to+\infty\},$$

$$G_m(0,T)=\{u\in L_{m,unif}(0,T):\ \int\limits_0^T\int\limits_{B(x_0,1)}|u(x,t)|^m dxdt\to 0$$

$$\text{as}\quad|x_0|\to+\infty\},$$

$$L_{m,unif}=\{u\in L_{m,loc}:\ \|u\|_{L_{m,unif}}=\sup_{x_0\in\mathbb{R}^3}\Big(\int\limits_{B(x_0,1)}|u(x)|^m dx\Big)^{1/m}<+\infty\},$$

$$L_{m,unif}(0,T)=\{u\in L_{m,loc}(Q_T):\ \|u\|_{L_{m,unif}(0,T)}=$$

$$=\sup_{x_0\in\mathbb{R}^3}\Big(\int\limits_0^T\int\limits_{B(x_0,1)}|u(x,t)|^m dxdt\Big)^{1/m}<+\infty\}.$$

As it has been shown in [Lemarie-Riesset (2002)] (see also references there), the space $\overset{\circ}{E}_m$ is in fact the closure of the set

$$\overset{\circ}{C}{}_0^\infty(\mathbb{R}^3)=\{u\in C_0^\infty(\mathbb{R}^3):\ \operatorname{div}u=0\quad\text{in}\quad\mathbb{R}^3\}$$

with respect to the norm of the space $L_{m,unif}$. For the readers' convenience, we give the proof of this fact in the last section of this chapter, see Lemma B.10.

In monograph [Lemarie-Riesset (2002)], P. G. Lemarie-Riesset proved that, for $g=0$, problem (B.1.1)–(B.1.3) has at least one weak solution

v with the following properties (see Definition 32.1 in [Lemarie-Riesset (2002)]): *for any* $T > 0$,

$$v \in L_\infty(0,T;L_{2,unif}), \qquad \sup_{x_0 \in \mathbb{R}^3} \int\limits_0^T \int\limits_{B(x_0,1)} |\nabla v|^2 dxdt < +\infty;$$

for any compact K,

$$\|v(\cdot,t) - a(\cdot)\|_{L_2(K)} \to 0 \quad \text{as} \quad t \to +0;$$

v is a suitable weak solution in the sense of Caffarelli-Kohn-Nirenberg

$$in \quad Q_T = \mathbb{R}^3 \times]0,T[.$$

This definition seems to be a bit weak and admits trivial non-uniqueness. Indeed, let a smooth vector-valued function $c(t)$ satisfy $c(0) = 0$. Then $v(x,t) = c(t)$ and $p(x,t) = -c'(t) \cdot x$ is also a weak solution for zero initial data. To avoid such type of uniqueness, one may add more restrictions on the velocity or on the pressure. Our definition involves the pressure in more explicit way and is follows.

Definition B.1. We call a pair of functions v and p defined in the space-time cylinder $Q_T = \mathbb{R}^3 \times]0,T[$ a local energy weak Leray-Hopf solution or just a local energy solution to the Cauchy problem (B.1.1)–(B.1.3) if they satisfy the following conditions:

$$v \in L_\infty(0,T;L_{2,unif}), \quad \nabla v \in L_{2,unif}(0,T),$$
$$p \in L_{\frac{3}{2}}(0,T;L_{\frac{3}{2},loc}(\mathbb{R}^3)); \tag{B.1.4}$$

$$v \; and \; p \; meet \; (B.1.1) \; in \; the \; sense \; of \; distributions; \tag{B.1.5}$$

$$the \, function \; t \mapsto \int\limits_{\mathbb{R}^3} v(x,t) \cdot w(x)\, dx \, is \; continuous \; on \; [0,T] \tag{B.1.6}$$

for any compactly supported function $w \in L_2(\mathbb{R}^3)$;

for any compact K,

$$\|v(\cdot,t) - a(\cdot)\|_{L_2(K)} \to 0 \quad as \quad t \to +0; \tag{B.1.7}$$

$$\int\limits_{\mathbb{R}^3} \varphi |v(x,t)|^2 \, dx + 2 \int\limits_0^t \int\limits_{\mathbb{R}^3} \varphi |\nabla v|^2 \, dxdt \leq \int\limits_0^t \int\limits_{\mathbb{R}^3} \Big(|v|^2 (\partial_t \varphi + \Delta \varphi)$$
$$+ v \cdot \nabla \varphi (|v|^2 + 2p) + 2\varphi g \cdot v \Big) \, dxdt \tag{B.1.8}$$

for a.a. $t \in]0,T[$ and for nonnegative smooth functions φ vanishing in a neighborhood of the parabolic boundary of the space-time cylinder $\mathbb{R}^3\times]0,T[$;

for any $x_0 \in \mathbb{R}^3$, there exists a function $c_{x_0} \in L_{\frac{3}{2}}(0,T)$ such that

$$p_{x_0}(x,t) \equiv p(x,t) - c_{x_0}(t) = p^1_{x_0}(x,t) + p^2_{x_0}(x,t), \tag{B.1.9}$$

for $(x,t) \in B(x_0,3/2)\times]0,T[$, where

$$p^1_{x_0}(x,t) = -\frac{1}{3}|v(x,t)|^2 + \frac{1}{4\pi}\int\limits_{B(x_0,2)} K(x-y) : v(y,t)\otimes v(y,t)\,dy,$$

$$p^2_{x_0}(x,t) = \frac{1}{4\pi}\int\limits_{\mathbb{R}^3\setminus B(x_0,2)} (K(x-y) - K(x_0-y)) : v(y,t)\otimes v(y,t)\,dy$$

and $K(x) = \nabla^2(1/|x|)$.

Remark B.3. It is easy to see that (B.1.4), (B.1.6)–(B.1.8) imply the following inequality:

$$\int\limits_{\mathbb{R}^3} \varphi(x)|v(x,t)|^2dx + 2\int\limits_{t_0}^{t}\int\limits_{\mathbb{R}^3} \varphi|\nabla v|^2dxds \leq \int\limits_{\mathbb{R}^3} \varphi(x)|v(x,t_0)|^2dx$$

$$+\int\limits_{t_0}^{t}\int\limits_{\mathbb{R}^3} \Big[|v|^2\Delta\varphi + \nabla\varphi\cdot v\Big(|v|^2+2p\Big) + 2\varphi g\cdot v\Big]\,dxds. \tag{B.1.10}$$

It is valid for any $t \in [0,T]$, for a.a. $t_0 \in [0,T]$, including $t_0 = 0$, and for any nonnegative function $\varphi \in C^\infty_0(\mathbb{R}^3)$.

Remark B.4. In turn, from (B.1.4), (B.1.6), and (B.1.10), it follows that if v and p are a local energy solution on the set $\mathbb{R}^3\times]0,T[$, then they are a local energy solution on the set $\mathbb{R}^3\times]t_0,T[$ for a.a. $t_0 \in [0,T]$, including $t_0 = 0$.

We are going to prove the following statements. The first of them shows that our information about pressure is sufficient to prove decay for both velocity v and p.

Theorem 1.6. *Assume that conditions (B.1.3) hold. Let v and p be a local energy solution to the Cauchy problem (B.1.1), (B.1.2). Then v and p satisfy the following additional properties:*

$$v(\cdot,t) \in \overset{\circ}{E}_2 \tag{B.1.11}$$

for all $t \in [0,T]$;

$$v(\cdot,t) \in \overset{\circ}{E}_3 \tag{B.1.12}$$

for a.a. $t \in [0,T]$;

$$\|v(\cdot,t) - a(\cdot)\|_{L_{2,unif}} \to 0 \quad as \quad t \to +0; \tag{B.1.13}$$

$$\sup_{x_0 \in \mathbb{R}^3} \int\limits_0^T \int\limits_{B(x_0,3/2)} |p(x,t) - c_{x_0}(t)|^{\frac{3}{2}}\, dxdt < +\infty,$$

$$\sup_{x_0 \in \mathbb{R}^3} \int\limits_0^T \int\limits_{B(x_0,3/2)} \mathbb{I}_{\{|x|>R\}} |p(x,t) - c_{x_0}(t)|^{\frac{3}{2}}\, dxdt \to 0 \tag{B.1.14}$$

as $R \to +\infty$, *where* $\mathbb{I}_{\{|x|>R\}}$ *is the characteristic function of the set* $\{x \in \mathbb{R}^3 : |x| > R\}$.

The main theorem of the chapter is Theorem 1.7 below.

Theorem 1.7. *Assume that conditions (B.1.3) hold. There exists at least one local energy solution to the Cauchy problem (B.1.1), (B.1.2).*

The substantial counterpart of the proof of Theorem 1.7 is the statement on the local in time existence of local energy weak solutions.

Proposition 1.8. *(local in time solvability) Assume that conditions (B.1.3) hold. There exist a number* $T_0 \in]0,T]$, *depending on* $\|a\|_{L_{2,unif}}$ *and on* $\|g\|_{L_{2,unif}(0,T)}$ *only, and two functions* v *and* p, *being a local energy solution to the Cauchy problem:*

$$\partial_t v(x,t) + div\, v(x,t) \otimes v(x,t) - \Delta\, v(x,t) = g(x,t) - \nabla\, p(x,t),$$
$$div\, v(x,t) = 0 \tag{B.1.15}$$

for $x \in \mathbb{R}^3$ *and* $0 < t < T_0$,

$$v(x,0) = a(x), \qquad x \in \mathbb{R}^3. \tag{B.1.16}$$

B.2 Proof of Theorem 1.6

Let us introduce the following decomposition:

$$p^2_{x_0}(x,t) = p_{x_0,R}(x,t) + \bar{p}_{x_0,R}(x,t) \qquad (x,t) \in B(x_0,3/2)\times]0,T[, \tag{B.2.1}$$

where

$$\bar{p}_{x_0,R}(x,t) = \frac{1}{4\pi} \int\limits_{\mathbb{R}^3\setminus B(x_0,2R)} (K(x-y) - K(x_0-y)) : v(y,t) \otimes v(y,t)\, dy.$$

Lemma B.5. *For any $x_0 \in \mathbb{R}^3$, for any $t \in]0,T[$, and for any $R \geq 1$, the following estimate is valid:*

$$\sup_{B(x_0,3/2)} |\bar{p}_{x_0,R}(x,t)| \leq \frac{c}{R} \|v(\cdot,t)\|^2_{L_{2,unif}}. \tag{B.2.2}$$

PROOF By our assumptions,

$$|K(x-y) - K(x_0-y)| \leq c\frac{|x-x_0|}{|x_0-y|^4}$$

for $x \in B(x_0, 3/2)$ and for $y \in \mathbb{R}^3 \setminus B(x_0, 2R)$. And then

$$|\bar{p}_{x_0,R}(x,t)| \leq c \int\limits_{\mathbb{R}^3\setminus B(x_0,2R)} \frac{1}{|x_0-y|^4} |v(y,t)|^2 dy$$

$$= c\sum_{i=0}^{\infty} \int\limits_{B(x_0,2^{i+2}R)\setminus B(x_0,2^{i+1}R)} \frac{1}{|x_0-y|^4} |v(y,t)|^2 dy$$

$$\leq c\sum_{i=0}^{\infty} \frac{1}{(2^{i+1}R)^4} \int\limits_{B(x_0,2^{i+2}R)} |v(y,t)|^2\, dy$$

$$\leq c\sum_{i=0}^{\infty} \frac{1}{(2^{i+1}R)^4} (2^{i+2}R)^3 \|v(\cdot,t)\|^2_{L_{2,unif}}.$$

Lemma B.5 is proved.

We let

$$\alpha(t) = \|v(\cdot,t)\|^2_{L_{2,unif}}, \qquad \beta(t) = \sup_{x_0\in\mathbb{R}^3} \int\limits_0^t \int\limits_{B(x_0,1)} |\nabla v|^2\, dxds,$$

$$\gamma(t) = \sup_{x_0\in\mathbb{R}^3} \int\limits_0^t \int\limits_{B(x_0,1)} |v|^3\, dxds.$$

By the known multiplicative inequality, we have

$$\gamma(t) \leq c\Big(\int\limits_0^t \alpha^3(s)\, ds\Big)^{\frac{1}{4}} \Big(\beta(t) + \int\limits_0^t \alpha(s)\, ds\Big)^{\frac{3}{4}}. \tag{B.2.3}$$

From our assumptions and from (B.2.3), it follows that:

$$\operatorname*{ess\,sup}_{0<t<T} \alpha(t) + \beta(T) + \gamma^{\frac{2}{3}}(T) \le A < +\infty. \tag{B.2.4}$$

Next, fix a smooth cut-off function χ so that

$$\chi(x) = 0, \quad x \in B(1), \qquad \chi(x) = 1, \quad x \notin B(2),$$

and then, for $\chi_R(x) = \chi(x/R)$, let

$$\alpha_R(t) = \|\chi_R v(\cdot,t)\|^2_{L_{2,unif}}, \qquad \beta_R(t) = \sup_{x_0\in\mathbb{R}^3} \int\limits_0^t \int\limits_{B(x_0,1)} |\chi_R \nabla v|^2\, dxds,$$

$$\gamma_R(t) = \sup_{x_0\in\mathbb{R}^3} \int\limits_0^t \int\limits_{B(x_0,1)} |\chi_R v|^3\, dxds, \quad G_R = \sup_{x_0\in\mathbb{R}^3} \int\limits_0^T \int\limits_{B(x_0,1)} |\chi_R g|^2\, dxds$$

$$\delta_R(t) = \sup_{x_0\in\mathbb{R}^3} \int\limits_0^t \int\limits_{B(x_0,3/2)} |\chi_R p_{x_0}|^{\frac{3}{2}} dxds.$$

An analogue of (B.2.3) is available with the form

$$\gamma_R(t) \le c\Big(\int\limits_0^t \alpha_R^3(s)\, ds\Big)^{\frac{1}{4}} \Big(\beta_R(t) + \int\limits_0^t \alpha_R(s)ds + \frac{1}{R^2}\int\limits_0^t \alpha(s)ds\Big)^{\frac{3}{4}}. \tag{B.2.5}$$

Lemma B.6. *Assume that v and p are a local energy weak Leray-Hopf solution to the Cauchy problem (B.1.1)–(B.1.3) in the space-time cylinder Q_T. Then we have the estimate*

$$\sup_{0<t<T} \alpha_R(t) + \beta_R(T) + \gamma_R^{\frac{2}{3}}(T) + \delta_R^{\frac{4}{3}}(T) \le C(T,A)\Big[\|\chi_R a\|^2_{L_{2,unif}} +$$

$$+ G_R + \frac{1}{R^{2/3}}\Big]. \tag{B.2.6}$$

Proof. To simplify our notation, we let $\hat{p} = p_{x_0}$.

We fix $x_0 \in \mathbb{R}^3$ and a smooth nonnegative function φ such that

$$\varphi = 1 \quad \text{in} \quad B(1), \qquad \operatorname{spt}\varphi \subset B(3/2)$$

and let $\varphi_{x_0}(x) = \varphi(x - x_0)$. For $\psi = \chi_R^2\varphi_{x_0}$, we find from inequality (B.1.10):

$$L \equiv \int\limits_{\mathbb{R}^3} \psi(x)|v(x,t)|^2\, dx + 2\int\limits_0^t \int\limits_{\mathbb{R}^3} \psi|\nabla v|^2 dxds = \sum_{i=1}^5 I_i, \tag{B.2.7}$$

where

$$I_1 = \int_{\mathbb{R}^3} \psi |a|^2 dx, \qquad I_2 = \int_0^t \int_{\mathbb{R}^3} |v|^2 \Delta \psi \, dxds,$$

$$I_3 = \int_0^t \int_{\mathbb{R}^3} \nabla \psi \cdot v |v|^2 \, dxds, \qquad I_4 = 2 \int_0^t \int_{\mathbb{R}^3} \nabla \psi \cdot v \hat{p} \, dxds,$$

$$I_5 = 2 \int_0^t \int_{\mathbb{R}^3} \psi g \cdot v \, dxds.$$

Obviously,

$$I_1 \leq c \|\chi_R a\|^2_{L_{2,unif}}, \tag{B.2.8}$$

$$I_2 \leq c \int_0^t \alpha_R(s) \, ds + C(T, A) \frac{1}{R}, \tag{B.2.9}$$

$$I_5 \leq c \Big(\int_0^t \alpha_R(s) \, ds + G_R \Big). \tag{B.2.10}$$

The term I_3 is evaluated with the help of Hölder inequality in the following way:

$$I_3 \leq c \gamma^{1/3}(t) \Big(\gamma_R^{2/3}(t) + \frac{c}{R} \gamma^{2/3}(t) \Big).$$

So, by (B.2.4),

$$I_3 \leq C(T, A) \Big(\gamma_R^{2/3}(t) + \frac{1}{R} \Big). \tag{B.2.11}$$

Next, we let

$$I_4 = I' + I'',$$

where

$$I'' = 4 \int_0^t \int_{B(x_0, 3/2)} \chi_R \varphi_{x_0} \nabla \chi_R \cdot v \hat{p} \, dxds$$

and, by Hölder inequality,

$$I'' \leq c\frac{1}{R}\Big(\int_0^t \int_{B(x_0,3/2)} |v|^3\,dxds\Big)^{\frac{1}{3}}\Big(\int_0^t \int_{B(x_0,3/2)} |\hat{p}|^{\frac{3}{2}}\,dxds\Big)^{\frac{2}{3}}.$$

From Lemma B.5 and the theory of singular integrals, we find

$$\int_0^t \int_{B(x_0,3/2)} |\hat{p}|^{\frac{3}{2}}\,dxds \leq c\int_0^t \int_{B(x_0,2)} |v|^3\,dxds + c\int_0^t \alpha^{\frac{3}{2}}(s)\,ds. \tag{B.2.12}$$

Now, (B.2.4) and (B.2.12) give us

$$I'' \leq C(T,A)\frac{1}{R}. \tag{B.2.13}$$

I' can be estimated with the help of Hölder inequality as follows:

$$I' \leq cJ\gamma_R^{\frac{1}{3}}(t), \tag{B.2.14}$$

where

$$J = \Big(\int_0^t \int_{B(x_0,3/2)} |\chi_R\hat{p}|^{\frac{3}{2}}\,dxds\Big)^{\frac{2}{3}}.$$

Obviously, $J \leq J_1 + J_2 + J_3$ with

$$J_1 = \Big(\int_0^t \int_{B(x_0,3/2)} |\chi_R p^1_{x_0}|^{\frac{3}{2}}\,dxds\Big)^{\frac{2}{3}}, \quad J_2 = \Big(\int_0^t \int_{B(x_0,3/2)} |\chi_R p_{x_0,\rho}|^{\frac{3}{2}}\,dxds\Big)^{\frac{2}{3}},$$

$$J_3 = \Big(\int_0^t \int_{B(x_0,3/2)} |\chi_R \bar{p}_{x_0,\rho}|^{\frac{3}{2}}\,dxds\Big)^{\frac{2}{3}}$$

for $\rho = \sqrt{R}$. We start with evaluation of J_1. Letting

$$\chi_R p^1_{x_0} = q_1 + q_2,$$

where

$$q_1(x,t) = -\frac{1}{3}\chi_R(x)|v(x,t)|^2$$

$$+\frac{1}{4\pi}\int_{B(x_0,2)} K(x-y)(\chi_R(x)-\chi_R(y)) : v(y,t)\otimes v(y,t)\,dy,$$

$$q_2(x,t) = \frac{1}{4\pi} \int\limits_{B(x_0,2)} K(x-y)\chi_R(y) : v(y,t) \otimes v(y,t)\, dy,$$

we use the theory of singular integrals and find the estimate for q_2:

$$\int\limits_0^t \int\limits_{B(x_0,3/2)} |q_2|^{\frac{3}{2}}\, dxds \leq c \int\limits_0^t \int\limits_{B(x_0,2)} |\chi_R|^{\frac{3}{2}} |v|^3\, dxds$$

$$\leq C(T,A)\gamma_R^{1/2}(t). \qquad \text{(B.2.15)}$$

Since

$$q_1(x,t) = -\frac{1}{3}\chi_R(x)|v(x,t)|^2$$

$$+\frac{1}{4\pi} \int\limits_{B(x_0,2)} K(x-y)(\chi_R(x) - \chi_R(x_0)) : v(y,t) \otimes v(y,t)\, dy$$

$$+\frac{1}{4\pi} \int\limits_{B(x_0,2)} K(x-y)(\chi_R(x_0) - \chi_R(y)) : v(y,t) \otimes v(y,t)\, dy,$$

the same arguments lead to the estimate

$$\int\limits_0^t \int\limits_{B(x_0,3/2)} |q_1|^{\frac{3}{2}}\, dxds \leq c\gamma^{1/2}(t)\gamma_R^{1/2}(t) + \frac{c}{R^{3/2}} \int\limits_0^t \int\limits_{B(x_0,2)} |v|^3\, dxds$$

$$+c \int\limits_0^t \int\limits_{B(x_0,2)} |\chi_R(x_0) - \chi_R(x))|^{\frac{3}{2}} |v(x,s)|^3\, dxds$$

$$\leq C(T,A)\Big(\frac{1}{R^{3/2}} + \gamma_R^{1/2}(t)\Big).$$

Combining the latter estimate with (B.2.15), we find

$$J_1 \leq C(T,A)\Big(\frac{1}{R} + \gamma_R^{1/3}(t)\Big). \qquad \text{(B.2.16)}$$

Next, we let

$$\chi_R p_{x_0,\rho} = q_3 + q_4,$$

where

$$q_3(x,t) = \frac{1}{4\pi} \int\limits_{B(x_0,2\rho)\setminus B(x_0,2)} (K(x-y) - K(x_0-y))(\chi_R(x)$$

$$-\chi_R(y)) : v(y,t) \otimes v(y,t)\, dy,$$

$$q_4(x,t) = \frac{1}{4\pi} \int\limits_{B(x_0,2\rho)\setminus B(x_0,2)} (K(x-y)-K(x_0-y))\chi_R(y) : v(y,t)\otimes v(y,t)\, dy.$$

For $x \in B(x_0, 3/2)$, we have

$$|q_3(x,t)| \le c \int\limits_{B(x_0,2\rho)\setminus B(x_0,2)} |K(x-y) - K(x_0-y)| \frac{|x-y|}{R} |v(y,t)|^2\, dy$$

$$\le c\frac{\rho}{R} \int\limits_{\mathbb{R}^3\setminus B(x_0,2)} |K(x-y) - K(x_0-y)||v(y,t)|^2\, dy.$$

The same arguments as in the proof of Lemma B.5 work here and show that

$$|q_3(x,t)| \le c\frac{\rho}{R}\alpha(t) = c\frac{1}{\sqrt{R}}\alpha(t) \le C(T,A)\frac{1}{\sqrt{R}}, \tag{B.2.17}$$

where we used $\rho = \sqrt{R}$. Similar arguments work for q_4:

$$|q_4(x,t)| \le c\alpha_R^{1/2}(t)\alpha^{1/2}(t) \le C(T,A)\alpha_R^{1/2}(t), \qquad x \in B(x_0, 3/2). \tag{B.2.18}$$

From (B.2.17) and (B.2.18), it follows that

$$J_2 \le C(T,A)\Big[\frac{1}{\sqrt{R}} + \Big(\int\limits_0^t \alpha_R^{3/4}(s)\, ds\Big)^{\frac{2}{3}}\Big]$$

$$\le C(T,A)\Big[\frac{1}{\sqrt{R}} + \Big(\int\limits_0^t \alpha_R^{3}(s)\, ds\Big)^{\frac{1}{6}}\Big]. \tag{B.2.19}$$

The term J_3 can be estimated with the help of (B.2.2):

$$J_3 \le c\frac{1}{\rho}\alpha(t) \le C(T,A)\frac{1}{\sqrt{R}}. \tag{B.2.20}$$

So, by (B.2.14), (B.2.16), (B.2.19), and (B.2.20), we have

$$J \le C(T,A)\Big[\frac{1}{\sqrt{R}} + \gamma_R^{1/3}(t) + \Big(\int\limits_0^t \alpha_R^{3}(s)\, ds\Big)^{\frac{1}{6}}\Big] \tag{B.2.21}$$

and

$$I' \le C(T,A)\gamma_R^{\frac{1}{3}}(t)\Big[\gamma_R^{1/3}(t) + \frac{1}{\sqrt{R}} + \Big(\int\limits_0^t \alpha_R^{3}(s)\, ds\Big)^{\frac{1}{6}}\Big]$$

$$\le C(T,A)\Big[\gamma_R^{2/3}(t) + \frac{1}{R} + \Big(\int\limits_0^t \alpha_R^{3}(s)\, ds\Big)^{\frac{1}{3}}\Big]. \tag{B.2.22}$$

Now, we can derive from (B.2.7)–(B.2.11), (B.2.13), and (B.2.22):

$$\alpha_R(t)+\beta_R(t)\le c\|\chi_R a\|^2_{L_{2,unif}}+cG_R+c\int\limits_0^t\alpha_R(s)\,ds$$

$$+C(T,A)\Big[\gamma_R^{2/3}(t)+\frac{1}{R}+\Big(\int\limits_0^t\alpha_R^3(s)\,ds\Big)^{\frac{1}{3}}\Big]$$

and, by (B.2.3) and Young inequality, we find the main inequality

$$\alpha_R(t)+\beta_R(t)\le c\|\chi_R a\|^2_{L_{2,unif}}+cG_R$$
$$+C(T,A)\Big[\frac{1}{R}+\int\limits_0^t\alpha_R(s)\,ds+\Big(\int\limits_0^t\alpha_R^3(s)\,ds\Big)^{\frac{1}{3}}\Big].\qquad\text{(B.2.23)}$$

The important consequence of (B.2.23) is as follows:

$$\alpha^3_R(t)\le c\|\chi_R a\|^6_{L_{2,unif}}+cG_R3+C(T,A)\Big[\frac{1}{R^3}+\int\limits_0^t\alpha^3_R(s)\,ds\Big].$$

The latter implies

$$\alpha_R(t)\le C(T,A)\Big[\|\chi_R a\|^2_{L_{2,unif}}+G_R+\frac{1}{R}\Big],$$

which, together with (B.2.23), (B.2.21), and (B.2.3), proves (B.2.6). Lemma B.6 is proved.

PROOF OF THEOREM 1.6 Now, the proof of Theorem 1.6 is easy. In particular, (B.1.11) follows from (B.1.4) and (B.1.6), while (B.1.12) is deduced from (B.2.4) and (B.2.6). In turn, (B.1.7) and (B.2.6) imply (B.1.13).

Regarding the pressure, we observe, by known results for singular integrals, that

$$\int\limits_{B(x_0,3/2)}|p^1_{x_0}(x,t)|^{\frac{3}{2}}\,dx\le c\int\limits_{B(x_0,2)}|v(x,t)|^3\,dx.$$

So, the first estimate in (B.1.14) follows from (B.2.2) and from (B.2.6). Finally, the second estimate in (B.1.14) is one of the statements of Lemma B.6, see (B.2.6). Theorem 1.6 is proved.

B.3 Regularized Problem

Assume that condition (B.1.3) holds. Then according to Lemma B.1.3, we can consider the following regularized problem:

$$\begin{gathered}\partial_t v^\varepsilon + F_\varepsilon(v^\varepsilon)\cdot\nabla v^\varepsilon - \Delta v^\varepsilon = g^\varepsilon - \nabla p^\varepsilon,\\ \operatorname{div} v^\varepsilon = 0\end{gathered}\tag{B.3.1}$$

in $\mathbb{R}^3\times]0,+\infty[$ with

$$g^\varepsilon \in L_2(Q_T), \qquad \operatorname{div} g^\varepsilon = 0, \tag{B.3.2}$$

$$v^\varepsilon|_{t=0} = a^\varepsilon \in \overset{\circ}{C}{}_0^\infty(\mathbb{R}^3) \tag{B.3.3}$$

in $\mathbb{R}^3$. Here,

$$F_\varepsilon(u)(x,t) \equiv \int\limits_{\mathbb{R}^3} \varrho_\varepsilon(x-\bar{x})u(\bar{x},t)\,d\bar{x},$$

ϱ_ε is a standard smoothing kernel,

$$\|a^\varepsilon - a\|_{L_{2,unif}} \to 0, \qquad \|g^\varepsilon - g\|_{L_{2,unif}(0,T)} \to 0 \tag{B.3.4}$$

as $\varepsilon \to 0$. And, we may assume that

$$\|a^\varepsilon\|_{L_{2,unif}} \leq 2\|a\|_{L_{2,unif}}, \qquad \|g^\varepsilon\|_{L_{2,unif}(0,T)} \leq 2\|g\|_{L_{2,unif}(0,T)} \tag{B.3.5}$$

for all ε. Moreover, we may assume also that g^ε is a function of class C^∞ in Q_T and, for each $\varepsilon > 0$, there exists $R_\varepsilon > 0$ such that the support of $g^\varepsilon(\cdot,t)$ lies in $B(R_\varepsilon)$ for all $t \in]0,T[$.

It is known that problem (B.3.1)–(B.3.3) has a unique smooth solution v^ε with finite energy. Moreover, we can define the pressure in the following way:

$$\begin{aligned}p^\varepsilon(x,t) = \frac{1}{4\pi}\int\limits_{\mathbb{R}^3}\frac{1}{|x-\bar{x}|}&\operatorname{div}\operatorname{div}\Big(v^\varepsilon(\bar{x},t)\otimes F_\varepsilon(v^\varepsilon)(\bar{x},t)\Big)\,d\bar{x}\\ &= -\frac{1}{3}v^\varepsilon(x,t)\cdot F_\varepsilon(v^\varepsilon)(x,t)\\ &+\frac{1}{4\pi}\int\limits_{\mathbb{R}^3} K(x-\bar{x}) : v^\varepsilon(\bar{x},t)\otimes F_\varepsilon(v^\varepsilon)(\bar{x},t)\,d\bar{x},\end{aligned}\tag{B.3.6}$$

where $K(x) = \nabla^2(1/|x|)$.

Our aim is to find estimates of v^ε and p^ε that are uniform with respect to ε.

In what follows, we shall use the following decomposition of the pressure. For any $x_0 \in \mathbb{R}^3$ and for any $0 < r \leq R$, we let

$$\hat{p}^{\varepsilon}_{x_0,r}(x,t) \equiv p^{\varepsilon}(x,t) - p^{\varepsilon}_{x_0,r}(t) = p^{1\varepsilon}_{x_0,r}(x,t) + p^{2\varepsilon}_{x_0,r,R}(x,t) + p^{3\varepsilon}_{x_0,R}(x,t), \tag{B.3.7}$$

where

$$p^{\varepsilon}_{x_0,r}(t) \equiv \frac{1}{4\pi} \int\limits_{\mathbb{R}^3 \setminus B(x_0,r)} K(x_0 - \bar{x}) : v^{\varepsilon}(\bar{x},t) \otimes F_{\varepsilon}(v^{\varepsilon})(\bar{x},t)\, d\bar{x},$$

$$p^{1\varepsilon}_{x_0,r}(x,t) \equiv -\frac{1}{3} v^{\varepsilon}(x,t) \cdot F_{\varepsilon}(v^{\varepsilon})(x,t) +$$

$$+\frac{1}{4\pi} \int\limits_{B(x_0,r)} K(x - \bar{x}) : v^{\varepsilon}(\bar{x},t) \otimes F_{\varepsilon}(v^{\varepsilon})(\bar{x},t)\, d\bar{x},$$

$$p^{2\varepsilon}_{x_0,r,R}(x,t) \equiv$$

$$\equiv \frac{1}{4\pi} \int\limits_{B(x_0,2R) \setminus B(x_0,r)} \Big(K(x - \bar{x}) - K(x_0 - \bar{x})\Big) : v^{\varepsilon}(\bar{x},t) \otimes F_{\varepsilon}(v^{\varepsilon})(\bar{x},t)\, d\bar{x},$$

$$p^{3\varepsilon}_{x_0,R}(x,t) \equiv$$

$$\equiv \frac{1}{4\pi} \int\limits_{\mathbb{R}^3 \setminus B(x_0,2R)} \Big(K(x - \bar{x}) - K(x_0 - \bar{x})\Big) : v^{\varepsilon}(\bar{x},t) \otimes F_{\varepsilon}(v^{\varepsilon})(\bar{x},t) d\bar{x}.$$

Using the same arguments as in the proof of Lemma B.5, we prove

Lemma B.7. *For any $x_0 \in \mathbb{R}^3$ and for any $R \geq 1$, we have the following estimate*

$$\sup_{x \in B(x_0,r)} |p^{3\varepsilon}_{x_0,R}(x,t)| \leq \frac{cr}{R} \|v^{\varepsilon}(\cdot,t)\|_{L_{2,unif}} \|F_{\varepsilon}(v^{\varepsilon})(\cdot,t)\|_{L_{2,unif}}. \tag{B.3.8}$$

Assuming that $0 < \varepsilon < 1$, we observe that

$$\|F_{\varepsilon}(v^{\varepsilon})(\cdot,t)\|_{L_{2,unif}} \leq c \|v^{\varepsilon}(\cdot,t)\|_{L_{2,unif}}. \tag{B.3.9}$$

Taking into account the standard estimates for singular integrals, Lemma B.7, and inequality (B.3.9), we find:

$$\|p^{1\varepsilon}_{x_0,r}(\cdot,t)\|_{L_{\frac{3}{2}}(B(x_0,r))} \leq c \|v^{\varepsilon}(\cdot,t)\|^2_{L_3((B(x_0,2))}, \tag{B.3.10}$$

$$\sup_{x\in B(x_0,3r/4)} |p^{2\varepsilon}_{x_0,r,R}(x,t)| \le C_1(r,R)\|v^\varepsilon(\cdot,t)\|^2_{L_{2,unif}}, \tag{B.3.11}$$

$$\sup_{x\in B(x_0,r)} |p^{3\varepsilon}_{x_0,R}(x,t)| \le c\frac{r}{R}\|v^\varepsilon(\cdot,t)\|^2_{L_{2,unif}}. \tag{B.3.12}$$

We let

$$\alpha_\varepsilon(t) = \|v^\varepsilon(\cdot,t)\|^2_{L_{2,unif}}, \qquad \beta_\varepsilon(t) = \sup_{x_0\in\mathbb{R}^3}\int_0^t\int_{B(x_0,1)} |\nabla v^\varepsilon|^2\, dxds,$$

$$\gamma_\varepsilon(t) = \sup_{x_0\in\mathbb{R}^3}\int_0^t\int_{B(x_0,1)} |v^\varepsilon|^3\, dxds, \quad G = \sup_{x_0\in\mathbb{R}^3}\int_0^T\int_{B(x_0,1)} |g(x,t)|^2\, dxdt.$$

By the known multiplicative inequality, we have

$$\gamma_\varepsilon(t) \le c\Big(\int_0^t \alpha^3_\varepsilon(s)\, ds\Big)^{\frac14}\Big(\beta_\varepsilon(t) + \int_0^t \alpha_\varepsilon(s)ds\Big)^{\frac34}. \tag{B.3.13}$$

Now, we can derive the energy estimate.

Lemma B.8. *For any $t \ge 0$, the following energy estimate is valid:*

$$\alpha_\varepsilon(t) + \beta_\varepsilon(t) \le c\Big[\|a\|^2_{L_{2,unif}} + G + \int_0^t (\alpha_\varepsilon(s) + \alpha^3_\varepsilon(s))\, ds\Big]. \tag{B.3.14}$$

PROOF We fix $x_0 \in \mathbb{R}^3$ and a smooth nonnegative function φ such that

$$\varphi = 1 \quad \text{in} \quad B(1), \qquad \operatorname{spt}\varphi \subset B(3/2)$$

and let $\varphi_{x_0}(x) = \varphi(x - x_0)$.

From system (B.3.1) and (B.3.2), it is easy to derive the identity

$$E \equiv \int_{\mathbb{R}^3} \varphi^2_{x_0}(x)|v^\varepsilon(x,t)|^2 dx + 2\int_0^t\int_{\mathbb{R}^3} \varphi^2_{x_0}|\nabla v^\varepsilon|^2 dxds +$$

$$= \int_{\mathbb{R}^3} \varphi^2_{x_0}|a^\varepsilon + \int_0^t\int_{\mathbb{R}^3} \Big[|v^\varepsilon|^2\Delta\varphi^2_{x_0} + \nabla\varphi^2_{x_0}\cdot F_\varepsilon(v^\varepsilon)\Big(|v^\varepsilon|^2 + 2\hat{p}^\varepsilon_{x_0,2}\Big) + \tag{B.3.15}$$

$$+2\varphi^2_{x_0} g^\varepsilon\cdot v^\varepsilon\Big]\, dxds.$$

By Hölder inequality and by (B.3.5), we show

$$E \leq c\Big[\|a\|^2_{L_{2,unif}} + G + \int\limits_0^t \alpha_\varepsilon(s)\,ds + \gamma_\varepsilon(t) + \int\limits_0^t \int\limits_{B(x_0,3/2)} |\hat{p}^\varepsilon_{x_0,2}|^{\frac{3}{2}}\,dxds\Big]. \tag{B.3.16}$$

On the other hand, (B.3.10)–(B.3.12) imply

$$\int\limits_0^t \int\limits_{B(x_0,3/2)} |\hat{p}^\varepsilon_{x_0,2}|^{\frac{3}{2}}\,dxds \leq c\Big(\gamma_\varepsilon(t) + \int\limits_0^t \alpha_\varepsilon^{\frac{3}{2}}(s)\,ds\Big). \tag{B.3.17}$$

Taking into account (B.3.13), we derive from (B.3.16) and (B.3.17) the following estimate

$$\alpha_\varepsilon(t) + \beta_\varepsilon(t) \leq c\Big[\|a\|^2_{L_{2,unif}} + G + \int\limits_0^t \alpha_\varepsilon(s)\,ds + \int\limits_0^t \alpha_\varepsilon^{\frac{3}{2}}(s)\,ds$$
$$+\Big(\int\limits_0^t \alpha_\varepsilon^3(s)\,ds\Big)^{\frac{1}{4}}\Big(\beta_\varepsilon(t) + \int\limits_0^t \alpha_\varepsilon(s)ds\Big)^{\frac{3}{4}}\Big].$$

Applying Young's inequality twice, we complete the proof of the lemma. Lemma B.8 is proved.

A simple consequence of Lemma B.8 is the following statement.

Lemma B.9. *There exist positive constants A and $T_0 \leq T$ depending on $\|a\|_{L_{2,unif}}$ and G only such that*

$$\sup_{0<t<T_0} \alpha_\varepsilon(t) + \beta_\varepsilon(T_0) + \gamma_\varepsilon^{\frac{2}{3}}(T_0) + \delta_\varepsilon^{\frac{4}{3}}(T_0) \leq A, \tag{B.3.18}$$

where

$$\delta_\varepsilon(t) = \sup_{x_0\in\mathbb{R}^3} \int\limits_0^t \int\limits_{B(x_0,3/2)} |\hat{p}^\varepsilon_{x_0,2}|^{\frac{3}{2}}\,dxds.$$

Indeed, let

$$T_0 := \min\Big\{T, \frac{\ln 2}{c(1 + (2c(\|a\|^2_{L_{2,unif}} + G))^2)}\Big\}. \tag{B.3.19}$$

We claim that if $0 \leq t < T_0$, then $\alpha_\varepsilon(t) < 2c(\|a\|^2_{L_{2,unif}} + G)$. Otherwise, there should exist $T' < T_0$ such that

$$\alpha_\varepsilon(t) < 2c(\|a\|^2_{L_{2,unif}} + G)$$

for $0 \leq t < T'$ and

$$\alpha_\varepsilon(T') = 2c(\|a\|^2_{L_{2,unif}} + G).$$

The main inequality implies the following estimate

$$\alpha_\varepsilon(t) \leq c\Big(\|a\|^2_{L_{2,unif}} + G + (1 + (2c(\|a\|^2_{L_{2,unif}} + G))^2) \int\limits_0^t \alpha_\varepsilon(\tau)d\tau\Big)$$

for $0 \leq t < T'$. In turn, this inequality implies

$$\alpha_\varepsilon(t) \leq c(\|a\|^2_{L_{2,unif}} + G)\exp\{c(1 + (2c(\|a\|^2_{L_{2,unif}} + G))^2)t\}$$

for the same t. And thus we find

$$2c(\|a\|^2_{L_{2,unif}} + G) \leq c(\|a\|^2_{L_{2,unif}} + G)\exp\{c(1 + (2c(\|a\|^2_{L_{2,unif}} + G))^2)T'\}.$$

But this is possible only if $T' \geq T_0$ and contradicts the above assumption.

B.4 Passing to Limit and Proof of Proposition 1.8

First, we fix $n \in \mathbb{N}$. From Lemma B.9, it follows that the following estimate is valid:

$$\sup_{0<t<T_0} \int\limits_{B(n)} |v^\varepsilon(x,t)|^2\,dx + \int\limits_0^{T_0}\int\limits_{B(n)} |\nabla v^\varepsilon|^2\,dxdt \leq cn^3 A. \qquad \text{(B.4.1)}$$

Using the known multiplicative inequality, we find from (B.4.1)

$$\int\limits_0^{T_0}\int\limits_{B(n)} |v^\varepsilon|^{\frac{10}{3}}\,dxdt \leq$$

$$\leq c\Big(\sup_{0<t<T_0} \int\limits_{B(n)} |v^\varepsilon(x,t)|^2\,dx\Big)^{\frac{2}{3}} \int\limits_0^{T_0}\int\limits_{B(n)} \Big(|\nabla v^\varepsilon|^2 + \frac{1}{n^2}|v^\varepsilon|^2\Big)\,dxdt$$

and thus

$$\int\limits_0^{T_0}\int\limits_{B(n)} |v^\varepsilon|^{\frac{10}{3}}\,dxdt \leq cn^5 A^{\frac{5}{3}}. \qquad \text{(B.4.2)}$$

To estimate the pressure, we use (B.3.10)–(B.3.12) with $x_0 = 0$ and $r = R = 2n$. So, we have

$$p^\varepsilon_n(x,t) \equiv \hat{p}^\varepsilon_{0,2n}(x,t),$$

where $\hat{p}^\varepsilon_{0,2n}(x,t) \equiv p^\varepsilon(x,t) - p^\varepsilon_{0,2n}(t)$, and

$$\int\limits_0^{T_0}\int\limits_{B(n)} |p^\varepsilon_n|^{\frac{3}{2}}\,dxdt \leq C(n,T_0,A). \tag{B.4.3}$$

The derivative in time can be estimated with the help of the Navier-Stokes equations in the following way:

$$\int\limits_0^{T_0}\int\limits_{B(n)} \partial_t v^\varepsilon \cdot w\,dxdt$$

$$= \int\limits_0^{T_0}\int\limits_{B(n)} \Big(v^\varepsilon \otimes F_\varepsilon(v^\varepsilon) : \nabla w - \nabla v^\varepsilon : \nabla w + p^\varepsilon_n \operatorname{div} w + g^\varepsilon \cdot w\Big)\,dxdt$$

$$\leq \Big(\int\limits_0^{T_0}\int\limits_{B(n)} |v^\varepsilon|^3\,dxdt\Big)^{\frac{1}{3}}\Big(\int\limits_0^{T_0}\int\limits_{B(n)} |F_\varepsilon(v^\varepsilon)|^3\,dxdt\Big)^{\frac{1}{3}}\Big(\int\limits_0^{T_0}\int\limits_{B(n)} |\nabla w|^3\,dxdt\Big)^{\frac{1}{3}}$$

$$+\Big(\int\limits_0^{T_0}\int\limits_{B(n)} |\nabla v^\varepsilon|^2\,dxdt\Big)^{\frac{1}{2}}\Big(\int\limits_0^{T_0}\int\limits_{B(n)} |\nabla w|^2\,dxdt\Big)^{\frac{1}{2}}$$

$$+\Big(\int\limits_0^{T_0}\int\limits_{B(n)} |p^\varepsilon_n|^{\frac{3}{2}}\,dxdt\Big)^{\frac{2}{3}}\Big(\int\limits_0^{T_0}\int\limits_{B(n)} |\nabla w|^3\,dxdt\Big)^{\frac{1}{3}}$$

$$+c\Big(\int\limits_0^{T_0}\int\limits_{B(n)} |g^\varepsilon|^2\,dxdt\Big)^{\frac{1}{2}}\Big(\int\limits_0^{T_0}\int\limits_{B(n)} |w|^2\,dxdt\Big)^{\frac{1}{2}}$$

for any $w \in C^\infty_0(B(n))$. Since

$$\int\limits_0^{T_0}\int\limits_{B(n)} |F_\varepsilon(v^\varepsilon)|^3\,dxdt \leq \int\limits_0^{T_0}\int\limits_{B(2n)} |v^\varepsilon|^3\,dxdt,$$

we have

$$\int\limits_0^{T_0}\int\limits_{B(n)} \partial_t v^\varepsilon \cdot w\,dxdt \leq C(n,T_0,A,G)\Big(\int\limits_0^{T_0}\int\limits_{B(n)} |\nabla w|^3\,dxdt\Big)^{\frac{1}{3}}$$

for any $w \in C_0^\infty(B(n))$. The latter estimates implies

$$\|\partial_t v^\varepsilon\|_{\mathcal{X}_n} \leq C(n, T_0, A, G), \tag{B.4.4}$$

where $\mathcal{X}_n$ is the space dual to $L_3(0, T_0; \overset{\circ}{W}{}_3^1(B(n)))$ and $\overset{\circ}{W}{}_3^1(B(n))$ is the closure of $C_0^\infty(B(n))$ in $W_3^1(B(n))$.

Now, we argue by induction in n. Let $n = 1$. Estimates (B.4.1)–(B.4.4) make it possible to apply the known compactness arguments and to find sequences $v^{1,k}$ and $p_1^{1,k}$ such that

$$\begin{array}{lll} v^{1,k} \overset{\star}{\rightharpoonup} v^1 & \text{in} & L_\infty(0, T_0; L_2(B(1))), \\ v^{1,k} \rightharpoonup v^1 & \text{in} & L_2(0, T_0; W_2^1(B(1))), \\ v^{1,k} \to v^1 & \text{in} & L_3(0, T_0; L_3(B(1))), \\ F_{1,k}(v^{1,k}) \to v^1 & \text{in} & L_3(0, T_0; L_3(B(\delta))), \qquad \forall \delta < 1, \\ p_1^{1,k} \rightharpoonup p_1 & \text{in} & L_{\frac{3}{2}}(0, T_0; L_{\frac{3}{2}}(B(1))) \end{array}$$

as $k \to +\infty$. We let $v = v^1$ and $p = p_1$ in $B(1)\times]0, T_0[$. Obviously, the pair v and p satisfies the Navier-Stokes equations in the sense of distributions and the local energy inequality in $B(1)\times]0, T_0[$. The latter means that

$$\int\limits_{B(1)} \varphi(x,t)|v(x,t)|^2\,dx + 2\int\limits_0^t\int\limits_{B(1)} \varphi|\nabla v|^2\,dxdt \leq \int\limits_0^t\int\limits_{B(1)} \Big(|v|^2(\partial_t\varphi + \Delta\varphi)$$

$$+v\cdot\nabla\varphi(|v|^2 + 2p) + 2\varphi g\cdot v\Big)\,dxdt$$

for a.a. $t \in]0, T_0[$ and for non-negative smooth functions φ vanishing in a neighborhood of the parabolic boundary of space-time cylinder $B\times]0, T_0[$.

Now, let $n = 2$. By the same arguments as above, we assert that there exists sequences $v^{2,k}$, which is a subsequence of $v^{1,k}$, and $p_2^{2,k}$ such that

$$\begin{array}{lll} v^{2,k} \overset{\star}{\rightharpoonup} v^2 & \text{in} & L_\infty(0, T_0; L_2(B(2))), \\ v^{2,k} \rightharpoonup v^2 & \text{in} & L_2(0, T_0; W_2^1(B(2))), \\ v^{2,k} \to v^2 & \text{in} & L_3(0, T_0; L_3(B(2))), \\ F_{2,k}(v^{2,k}) \to v^2 & \text{in} & L_3(0, T_0; L_3(B(\delta))), \qquad \forall \delta < 2, \\ p_2^{2,k} \rightharpoonup p_2 & \text{in} & L_{\frac{3}{2}}(0, T_0; L_{\frac{3}{2}}(B(2))). \end{array}$$

The functions v^2 and p_2 satisfy the Navier-Stokes equations and the local energy inequality in the space-time cylinder $B(2)\times]0, T_0[$. Obviously, that $v^2 = v$ on $B\times]0, T_0[$. So, we may extend v by letting $v = v^2$ on

$B(2)\times]0,T_0[$. As to the function p_2, it follows from the Navier-Stokes equations that $\nabla p_2 = \nabla p$ on $B\times]0,T_0[$. This means that $p_2(x,t) - h_2(t) = p(x,t)$ for $x \in B$ and for $t \in]0,T_0[$. Since both p_2 and p belong to $L_{\frac{3}{2}}(0,T_0;L_{\frac{3}{2}}(B))$, we conclude the $h_2 \in L_{\frac{3}{2}}(0,T_0)$. This allows to extend the function p to $B(2)\times]0,T_0[$ so that $p = p_2 - h_2$ on $B(2)\times]0,T_0[$. Clearly, $p \in L_{\frac{3}{2}}(0,T_0;L_{\frac{3}{2}}(B(2)))$ and the functions v and p satisfies the Navier-Stokes equations and the local energy inequality on the space-time cylinder $B(2)\times]0,T_0[$.

In the case $n = 3$, we repeat the above arguments choosing a subsequence of the sequence $v^{2,k}$ and replacing balls B and $B(2)$ with balls $B(2)$ and $B(3)$, respectively. Continuing this process, we arrive at the following result. There exist two functions v and p defined on $\mathbb{R}^3\times]0,T_0[$ such that

$$v \in L_\infty(0,T_0;L_{2,loc}(\mathbb{R}^3)) \cap L_2(0,T_0;W^1_{2,loc}(\mathbb{R}^3)),$$
$$p \in L_{\frac{3}{2}}(0,T_0;L_{\frac{3}{2},loc}(\mathbb{R}^3)). \tag{B.4.5}$$

Next, letting $v^{\{k\}} = v^{k,k}$, we observe that $v^{\{k\}}$ is a subsequence of the sequence $\{v^{n,k}\}_{k=n}^\infty$, i.e., there exists a sequence of natural numbers $\{r_k\}_{k=n}^\infty$ having the following properties:

$$v^{\{k\}} = v^{n,r_k}, \qquad k = n, n+1, ..., \qquad r_n = n, \qquad r_k \to \infty$$

as $k \to \infty$. Then we may let

$$p_n^{\{k\}} = p_n^{n,r_k}.$$

Obviously, $p_n^{\{k\}}$ is a subsequence of the sequence $\{p_n^{n,k}\}_{k=n}^\infty$. For these new sequences and for any $n \in \mathbb{N}$, we have

$$\begin{aligned}
&v^{\{k\}} \overset{\star}{\rightharpoonup} v && \text{in} \quad L_\infty(0,T_0;L_2(B(n))),\\
&v^{\{k\}} \rightharpoonup v && \text{in} \quad L_2(0,T_0;W^1_2(B(n))),\\
&v^{\{k\}} \to v && \text{in} \quad L_3(0,T_0;L_3(B(n))),\\
&F_{\{k\}}(v^{\{k\}}) \to v && \text{in} \quad L_3(0,T_0;L_3(B(\delta))), \qquad \forall \delta < n,\\
&p_n^{\{k\}} \rightharpoonup p_n && \text{in} \quad L_{\frac{3}{2}}(0,T_0;L_{\frac{3}{2}}(B(n)))
\end{aligned} \tag{B.4.6}$$

and

$$\|\partial_t v\|_{\mathcal{X}_n} \leq C(n,T_0,A,G), \tag{B.4.7}$$

$$p_n(x,t) = p(x,t) - c_n(t), \qquad x \in B(n), \quad t \in]0,T_0[\tag{B.4.8}$$

for some $c_n \in L_{\frac{3}{2}}(0,T_0)$. From (B.3.18) and (B.4.6), it is easy to derive the estimate

$$\operatorname*{ess\,sup}_{0<t<T_0} \|v(\cdot,t)\|^2_{L_{2,unif}} + \sup_{x_0\in\mathbb{R}^3} \int\limits_0^{T_0} \int\limits_{B(x_0,1)} |\nabla v|^2\,dxdt \leq 2A. \tag{B.4.9}$$

Now, by (B.4.7) and (B.4.9), we see that, for each $n \in \mathbb{N}$,

$$the\, function\; t \mapsto \int\limits_{B(n)} v(x,t)\cdot w(x)\,dx\, is\; continuous\; on\; [0,T_0] \qquad \text{(B.4.10)}$$

for any $w \in L_2(B(n))$.

We notice further that for the solution of the regularized problem we have the following identity:

$$\int\limits_{\mathbb{R}^3} \varphi(x)|v^{\{k\}}(x,t)|^2 dx + 2\int\limits_0^t\int\limits_{\mathbb{R}^3} \varphi|\nabla v^{\{k\}}|^2 dxds = \int\limits_{\mathbb{R}^3}\varphi|a^{\{k\}}|^2 dx+$$

$$+\int\limits_0^t\int\limits_{\mathbb{R}^3}\Big[|v^{\{k\}}|^2\Delta\varphi + \nabla\varphi\cdot F_{\{k\}}(v^{\{k\}})\Big(|v^{\{k\}}|^2 + 2p_n^{\{k\}}\Big)+ \qquad \text{(B.4.11)}$$

$$+2\varphi g^{\{k\}}\cdot v^{\{k\}}\Big]\,dxds,$$

which is valid for any function $\varphi \in C_0^\infty(\mathbb{R}^3)$. Taking into account (B.3.4), (B.4.6)–(B.4.8), and (B.4.10), we deduce from (B.4.11) the inequality

$$\int\limits_{\mathbb{R}^3}\varphi(x)|v(x,t)|^2 dx + 2\int\limits_0^t\int\limits_{\mathbb{R}^3}\varphi|\nabla v|^2 dxds \le \int\limits_{\mathbb{R}^3}\varphi|a|^2 dx$$

$$+\int\limits_0^t\int\limits_{\mathbb{R}^3}\Big[|v|^2\Delta\varphi + \nabla\varphi\cdot v\Big(|v|^2+2p\Big) \qquad \text{(B.4.12)}$$

$$+2\varphi f\cdot v\Big]\,dxds.$$

The latter holds for any $t \in [0,T_0]$ and for any nonnegative function $\varphi \in C_0^\infty(\mathbb{R}^3)$. On the other hand, from (B.4.10) and from (B.4.12) it follows that

$$\int\limits_{\mathbb{R}^3}\varphi|v(x,t)-a(x)|^2\,dx \to 0 \quad as \quad t\to +0 \qquad \text{(B.4.13)}$$

for all $\varphi \in C_0^\infty(\mathbb{R}^3)$. So, v meets (B.1.7). The validity of (B.1.8) follows from (B.4.6). It remains to establish decomposition (B.1.9).

Thanks to (B.3.18), we have

$$\int_0^{T_0}\int_{B(x_0,3/2)} |\hat{p}^{\varepsilon}_{x_0,2}|^{\frac{3}{2}}\,dxdt \leq A. \tag{B.4.14}$$

We would like to emphasize that the constant on the right-hand sides of (B.4.14) is independent of ε and x_0. Let $\hat{p}^{\{k\}}_{x_0,2}$ be the sequence generated by $v^{\{k\}}$ via (B.3.7). For each $x_0 \in \mathbb{R}^3$, we can find subsequences $\hat{p}^{\{k\},x_0}_{x_0,2}$ and $v^{\{k\},x_0}$ such that

$$\hat{p}^{\{k\},x_0}_{x_0,2} \rightharpoonup p_{x_0} \quad \text{in} \quad L_{\frac{3}{2}}(B(x_0,3/2)\times]0,T_0[).$$

So, it follows from (B.4.14) that

$$\int_0^{T_0}\int_{B(x_0,3/2)} |p_{x_0}|^{\frac{3}{2}}\,dxdt \leq A \tag{B.4.15}$$

for each $x_0 \in \mathbb{R}^3$. Passing to the limit in the Navier-Stokes equations on the set $B(x_0,3/2)\times]0,T_0[$, we show that

$$\nabla(p - p_{x_0}) = 0$$

on $B(x_0,3/2)\times]0,T_0[$ in the sense of distributions. So, we state that, for any $x_0 \in \mathbb{R}^3$, there exists a function $c_{x_0} \in L_{\frac{3}{2}}(0,T_0)$ having the property

$$p(x,t) - p_{x_0}(x,t) = c_{x_0}(t) \tag{B.4.16}$$

for $x \in B(x_0,3/2)$ and $t \in]0,T_0[$. Now, let us show the validity of (B.1.9). Using decomposition (B.3.7) and the theory of singular integrals, we observe

$$p^{1\{k\},x_0}_{x_0,2} \to p^1_{x_0} \quad in \quad L_{\frac{3}{2}}(B(x_0,3/2)\times]0,T_0[).$$

Obviously,

$$p^{2\{k\},x_0}_{x_0,2,R} \to p^2_{x_0,R} \quad in \quad L_{\frac{3}{2}}(B(x_0,3/2)\times]0,T_0[),$$

where

$$p^2_{x_0,R}(x,t) = \frac{1}{4\pi}\int_{B(x_0,2R)\setminus B(x_0,2)} (K(x-y) - K(x_0-y)) : v(y,t)\otimes v(y,t)\,dx.$$

By (B.3.12), we also have

$$\sup_{x\in B(x_0,3/2)} |p^{3\{k\},x_0}_{x_0,R}(x,t)| \leq \frac{1}{R}C(A,T_0)$$

on $]0,T_0[$. On the other hand, the integral in the definition of $p^2_{x_0,R}$ converges to $p^2_{x_0}$ as $R \to +\infty$. This follows from the Lemma B.7 and the inequality

$$\int_{B(x_0,2R)\setminus B(x_0,2)} |K(x-y) - K(x_0-y)||v(y,t)|^2\,dx \leq c\|v(\cdot,t)\|^2_{L_{2,unif}}$$

which is valid on $B(x_0,3/2)\times]0,T_0[$. Passing to the limit as $R \to +\infty$, we show (B.1.11). Proposition 1.8 is proved.

B.5 Proof of Theorem 1.7

The idea of the proof of the main theorem is the same in [Lemarie-Riesset (2002)]. It is based on the theory of solvability of the Cauchy problems for the Stokes and Navier-Stokes systems with initial data from $\overset{\circ}{E}_3$. A proof of them can be done along standard lines and is omitted here.

We start with the most important part: the linear theory. Consider the Cauchy problem for the Stokes system:

$$\partial_t v(x,t) - \Delta v(x,t) + \nabla p(x,t) = -\operatorname{div} f(x,t) + g(x,t),$$
$$\operatorname{div} v(x,t) = 0 \tag{B.5.1}$$

for $x \in \mathbb{R}^3$ and $0 < t < T$,

$$v(x,0) = a(x), \qquad x \in \mathbb{R}^3. \tag{B.5.2}$$

It is supposed that

$$a \in \overset{\circ}{E}_3, \qquad g \in \overset{\circ}{G}_3(0,T), \qquad f \in G_{\frac{5}{2}}(0,T). \tag{B.5.3}$$

Theorem 5.9. *Assume that conditions (B.5.3) hold. There exists a unique pair of functions v and p having the following properties:*

$$v \in L_\infty(0,T;L_{3,unif}), \qquad (1+\sqrt{|v|})|\nabla v| \in L_{2,unif}(0,T),$$

$$p \in L_{\frac{5}{2}}(0,T;L_{\frac{5}{2},loc}(\mathbb{R}^3));$$

v and p satisfy (B.5.1) in the sense of distributions;

$$\|v\|_{L_\infty(0,T;L_{3,unif})} \le c\Big[\|f\|_{L_{\frac{5}{2},unif}(0,T)} + \|g\|_{L_{3,unif}(0,T)} + \|a\|_{L_{3,unif}}\Big];$$

the $t \mapsto \int\limits_{\mathbb{R}^3} v(x,t)\cdot u(x)\,dx$ is continuous on $[0,T]$ for any compactly supported functions $u \in L_{\frac{3}{2}}(\mathbb{R}^3)$;

$$v \in \overset{\circ}{G}_5(0,T), \qquad v(\cdot,t) \in \overset{\circ}{E}_3, \qquad \forall t \in [0,T];$$

$$\|v(\cdot,t) - v(\cdot,t_0)\|_{L_{3,unif}} \to 0 \qquad as\ t \to t_0+0, \qquad \forall t_0 \in [0,T],$$

$$v(\cdot,0) = a(\cdot);$$

for any $x_0 \in \mathbb{R}^3$*, there exists* $c_{x_0} \in L_{\frac{5}{2}}(0,T)$ *such that*

$$p(x,t) - c_{x_0}(t) = p^1_{x_0}(x,t) + p^2_{x_0}(x,t)$$

for any $x \in B(x_0, 3/2)$ *and for any* $t \in]0,T[$*, where*

$$p^1_{x_0}(x,t) = -\frac{1}{3} trf(x,t) + \frac{1}{4\pi} \int\limits_{B(x_0,2)} K(x-y) : f(y,t)\, dy,$$

$$p^2_{x_0}(x,t) = \frac{1}{4\pi} \int\limits_{\mathbb{R}^3 \setminus B(x_0,2)} (K(x-y) - K(x_0-y)) : f(y,t)\, dy$$

and

$$\sup_{x_0 \in \mathbb{R}^3} \int\limits_0^T \int\limits_{B(x_0,3/2)} |p(x,t) - c_{x_0}(t)|^{\frac{5}{2}}\, dxdt < +\infty,$$

$$\sup_{x_0 \in \mathbb{R}^3} \int\limits_0^T \int\limits_{B(x_0,3/2)} \mathbb{I}_{\{|x|>R\}} |p(x,t) - c_{x_0}(t)|^{\frac{5}{2}}\, dxdt \to 0$$

as $R \to +\infty$.

Using Theorem 5.9 and successive approximations, see, for example, [Escauriaza *et al.* (2003)] and [Galdi (2000)], we can prove the following theorems about solvability of the Cauchy problem:

$$\partial_t v(x,t) + \operatorname{div} v(x,t) \otimes v(x,t) - \Delta v(x,t) + \nabla p(x,t) = g(x,t), \tag{B.5.4}$$

$$\operatorname{div} v(x,t) = 0$$

for $x \in \mathbb{R}^3$ and $0 < t < T$,

$$v(x,0) = a(x), \qquad x \in \mathbb{R}^3, \tag{B.5.5}$$

under assumptions that

$$a \in \overset{\circ}{E}_3, \qquad g \in \overset{\circ}{G}_3(0,T). \tag{B.5.6}$$

Theorem 5.10. *Suppose that conditions (B.5.6) hold. There exists a number* $T_0 \in]0,T[$ *with the following property. Given* a *and* g*, there exists a pair of functions* v *and* p *that is a local energy solution in the space-time cylinder* $Q_{T_0} = \mathbb{R}^3 \times]0,T_0[$*, such that*

$$v \in C([0,T_0]; \overset{\circ}{E}_3) \cap \overset{\circ}{G}_5(0,T_0),$$

$$\sqrt{|v|}|\nabla v| \in L_{2,unif}(0,T_0), \qquad p \in L_{\frac{5}{2}}(0,T_0;L_{\frac{5}{2},loc}(\mathbb{R}^3));$$

$$\sup_{x_0\in\mathbb{R}^3}\int\limits_0^{T_0}\int\limits_{B(x_0,3/2)}|p(x,t)-c_{x_0}(t)|^{\frac{5}{2}}\,dxdt < +\infty,$$

$$\sup_{x_0\in\mathbb{R}^3}\int\limits_0^{T_0}\int\limits_{B(x_0,3/2)}\mathbb{I}_{\{|x|>R\}}|p(x,t)-c_{x_0}(t)|^{\frac{5}{2}}\,dxdt \to 0$$

as $R\to+\infty$.

Moreover, assume that a pair u *and* q *is a local energy solution to the Cauchy problem*

$$\begin{gathered}\partial_t u(x,t) + div\, u(x,t)\otimes u(x,t) - \Delta\, u(x,t) + \nabla\, q(x,t) = g(x,t),\\ div\, u(x,t) = 0\end{gathered} \tag{B.5.7}$$

for $x\in\mathbb{R}^3$ *and* $0<t<T_0$,

$$u(x,0) = a(x), \qquad x\in\mathbb{R}^3. \tag{B.5.8}$$

Then, $u=v$.

Theorem 5.11. *Suppose that conditions (B.5.6) hold. Given* $T>0$, *there exists a constant* $\varepsilon(T)$ *with the following property. If*

$$\|g\|_{L_{3,unif}(0,T)} + \|a\|_{L_{3,unif}} \le \varepsilon(T),$$

then there exists a pair of functions v *and* p *that is a local energy solution in the space-time cylinder* Q_T, *such that:*

$$v\in C([0,T];\overset{\circ}{E}_3)\cap \overset{\circ}{G}_5(0,T),$$

$$\sqrt{|v|}|\nabla v| \in L_{2,unif}(0,T), \qquad p \in L_{\frac{5}{2}}(0,T;L_{\frac{5}{2},loc}(\mathbb{R}^3));$$

$$\|v\|_{L_\infty(0,T;L_{3,unif})} \le c\varepsilon(T);$$

$$\sup_{x_0\in\mathbb{R}^3}\int\limits_0^{T}\int\limits_{B(x_0,3/2)}|p(x,t)-c_{x_0}(t)|^{\frac{5}{2}}\,dxdt < +\infty,$$

$$\sup_{x_0\in\mathbb{R}^3}\int\limits_0^{T}\int\limits_{B(x_0,3/2)}\mathbb{I}_{\{|x|>R\}}|p(x,t)-c_{x_0}(t)|^{\frac{5}{2}}\,dxdt \to 0$$

as $R\to+\infty$.

Moreover, assume that a pair u *and* q *is a local energy solution to the Cauchy problem (B.5.4), (B.5.5). Then,* $u=v$.

Now, let us outline the proof of Theorem 1.7.

So, according to Proposition 1.8, we can find a number $T_0 \in]0,T]$ and a pair of functions v and p that are a local energy solution in the space-time cylinder Q_{T_0}. If $T_0 = T$, then we are done. Assume that it is not. By Theorem 1.6, we can find $t_0 \in]0,T_0[$ so that

$$v(\cdot,t_0) \in \overset{\circ}{E}_3,$$

$$v \textit{ and } p \textit{ are a local energy solution in } \mathbb{R}^3\times]t_0,T_0[.$$

Next, there exist T_1 and a pair of functions u and q that is a local energy solution in $\mathbb{R}^3\times]t_0,T_1[$ and $u \in C([t_0,T_1];\overset{\circ}{E}_3)$ with $u(\cdot,t_0) = v(\cdot,t_0)$. However, we know that there must be $v = u$ in $\mathbb{R}^3\times]t_0,T_1[$. Without loss of generality, we may assume that $T_1 < T$. Using density of smooth functions, let us decompose $v(\cdot,t_0) = a_1 + a_2$ and $g = g_1 + g_2$ so that

$$\|g_1\|_{L_{3,unif}(0,T)} + \|a_1\|_{L_{3,unif}} \leq \varepsilon(T-t_0),$$

$$a_2 \in \overset{\circ}{C}{}^\infty_0(\mathbb{R}^3),$$

and g_2 is a function of class C^∞ in Q_T and there exists $R_2 > 0$ such that the support of $g_2(\cdot,t)$ lies in $B(R_2)$ for all $t \in]t_0,T[$. According to Theorem 5.11, there exists a pair u_1 and q_1, which is a local energy solutions to the Cauchy problem:

$$\partial_t u_1(x,t) + \operatorname{div} u_1(x,t)\otimes u_1(x,t) - \Delta u_1(x,t) + \nabla q_1(x,t) = g_1(x,t),$$

$$\operatorname{div} u_1(x,t) = 0$$

for $x \in \mathbb{R}^3$ and $t_0 < t < T$,

$$u_1(x,t_0) = a_1(x), \qquad x \in \mathbb{R}^3.$$

Moreover,

$$\|u_1\|_{L_\infty(t_0,T;L_{3,unif})} \leq c\varepsilon(T-t_0). \tag{B.5.9}$$

We seek functions u_2 and q_2, solving the following Cauchy problem:

$$\partial_t u_2(x,t) + \operatorname{div}\,(u_2(x,t)\otimes u_2(x,t) + u_1(x,t)\otimes u_2(x,t)+$$

$$+u_2(x,t)\otimes u_1(x,t)) - \Delta u_2(x,t) + \nabla q_2(x,t) = g_2(x,t), \quad \operatorname{div} u_2(x,t) = 0$$

for $x \in \mathbb{R}^3$ and $t_0 < t < T$,

$$u_2(x,t_0) = a_2(x), \qquad x \in \mathbb{R}^3.$$

We state that this problem has a weak Leray-Hopf solution with the finite global energy satisfying the local energy inequality. To see that it is really possible, let us comment the crucial term in proving *a priori* global energy estimate. This term has the form

$$I_0 = \int_{t_0}^{t}\int_{\mathbb{R}^3} (u_2 \otimes u_1 + u_1 \otimes u_2) : \nabla u_2 dxds.$$

So, we need to estimate the integral

$$I = \int_{t_0}^{t}\int_{\mathbb{R}^3} |u_1|^2|u_2|^2 dxds.$$

To this end, we fix $x_0 \in \mathbb{R}^3$ and apply successively Hölder's and Gagliardo's inequalities, and estimate (B.5.9):

$$\int_{t_0}^{t}\int_{B(x_0,1)} |u_1|^2|u_2|^2 dxds \leq \int_{t_0}^{t}\Big(\int_{B(x_0,1)} |u_1|^3 dx\Big)^{\frac{2}{3}}\Big(\int_{B(x_0,1)} |u_2|^6 dx\Big)^{\frac{1}{3}} ds$$

$$\leq c\|u_1\|^2_{L_\infty(t_0,T;L_{3,unif})}\int_{t_0}^{t}\int_{B(x_0,1)} \Big(|\nabla u_2|^2 + |u_2|^2\Big) dxds$$

$$\leq c\varepsilon^2(T-t_0)\int_{t_0}^{t}\int_{B(x_0,1)} \Big(|\nabla u_2|^2 + |u_2|^2\Big) dxds.$$

Using Besicovitch covering lemma, we can easily show

$$I \leq c\varepsilon^2(T-t_0)\int_{t_0}^{t}\int_{\mathbb{R}^3} \Big(|\nabla u_2|^2 + |u_2|^2\Big) dxds$$

and therefore

$$I_0 \leq c\varepsilon(T-t_0)\int_{t_0}^{t}\int_{\mathbb{R}^3} \Big(|\nabla u_2|^2 + |u_2|^2\Big) dxds.$$

The latter allows us to hide I_0 into the left-hand side of the global energy inequality by choosing $\varepsilon(T-t_0)$ sufficiently small and to find

$$\int_{\mathbb{R}^3} |u_2(x,t)|^2 dx + \int_{t_0}^{t}\int_{\mathbb{R}^3} |\nabla u_2|^2 dxds \leq \int_{\mathbb{R}^3} |a_2(x)|^2 dx + 2\int_{t_0}^{t}\int_{\mathbb{R}^3} g_2 \cdot u_2 dxds.$$

Using this estimate and suitable approximations, we can easily prove our statements about u_2. In addition, all above arguments show that pressure q_2 may be taken in the form

$$q_2 = \frac{1}{4\pi} K * (u_2 \otimes u_2 + u_1 \otimes u_2 + u_2 \otimes u_1)$$

and, moreover,

$$q_2 \in L_{\frac{3}{2}}(\mathbb{R}^3 \times]t_1, T[)$$

and we have representation:

$$q_2(x,t) - c_{2x_0}(t) = q^1_{2x_0}(x,t) + q^2_{2x_0}(x,t)$$

for $x_0 \in B(x_0, 3/2)$ and $t \in]t_0, T[$, where

$$q^1_{2x_0} = -\frac{1}{3}(|u_2(x,t)|^2 + 2u_1(x,t) \cdot u_2(x,t)) +$$

$$+\frac{1}{4\pi} \int\limits_{B(x_0,2)} K(x-y) : (u_2 \otimes u_2 + u_1 \otimes u_2 + u_2 \otimes u_1)(y,t)dy,$$

$$q^2_{2x_0} = \frac{1}{4\pi} \int\limits_{\mathbb{R}^3 \setminus B(x_0,2)} (K(x-y) - K(x_0-y)) : (u_2 \otimes u_2 + u_1 \otimes u_2 + u_2 \otimes u_1)(y,t)dy,$$

$$c_{2x_0}(t) = \frac{1}{4\pi} \int\limits_{\mathbb{R}^3 \setminus B(x_0,2)} K(x_0 - y) : (u_2 \otimes u_2 + u_1 \otimes u_2 + u_2 \otimes u_1)(y,t)dy.$$

Now, we let

$$u = u_1 + u_2, \qquad q = q_1 + q_2.$$

Our task is to verify that this new pair is a local energy solution to the Cauchy problem:

$$\partial_t u(x,t) + \operatorname{div} u(x,t) \otimes u(x,t) - \Delta u(x,t) + \nabla q(x,t) = g(x,t),$$

$$\operatorname{div} u(x,t) = 0$$

for $x \in \mathbb{R}^3$ and $t_0 < t < T$,

$$u(x,t_0) = a(x), \qquad x \in \mathbb{R}^3.$$

The most difficult part of this task is to show that u and q satisfy the local energy inequality. It can be done essentially in the same way as the corresponding part of the proof of the uniqueness for $C([0,T]; \overset{\circ}{E}_3)$-solutions. And this immediately implies that $u = v$ in the $\mathbb{R}^3 \times]t_0, T_1[$. Since $p(x,t) - q(x,t) = c(t) \in L_{\frac{3}{2}}(t_0, T_1)$, we can change function c_{x_0} in a suitable way and assume that $q = p$ in the $\mathbb{R}^3 \times]t_0, T_1[$. So, the pair u and q can be regarded as a required extension of v and p to the whole space-time cylinder Q_T. Theorem 1.7 is proved.

B.6 Density

Lemma B.10. *For any* $f \in \overset{\circ}{E}_m$ *and for any* $\varepsilon > 0$*, there exists* $f_\varepsilon \in \overset{\circ}{C}{}_0^\infty(\mathbb{R}^3)$ *such that*

$$\|f - f_\varepsilon\|_{L_{m,unif}} < \varepsilon. \tag{B.6.1}$$

PROOF Let

$$B_k = B(x_k, 2), \qquad x_k \in \mathbb{Z}^3.$$

Clearly $\bigcup\limits_k B_k = \mathbb{R}^3$. Moreover, there is a universal constant N_2 with the following property:

for any $x_0 \in \mathbb{R}^3$*, the ball* $B(x_0, 1)$ *is covered by not more*

than N_2 *balls* B_k*.*

For this covering, we can find a partition of unity such that

$$\varphi_k \in C_0^\infty(\mathbb{R}^3), \qquad \operatorname{spt}\varphi_k \subset B_k, \qquad \varphi_k \geq 0, \qquad \sum_k \varphi_k = 1.$$

Now, given $R > 0$, we introduce two smooth cut-off functions

$$\chi(x) = 1, \quad x \in B(1), \qquad \operatorname{spt}\chi \subset B(2), \qquad \chi_R(x) = \chi(x/R).$$

We fix a ball B_k. There exists a function $v^k \in \overset{\circ}{W}{}_2^1(B_k; \mathbb{R}^3)$ that is a solution of the equation

$$\operatorname{div} v^k = f \cdot \nabla\, \chi_R \varphi_k - \frac{1}{|B(2)|} \int\limits_{B_k} f \cdot \nabla\, \chi_R \varphi_k \, dx$$

and satisfies the estimate

$$\|v^k\|_{B_k} \leq \frac{c}{R} \|f\|_{B_k} \leq \frac{c}{R} \|f\|_{L_{m,unif}}$$

with a universal constant c. Extending v^k by zero to the whole $\mathbb{R}^3$, we set

$$v^R = \sum_k v^k$$

and observe that, for each $R > 0$, the function v^R has a compact support and, moreover,

$$\operatorname{div} v^R = f \cdot \nabla\, \chi_R \qquad \text{in } \mathbb{R}^3,$$

$$\|v^R\|_{B(x_0,1)} \leq \sum_k \|v^k\|_{B(x_0,1)} \leq \frac{cN_2}{R} \|f\|_{L_{m,unif}}, \qquad \forall x_0 \in \mathbb{R}^3.$$

Next, we let

$$u^R = f\chi_R - v^R.$$

Obviously, we have

$$\operatorname{div} u^R = 0 \qquad \text{in } \mathbb{R}^3,$$

and u^R has a compact support. Since $f \in E_2$, we see that, for an arbitrary $\varepsilon > 0$, we can find $R > 0$ such that

$$\|f - u^R\|_{L_{m,unif}} \leq \|f - f\chi_R\|_{L_{m,unif}} + \|v^R\|_{L_{m,unif}}$$

$$\leq \|f - f\chi_R\|_{L_{m,unif}} + \frac{cN_2}{R}\|f\|_{L_{m,unif}} < \varepsilon.$$

To complete the proof of the lemma, it is enough to smooth u^R which is easy. Lemma B.10 is proved.

B.7 Comments

The main source for the content of Appendix B is the monograph of P.-G. Lemarie-Riesset [Lemarie-Riesset (2002)]. Our interpretation of his results is given in the paper [Kikuchi and Seregin (2007)] and we follow it here. We wish to emphasize that the Lemarie-Riesset conception of local energy weak Leray-Hopf solution[1] is heavily used in Chapter 7.

[1] In fact, G.-P. Lemarie-Riesset himself calls them simply local Leray solutions

Bibliography

Bogovskii, M. E. (1980). Solution of some problems of vector analysis with the operators Div and Grad, *in: Trudy S. L. Sobolevs Seminar* **80**, 540.

Caffarelli, L., Kohn, R. V. and Nirenberg, L. (1982). Partial regularity of suitable weak solutions of the Navier-Stokes equations, *Comm. Pure Appl. Math.* **XXXV**, 771–831.

Cannone, M. (2002). *Harmonic analysis tools for solving the incompressible Navier-Stokes equations*, Handbook of Mathematical Fluid Dynamics, vol. 3.

Chen, C. C., Strain, R. M., Tsai, T. P. and Yau, H. T. (2008). Lower bound on the blow-up rate of the axisymmetric Navier-Stokes equations, *Int. Math. Res. Not.*: article ID rnn016, 31 pages, doi:10.1093/imrn/rnn016.

Chen, C. C., Strain, R. M., Tsai, T. P. and Yau, H. T. (2009). Lower bounds on the blow-up rate of the axisymmetric Navier-Stokes equations II, *Communications in Partial Differential Equations* **34**, 203–232.

Chen Z. M. and Price W. G. (2001). Blow-up rate estimates for weak solutions of the Navier–Stokes equations, *R. Soc. Lond. Proc. Ser. A Math. Phys. Eng. Sci.* **457**, 2625–2642.

Choe, H. L., Lewis, J. L. (2000). On the singular set in the Navier-Stokes equations, *J. Functional Anal.* **175**, 348–369.

Constantin, P. and Foias, C. (1988). *Navier-Stokes Equations*, Chicago Lecture in Mathematics.

Escauriaza, L. (2000). Carleman inequalities and the heat operator, *Duke Mathematical Journal* **104**(2000), 113 - 126.

Escauriaza, L. E. Kenig, C. E, Ponce, G. and Vega, L. (2006). Decay at infinity of caloric functions within characteristic hyperplanes, *Math. Res. Lett.* **13**, 441–453.

Escauriaza, L., Seregin, G. and Šverák, V. (2002). On backward uniqueness for parabolic equations, *Zap. Nauchn. Seminar. POMI* **288**, 100–103.

Escauriaza, L., Seregin, G., Šverák, V. (2003). Backward uniqueness for the heat operator in a half-space, *Algebra i Analiz* **15**, 201–214.

Escauriaza, L., Seregin, G., Šverák, V. (2003). On backward uniqueness for parabolic equations, *Arch. Rational Mech. Anal.* **169**, 147–157.

Escauriaza, L., Seregin, G. and Šverák, V. (2003). $L_{3,\infty}$-solutions to the Navier-Stokes equations and backward uniqueness, *Russian Mathematical Surveys* **58**, 211–250.

Escauriaza, L. and Vega, L. (2001). Carleman inequalities and the heat operator II, *Indiana university Mathematics Journal* **50**, 1149 – 1169.

Fefferman, C. (2000). http://www.claymath.org/millennium/Navier-Stokes equations.

Galdi, G. P. (2011). *An Introduction to the Mathematical Theory of the Navier-Stokes Equations: Steady-State Problems*, Springer Monographs in Mathematics, Second edition.

Galdi, G. P. (2000). *An introduction to the Navier-Stokes initial-boundary value problem*, Fundamental directions in mathematical fluid mechanics/Giovanni P. Galdi...ed.,-Basel; Boston; Berlin: Birkhäuser 2000, 1-70.

Giga, Y. (1986). Solutions for semilinear parabolic equations in L^p and regularity of weak solutions of the Navier-Stokes equations, *J. of Diff. Equations* **62**, 186–212.

Gustafson, S., Kang, K. and Tsai, T.-P. (2007). Interior regularity criteria for suitable weak solutions of the Navier-Stokes equations, *Commun. Math. Phys.* **273**, 161–176.

Hopf, E. (1950-51). Über die Anfangswertaufgabe für die hydrodynamischen Grundgleichungen, *Math. Nachrichten*, **4**, 213–231.

Hörmander, L. (1963). *Linear Partial Differential Operators*, Springer Verlag.

Kang, K. (2005). Unbounded normal derivative for the Stokes system near boundary, *Math. Ann.* **331** (2005), 87–109.

Kapitanskii, L. V. and Pileckas, K. I. (1984). Some problems of vector analysis, *Zap. Nauchn. Semin. LOMI* **138**, 65-85.

Kato, T. (1984). Strong L_p-solutions of the Navier-Stokes equation in R^m, with applications to weak solutions, *Math. Z.* **187**, 471–480.

Kikuchi, N. and Seregin, G. (2007). *Weak solutions to the Cauchy problem for the Navier-Stokes equations satisfying the local energy inequality.* AMS translations, Series 2, **220**, 141–164.

Kim H. and Kozono H. (2004). Interior regularity criteria in weak spaces for the Navier-Stokes equations, *manuscripta math.* **115**, 85–100.

Kiselev, A. A. and Ladyzhenskaya, O. L. (1957). On the existence and uniqueness of the solution of the non-stationary problem for a viscous, incompressible fluid, *Izv. Akad. Nauk SSSR Ser. Mat.*, **21**, 655–680.

Koch, G., Nadirashvili, N., Seregin, G. and Šverák, V. (2009). Liouville theorems for the Navier-Stokes equations and applications, *Acta Mathematica* **203**, 83–105.

Koch, H. and Tataru, D. (2001). Well-posedness for the Navier-Stokes equations. *Adv. Math.* **157**, 22–35.

Ladyzhenskaya, O. A., Solution "in the large" of the boundary-value problem for the Navier-Stokes equations in the case of two space variables. *Dokl. Akad. Nauk SSSR*, **123**, 427–429.

Ladyzhenskaya, O. A. (1967). On the uniqueness and on the smoothness of weak solutions of the NavierStokes equations. *Zap. Nauchn. Sem. LOMI*, **5**, 169–185.

Ladyzhenskaya, O. A. (1970). *Mathematical problems of the dynamics of viscous incompressible fluids*, 2nd edition, Nauka, Moscow.

Ladyzhenskaya, O. A. (1968). On unique solvability of the three-dimensional Cauchy problem for the Navier-Stokes equations under the axial symmetry, *Zap. Nauchn. Sem. LOMI* **7**, 155–177.

Ladyzhenskaya, O. A. and Seregin, G. A. (1999). On partial regularity of suitable weak solutions to the three-dimensional Navier-Stokes equations, *J. math. fluid mech.* **1**, 356–387.

Ladyzhenskaya, O. A. and Solonnikov, V. A. (1976). Some problems of vector analysis and generalized settings of boundary-value problems for the Navier-Stokes equations, *Zap. Nauchn. Semin. LOMI* **59**, 81-116.

Ladyzhenskaya, O. A., Solonnikov, V. A. and Uralt'seva, N. N.(1967). *Linear and quasi-linear equations of parabolic type*, Moscow, 1967; English translation, American Math. Soc., Providence 1968.

Ladyzhenskaya, O. A. and Uraltseva, N. N. (1973). *Linear and quasilinear equations of elliptic type*, Nauka, Moscow, 1973.

Lemarie-Riesset, P. G. (2002). *Recent developments in the Navier-Stokes problem*, Chapman&Hall/CRC reseacrh notes in mathematics series.

Leray, J. (1934). Sur le mouvement d'un liquide visqueux emplissant l'espace, *Acta Math.* **63**, 193–248.

Lions, P. L. (1996). *Mathematical Topics in Fluid Mechanics: Volume 1: Incompressible Models*, Oxford University Press.

Lin, F.-H. (1998). A new proof of the Caffarelly-Kohn-Nirenberg theorem, *Comm. Pure Appl. Math.* **51**, 241–257.

Necas, J. (1966). Sur les normes equivalentes dans $W_p^{(k)}(\Omega)$ et sur la coercitivite des formes formellement positive, *in Equations aux derivees partielles, Presses de l'Universite de Montreal*, 1966.

Necas, J., Ruzicka, M. and Šverák, V. (1996). On Leray's self-similar solutions of the Navier-Stokes equations. *Acta Math.* **176**, 283–294.

Prodi, G. (1959) Un teorema di unicitaper le equazioni di Navier-Stokes, *Annali di Mat.* **48**,173–182.

Rusin, W., Sverak, V. (2007). Miminimal initial data for potential Navier-Stokes singularities, arXiv:0911.0500.

Scheffer, V. (1976). Partial regularity of solutions to the Navier-Stokes equations, *Pacific J. Math.* **66**, 535–552.

Scheffer, V. (1977). Hausdorff measure and the Navier-Stokes equations, *Commun. Math. Phys.* **55**, 97–112.

Scheffer, V. (1980). The Navier-Stokes equations in a bounded domain, *Commun. Math. Phys.* **73**, 1–42.

Scheffer, V. (1982). Boundary regularity for the Navier-Stokes equations in a half-space, *Commun. Math. Phys.* **85**, 275–299.

Seregin, G. A. (1999). Interior regularity for solutions to the modified Navier-Stokes equations, *J. of math. fluid mech.* **1**, 235–281.

Seregin, G. (2007). *Local regularity theory of the Navier-Stokes equations*, Handbook of Mathematical Fluid Mechanics, Vol. 4, Edited by Friedlander, D. Serre, 159–200.

Seregin, G. A. (2007). Estimates of suitable weak solutions to the Navier-Stokes equations in critical Morrey spaces, *J. Math. Sci.* **143:2**, 2961–2968.

Seregin, G. (2009). A note on local boundary regularity for the Stokes system, *Zapiski Nauchn. Seminar. POMI*, **370**, 151–159.

Seregin, G. (2007). Local regularity for suitable weak solutions of the Navier-Stokes equations, *Russian Math. Surveys* **62:3**, 595–614.

Seregin, G. A. (2001). On the number of singular points of weak solutions to the Navier-Stokes equations, *Comm. Pure Appl. Math.*, **54**, 1019-1028.

Seregin, G. A. Local regularity of suitable weak solutions to the Navier-Stokes equations near the boundary, *J. math. fluid mech.*, **4**, 1–29.

Seregin, G. A. (2002) Differentiability properties of weak solutions to the Navier-Stokes equations, *Algebra and Analysis* **14**, 193–237.

Seregin, G. A. (2007). Navier-Stokes equations: almost $L_{3,\infty}$-cases, *Journal of mathematical fluid mechanics* **9**, 34–43.

Seregin, G. (2011). A note on necessary conditions for blow-up of energy solutions to the Navier-Stokes equations, *Progress in Nonlinear Differential Equations and Their Applications, Springer Basel AG* **60**, 631-645. 98.

Seregin, G. (2010). Necessary conditions of potential blow up for the Navier-Stokes equations, *Zapiski Nauchn.Seminar. POMI* **385**, 187-199.

Seregin, G. (2012). A certain necessary condition of potential blow up for Navier-Stokes equations, *Commun. Math. Phys.* **312**, 833-845.

Seregin, G. (2013). *Selected topics of local regularity theory for the Navier-Stokes equations.* Topics in mathematical fluid mechanics, 239313, Lecture Notes in Math., 2073, Springer, Heidelberg, 2013.

Seregin, G. and Šverák, V. (2009). On Type I singularities of the local axi-symmetric solutions of the Navier-Stokes equations, *Communications in PDE's* **34**, 171-201.

Seregin, G. and Šverák, V. (2010). On a bounded shear flow in half space, *Zapiski Nauchn. Seminar. POMI* **385**, 200–205.

Seregin, G. and Zajaczkowski, W. (2006). A sufficient condition of local regularity for the Navier-Stokes equations, *Zapiski Nauchn. Seminar. POMI* **336**, 46-54.

Seregin, G. and Zajaczkowski, W. (2007). A sufficient condition of regularity for axially symmetric solutions to the Navier-Stokes equations, *SIMA J. Math. Anal.* **39**, 669–685.

Serrin, J. (1962). On the interior regularity of weak solutions of the Navier-Stokes equations, *Arch. Ration. Mech. Anal.* **9**, 187–195.

Stein, E. (1970). *Singular Integrals and Differentiabilty Properties of Functions*, Princeton University Press, Princeton, New Jersey.

Struwe, M. (1988). On partial regularity results for the Navier-Stokes equations, *Comm. Pure Appl. Math.* **41**, 434–458.

Takahashi, S. (1990). On interior regularity criteria for weak solutions of the Navier-Stokes equations, *Manuscripta Math.* **69**, 237–254.

Tataru, D. (2000). Carleman estimates, unique continuation, and applications,

notes downloadable from `http://math.berkeley.edu/~tataru/ucp.html`
Temam, R. (2001). *Navier-Stokes Equations: Theory and Numerical Analysis*, American Mathematical Soc., 408 p.
Ukhovskij, M. R. and Yudovich, V. L. (1968). Axially symmetric motions of ideal and viscous fluids filling all space, *Prikl. Mat. Mech.* **32**, 59–69.

Index

Printed in the United States
By Bookmasters